Mass Spectrometry

WILEY SERIES ON MASS SPECTROMETRY

Series Editors
Dominic M. Desiderio
Departments of Neurology and Biochemistry
University of Tennessee Health Science Center
Joseph A. Loo
Department of Chemistry and Biochemistry UCLA

Founding Editors
Nico M. M. Nibbering (1938–2014)
Dominic M. Desiderio

John R. de Laeter *Applications of Inorganic Mass Spectrometry*
Michael Kinter and Nicholas E. Sherman *Protein Sequencing and Identification Using Tandem Mass Spectrometry*
Chhabil Dass *Principles and Practice of Biological Mass Spectrometry*
Mike S. Lee *LC/MS Applications in Drug Development*
Jerzy Silberring and Rolf Eckman *Mass Spectrometry and Hyphenated Techniques in Neuropeptide Research*
J. Wayne Rabalais *Principles and Applications of Ion Scattering Spectrometry: Surface Chemical and Structural Analysis*
Mahmoud Hamdan and Pier Giorgio Righetti *Proteomics Today: Protein Assessment and Biomarkers Using Mass Spectrometry, 2D Electrophoresis, and Microarray Technology*
Igor A. Kaltashov and Stephen J. Eyles *Mass Spectrometry in Structural Biology and Biophysics: Architecture, Dynamics, and Interaction of Biomolecules, Second Edition*
Isabella Dalle-Donne, Andrea Scaloni, and D. Allan Butterfield *Redox Proteomics: From Protein Modifications to Cellular Dysfunction and Diseases*
Silas G. Villas-Boas, Ute Roessner, Michael A.E. Hansen, Jorn Smedsgaard, and Jens Nielsen *Metabolome Analysis: An Introduction*
Mahmoud H. Hamdan *Cancer Biomarkers: Analytical Techniques for Discovery*
Chabbil Dass *Fundamentals of Contemporary Mass Spectrometry*
Kevin M. Downard (Editor) *Mass Spectrometry of Protein Interactions*
Nobuhiro Takahashi and Toshiaki Isobe *Proteomic Biology Using LC-MS: Large Scale Analysis of Cellular Dynamics and Function*
Agnieszka Kraj and Jerzy Silberring (Editors) *Proteomics: Introduction to Methods and Applications*
Ganesh Kumar Agrawal and Randeep Rakwal (Editors) *Plant Proteomics: Technologies, Strategies, and Applications*
Rolf Ekman, Jerzy Silberring, Ann M. Westman-Brinkmalm, and Agnieszka Kraj (Editors) *Mass Spectrometry: Instrumentation, Interpretation, and Applications*
Christoph A. Schalley and Andreas Springer *Mass Spectrometry and Gas-Phase Chemistry of Non-Covalent Complexes*
Riccardo Flamini and Pietro Traldi *Mass Spectrometry in Grape and Wine Chemistry*
Mario Thevis *Mass Spectrometry in Sports Drug Testing: Characterization of Prohibited Substances and Doping Control Analytical Assays*
Sara Castiglioni, Ettore Zuccato, and Roberto Fanelli *Illicit Drugs in the Environment: Occurrence, Analysis, and Fate Using Mass Spectrometry*
Ángel Garciá and Yotis A. Senis (Editors) *Platelet Proteomics: Principles, Analysis, and Applications*
Luigi Mondello *Comprehensive Chromatography in Combination with Mass Spectrometry*
Jian Wang, James MacNeil, and Jack F. Kay *Chemical Analysis of Antibiotic Residues in Food*
Walter A. Korfmacher (Editor) *Mass Spectrometry for Drug Discovery and Drug Development*
Alejandro Cifuentes (Editor) *Foodomics: Advanced Mass Spectrometry in Modern Food Science and Nutrition*
Christine M. Mahoney (Editor) *Cluster Secondary Ion Mass Spectrometry: Principles and Applications*
Despina Tsipi, Helen Botitsi, and Anastasios Economou *Mass Spectrometry for the Analysis of Pesticide Residues and their Metabolites*
Xianlin Han *Lipidomics: Comprehensive Mass Spectrometry of Lipids*
Marek Smoluch, Giuseppe Grasso, Piotr Suder, and Jerzy Silberring (Editors) *Mass Spectrometry: An Applied Approach, Second Edition*

Mass Spectrometry

An Applied Approach

Edited by
Marek Smoluch, Giuseppe Grasso, Piotr Suder,
and Jerzy Silberring

Second Edition

This edition first published 2019
© 2019 John Wiley & Sons, Inc.

Edition History
John Wiley & Sons Inc. (1e, 2008)

Registered Office
John Wiley & Sons, Inc., 111 River Street, Hoboken, NJ 07030, USA

Editorial Office
111 River Street, Hoboken, NJ 07030, USA

For details of our global editorial offices, customer services, and more information about Wiley products visit us at www.wiley.com.

Wiley also publishes its books in a variety of electronic formats and by print-on-demand. Some content that appears in standard print versions of this book may not be available in other formats.

Library of Congress Cataloging-in-Publication Data

Names: Smoluch, Marek, 1977– editor. | Grasso, Giuseppe, 1974– editor. | Suder, Piotr, 1949– editor. | Silberring, Jerzy, 1949– editor.
Title: Mass spectrometry : an applied approach.
Description: Second edition / edited by Marek Smoluch, Giuseppe Grasso, Piotr Suder, Jerzy Silberring. | Hoboken, NJ : Wiley, 2019. | Series: Wiley series on mass spectrometry | Includes index. |
Identifiers: LCCN 2019011643 (print) | LCCN 2019012925 (ebook) | ISBN 9781119377337 (Adobe PDF) | ISBN 9781119377344 (ePub) | ISBN 9781119377306 (hardcover)
Subjects: LCSH: Mass spectrometry. | Spectrum analysis.
Classification: LCC QP519.9.M3 (ebook) | LCC QP519.9.M3 M314945 2019 (print) | DDC 572/.36–dc23
LC record available at https://lccn.loc.gov/2019011643

Cover design: Wiley
Cover image: Courtesy of Marian Hanik

Set in 10/12pt Warnock by SPi Global, Pondicherry, India

Printed in United States of America

V10010664_060319

In memory of Nicolaas Martinus Maria Nibbering (Nico)
29.05.1938 – 29.08.2014

Contents

List of Contributors

Kathrin Altweg
Physikalisches Institut
Universität Bern
Bern
Switzerland

Robert Anczkiewicz
Institute of Geological Sciences
Polish Academy of Sciences
Krakow
Poland

Anna Antolak
Department of Biochemistry and
Neurobiology
Faculty of Materials Science and
Ceramics, AGH University of Science
and Technology
Kraków
Poland

Francesco Bellia
CNR Institute of Biostructures and
Bioimaging
Catania
Italy

Anna Bodzon-Kulakowska
Department of Biochemistry and
Neurobiology
Faculty of Materials Science and
Ceramics, AGH University of Science

and Technology
Kraków
Poland

Pawel Ciborowski
Department of Pharmacology and
Experimental Neuroscience
University of Nebraska Medical
Center
Omaha
NE, USA

Vincenzo Cunsolo
Dipartimento di Scienze Chimiche
Università di Catania
Catania
Italy

Giuseppe Di Natale
CNR Institute of Crystallography
(IC), Secondary Site
Catania
Italy

Anna Drabik
Department of Biochemistry and
Neurobiology
Faculty of Materials Science and
Ceramics, AGH University of Science
and Technology
Kraków
Poland

Salvatore Foti
Dipartimento di Scienze Chimiche
Università di Catania
Catania
Italy

Giuseppe Grasso
Dipartimento di Scienze Chimiche
Università di Catania
Catania
Italy

Emma Harwood
Department of Pharmacology and
Experimental Neuroscience
University of Nebraska Medical
Center
Omaha
NE, USA

Claudio Iacobucci
Institute of Pharmacy
Martin Luther University
Halle-Wittenberg
Halle (Saale)
Germany

Dorota Kwiatkowska
Polish Anti-Doping Laboratory
Warsaw
Poland

Przemyslaw Mielczarek
Department of Biochemistry and
Neurobiology
Faculty of Materials Science and
Ceramics, AGH University of Science
and Technology
Kraków
Poland

Vera Muccilli
Dipartimento di Scienze Chimiche
Università di Catania
Catania
Italy

Joanna Ner-Kluza
Department of Biochemistry and
Neurobiology
Faculty of Materials Science and
Ceramics, AGH University of Science
and Technology
Kraków
Poland

Aleksandra Pawlaczyk
Faculty of Chemistry
Institute of General and Ecological
Chemistry
Lodz University of Technology
Łódź
Poland

Katarzyna Pawlak
Department of Analytical Chemistry
Faculty of Chemistry
Warsaw University of Technology
Warsaw
Poland

Kinga Piechura
Department of Biochemistry and
Neurobiology
Faculty of Materials Science and
Ceramics, AGH University of Science
and Technology
Kraków
Poland

Rosaria Saletti
Dipartimento di Scienze Chimiche
Università di Catania
Catania
Italy

Jerzy Silberring
Department of Biochemistry and
Neurobiology
Faculty of Materials Science and
Ceramics, AGH University of Science
and Technology
Kraków
Poland
and
Centre of Polymer and Carbon
Materials
Polish Academy of Sciences
Zabrze
Poland

Marek Smoluch
Department of Biochemistry and
Neurobiology
Faculty of Materials Science and
Ceramics, AGH University of Science
and Technology
Kraków
Poland

Giuseppe Spoto
Dipartimento di Scienze Chimiche
Università di Catania
Catania
Italy

Piotr Stefanowicz
Faculty of Chemistry
University of Wrocław

Wrocław
Poland

Piotr Suder
Department of Biochemistry and
Neurobiology
Faculty of Materials Science and
Ceramics, AGH University of Science
and Technology
Kraków
Poland

Zbigniew Szewczuk
Faculty of Chemistry
University of Wrocław
Wrocław
Poland

Małgorzata Iwona Szynkowska
Faculty of Chemistry
Institute of General and Ecological
Chemistry
Lodz University of Technology
Łódź
Poland

Nunzio Tuccitto
Dipartimento di Scienze Chimiche
Università di Catania
Catania
Italy

Fang Yu
Department of Biostatistics
University of Nebraska Medical
Center
Omaha
NE, USA

Preface

Rapid development of genomics, proteomics, combinatorial chemistry, and medical/toxicological diagnostics triggered the rapid development of various mass spectrometry techniques to fulfill requirements of many disciplines, such as biomedical sciences, toxicology, forensic research, and pharmacology. Mass spectrometry (MS) is a unique method that not only allows mass measurement but also provides detailed identification of molecules and traces the fate of compounds in vivo and in vitro. Among others, mass spectrometry may, at least partially, identify amino acid sequence of peptides and proteins, assign sites of posttranslational modifications, identify bacterial strains, and verify structures of organic compounds. The latter is particularly useful for detection of novel drugs of abuse, explosives, etc. A yet another challenge is rapid selection of combinatorial libraries, containing vast number of elements, and a novel place of mass spectrometry in nanomedicine, being a combination of diagnostics and therapy (theranostics).

MS has been proven as an efficient tool to analyze complex biological mixtures by applying hyphenated techniques, such as GC/MS, LC/MS, CE/MS, and TLC/MS, where mass spectrometer acts as a sensitive and highly specific detector. Such approaches may find their applications in, e.g. genomics, functional proteomics to reveal the role of entire pathways in biological systems (systems biology). Another interesting capability of MS is identification of the low molecular mass compounds that are not coded by genes. This aspect is a basis of metabolomics and remains, together with proteomics, a complementary way to study functions of the genes (transcriptomics).

Our goal was to offer you a book, which is written in an understandable language, avoiding complex equations and advanced physics, bearing in mind that the most important aspect for the readers are practical aspects, potential applications, and selection of proper methodologies to solve their analytical and scientific problems.

1

Introduction

Jerzy Silberring[1,2] and Marek Smoluch[1]

[1] Department of Biochemistry and Neurobiology, Faculty of Materials Science and Ceramics, AGH University of Science and Technology, Kraków, Poland
[2] Centre of Polymer and Carbon Materials, Polish Academy of Sciences, Zabrze, Poland

Mass spectrometry underwent a rapid and dynamic development during recent years. Innovative solutions brought highly advanced instruments that fulfill user demands with respect to sensitivity, speed, and simplicity of operation.

Mass spectrometer, independently on its construction, measures the ratio of mass of a molecule to its charge, m/z. While interpreting data obtained during analysis, it should be carefully noted that not always the m/z value can be directly related to the molecular mass of the analyzed compound. This happens when multiple charges are being attached to the molecule (multiple ionization), which results from the attachment or depletion of a proton or several protons. Even a popular electron impact ionization generates radicals, depleted of one electron. Typically, we tend to neglect this electron while analyzing spectra at low resolution. However, this lack of one electron will be clearly seen during high-resolution analysis using Fourier transform ion cyclotron resonance (FT-ICR) instrument. The multiplicity of ionization depends on the ion source used and their different types. These are described in the following chapters of this book.

The principle of operation of the apparatus can be compared to the sensitive balance, by which we weigh mass of molecules. Another association implies a comparison of a mass spectrometer with electrophoresis in a vacuum because the analyzed molecules, in the form of ions, are accelerated in the device under the influence of applied potential.

Until recently, the mass spectrometer consisted of elements traditionally associated with various ionization methods. For example, matrix-assisted laser desorption/ionization (MALDI) was combined with the time-of-flight (TOF) analyzer, and electron impact/chemical ionization (EI/CI) was typically used

Mass Spectrometry: An Applied Approach, Second Edition. Edited by Marek Smoluch, Giuseppe Grasso, Piotr Suder, and Jerzy Silberring.
© 2019 John Wiley & Sons, Inc. Published 2019 by John Wiley & Sons, Inc.

with quadrupole or magnetic and electrostatic analyzers. Today's constructions are built of "blocks" that can create combinations not yet very common, such as MALDI with ion trap, electron ionization with TOF, inductively coupled plasma (ICP) with TOF, and electrospray with TOF (qTOF). New constructions like Orbitrap are relatively cheap and affordable by many laboratories.

The mass spectrometer consists of several basic elements, presented schematically in Figure 1.1.

The basic requirement for a substance to be analyzed in a mass spectrometer is its ability to ionize. Ions can move in a vacuum under the influence of an applied electric field. It is important to note that vacuum is necessary inside mass spectrometer, where ions are analyzed. Ion sources, in many cases, do not require vacuum at all. The heterogeneous ion beam is separated in the analyzer depending on the *m/z* values for the individual ions. Separated ions are then introduced into a detector that converts quantum ion current into electrical current. The system control software transcribes the intensity of these signals as a function of the *m/z* value and presents these data as a mass spectrum, as shown in Figure 1.2.

Spectra are derived from the substances that are present in the sample. The mass spectrometer can simultaneously analyze the mixture (to some extent), which is extremely important in the study of complex biological material or other unknown samples. It is also possible to analyze selected substances only in the mixture. This saves analysis time, reduces the amount of data on the hard drive, and improves the signal-to-noise ratio. This method is referred to as the single ion monitoring (SIM) or multiple ion monitoring (MIM) and is mainly used for quantitative analysis of compounds and their fragment ions.

The main advantages of a mass spectrometer, compared with other techniques, are as follows:

- Speed of analysis.
- High sensitivity, reaching femto-/attomolar level.
- Simultaneous analysis of many components of mixtures.
- Ability to obtain information on the structure of the compounds (including amino acid sequence) and posttranslational modifications.
- Possibility of combining with separation techniques (e.g. gas and liquid chromatography, capillary electrophoresis, isotachophoresis).
- Quantitative analysis.

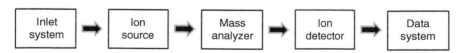

Figure 1.1 Components of the mass spectrometer.

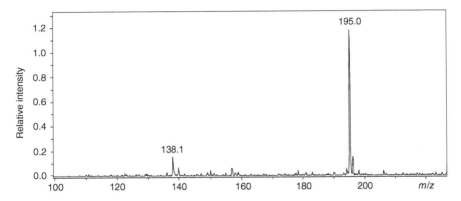

Figure 1.2 Exemplary mass spectrum of caffeine obtained by electrospray ionization mass spectrometry (ESI-MS) technique. Signal at *m/z* 195.0 corresponds to the protonated molecule of caffeine.

- Analysis of the elemental composition.
- Analysis of isotopic composition.
- Unambiguous identification of the substance.

This latter feature distinguishes mass spectrometry from other detection techniques encountered in chromatographic or electrophoretic methods. It is worth emphasizing here that the chromatogram obtained by ultraviolet–visible (UV/Vis) detection, electrochemical or other, generates only the signal intensity for the eluted fraction and the retention time. However, this is insufficient to obtain detailed information about the nature of the substance eluted from the column. The overlap of several components additionally complicates such analysis. Simply put, retention time is not a sufficient method of identification even if standards are provided. We cannot assume in such case that there is no other component in the mixture having the same retention time. The mass spectrometer, used as a detector, gives the exact mass of the substance, together with the information on its structure, hence eliminating the above problems. By analogy to the UV/Vis chromatogram, the mass spectrometer generates a mass chromatogram (Figure 1.3) having several features:

- It provides the relationship between the retention time of the substance and the intensity of the peaks on the spectrum (quantitative analysis).
- It also provides information on substances eluted from the column at the same time.
- It moreover provides detailed information about the structure of the compounds (identification of unknown components).

Figure 1.3 shows the retention time on the horizontal axis and the absolute intensity of the signals (a.i.) on the vertical axis. Individual components eluted

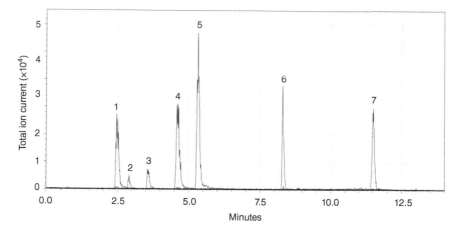

Figure 1.3 Mass chromatogram of several designer drugs separated by the reversed-phase liquid chromatography (LC)-ESI-MS.

from the chromatographic column are recorded by a mass spectrometer. Each of the detected substances is simultaneously characterized by mass spectrum and fragmentation spectrum. Surface under the peaks on a mass chromatogram is a measure of the concentration of individual elements in a sample.

An important feature of modern equipment is also very high accuracy in determining the weight of the substance, reaching the sixth decimal place. This allows to perform measurements with a resolution much higher than that necessary to determine the loss of one electron by a molecule! The following chapters of this book describe, in more details, the variety of techniques used in mass spectrometry and their applications in various areas of our life.

2

A Brief History of Mass Spectrometry

Marek Smoluch[1] and Jerzy Silberring[1,2]

[1] Department of Biochemistry and Neurobiology, Faculty of Materials Science and Ceramics,
AGH University of Science and Technology, Kraków, Poland
[2] Centre of Polymer and Carbon Materials, Polish Academy of Sciences, Zabrze, Poland

Mass spectrometry (MS) is almost 120 years old, but despite its age, it is still an extremely attractive technique. Let us compare here its birth and development to the fate of the title character of the novel *The Curious Case of Benjamin Button* by F.S. Fitzgerald. The main character becomes younger and younger over time. This is somehow similar to MS, still showing new faces and possibilities.

The origin of development of this technique is considered to be 1897, when the British physicist J.J. Thomson conducted research on tubular radiation and, at that time, he experimentally confirmed the existence of an electron by estimating its m/z value. This resulted in the construction, in 1912, of a device called *the parabola spectrograph* for spectra measurements of O_2, N_2, CO, and CO_2. F.W. Aston, Thomson's assistant, applied this technique for studying isotopic compositions and, in 1919, presented another construction called a *mass spectrograph*. From this time, there was a significant increase in the awareness of the equipment, which can quickly determine isotopic composition of the elements. The breakthrough in MS dates back to the Second World War. In 1939, the Manhattan Project management asked A. Nier for help to solve an important issue: which uranium isotope is responsible for the fission reaction and how to gather the necessary material for further work. The answer to this question was crucial to enable a possibility to construct the atomic bomb. Nier had a machine built by E. Lawrence, which was based on a magnetic analyzer, a technique that was not efficient in separation of isotopes for military purposes. Around the same time, MS was used in petrochemical industry to evaluate the components of crude oil. This branch of industry had a lot of money, which was always driving the advancement of technology. The first

Mass Spectrometry: An Applied Approach, Second Edition. Edited by Marek Smoluch,
Giuseppe Grasso, Piotr Suder, and Jerzy Silberring.

spectrometers offered relatively small measuring capacities, the *m/z* range was about 70, and the spectrum recording time was about 20 minutes. During this period, the instruments were complicated, and only the chosen were able to manage those "black boxes." With the development of computers driven by the very complex operating systems (e.g. UNIX), they became increasingly incomprehensible to scientists. The basis for further and rapid development became commercialization of the production. The apparatus no longer had to be built on its own, but it could simply be purchased.

Postwar applications of MS were focused on the analysis of the low molecular weight compounds due to the lack of ionization techniques suitable for the studies of higher molecular masses, predominantly peptides and proteins. The primary source of ions was electron ionization (EI). However, it is worth mentioning here that gas chromatography–mass spectrometry (GC-MS) systems were also utilized to analyze amino acids and peptides.

The breakthrough came in 1981 when Michael Barber from Manchester developed the *fast atom bombardment* (FAB). For the first time, scientists could analyze biological compounds (including peptides and lipids) in solutions and not only in the gas phase. An additional advantage of this technique was the spontaneous fragmentation leading to assignment of the amino acid sequence. The problem was the presence of glycerol, effectively contaminating the ion source that had to be thoroughly cleaned at least once a week, and in the case of a source connected to the liquid chromatography (LC) (continuous-flow FAB), cleaning routine had to be carried out daily. In parallel, the analysts had at their disposal the thermospray ionization, the protoplast of the electrospray method, but the source was operated under high vacuum.

Another breakthrough in the development of ionization techniques was made in the mid-1980s of the previous century by the introduction of electrospray technique (J. B. Fenn with the team, 1984) and matrix-assisted laser desorption/ionization (MALDI) in 1985, with the name given by the creators of the source (M. Karas and F. Hillenkamp). It was further developed by K. Tanaka, who applied it to the analysis of higher molecular compounds and obtained the Nobel Prize for his achievements, together with J.B. Fenn (for electrospray ionization [ESI]) and K. Wüthrich (for NMR). ESI and MALDI are complementary techniques, and their main advantage is the ability to analyze compounds in a very wide range of masses. ESI was the first method operating at atmospheric pressure, enabling direct coupling of separation techniques and introduction of the sample in solution.

Initially, quadrupole or sector analyzers were used, and ion traps were introduced in 1983. Ion traps were unwillingly accepted by the world of scientists, because of the very low resolution, which then reached the value of only 50–100! The promising designs included the time-of-flight (TOF) analyzers most commonly linked to the MALDI source. The initial TOF constructions were also characterized by a low resolution on the order of 50; however, the rapid development of electronics and introduction of delayed extraction and ion reflectron have already made it possible to reach a resolution of 10–15 000 by the end of the 1980s. Interestingly, the TOF was the first to be used in conjunction with the gas chromatograph in the 1950s of the twentieth century. Much later TOF was replaced with quadrupoles and traps. The problem was the lack of sufficiently fast detectors capable of counting the rapidly passing ions. It is also worth mentioning the systems for controlling the spectrometers and methods of mass spectra acquisition. Initially the photographic plates were applied, later thermosensitive paper printouts, and next computer systems controlled by software incomprehensible for the ordinary users. Anyone who controlled mass spectrometer with the PDP-11 or used UNIX commands knows exactly what we are talking about.

In the mid-1980s of the twentieth century, alongside the powerful machines weighing several hundred kilograms, miniaturization of the equipment began. One of the first constructions was the MALDI-TOF spectrometer, which could be placed on a laboratory bench (*benchtop*). This machine was designed by Vestec and led by Marvin Vestal, a genius in this field.

In parallel with the development of the construction, MS applications have been published in various fields of science and technology. One of the pioneers was Fred McLafferty, who described the gas-phase rearrangements, named after him. McLafferty, along with F. Turecek, has made a history as authors of a book describing the fragmentation pathways that are the basis of every MS operator. A great contribution should be attributed to Howard Morris for his work on the analysis of peptides and methods of their sequencing and identification. This is one of the few authors whose works were written in a way understandable even for the layman. The basis of nomenclature for the resulting peptide fragments is owed to Klaus Biemann and Peter Roepstorff. The latter, along with his assistant M. Mann, showed that protein digestion with a proteolytic enzyme results in the unique fragments for a given protein, which was the basis for identifying proteins using knowledge of the masses of several peptides (peptide mapping). Mann, together with his coworker M. Wilm, later became famous for introducing the nanospray, a technique commonly used in modern analytics.

It is impossible to list here all who contributed to the development of both the technique itself and the applications. The pioneers of the 1980s and even the early 1990s had to intensely convince potential users of advantages of MS over high performance liquid chromatography (HPLC). Many researchers deeply believed that MS would not solve their problems and, in addition, was much more expensive and more complex than the previously used methods, such as HPLC.[1]

---

1 Those interested in further details, please visit the homepage http://www.chemheritage.org/ research/policy-center/oral-history-program/projects/critical-mass/index.aspx which we used. The reader will find fascinating descriptions of the development of mass spectrometry and interviews with scientists and constructors of these devices.

Major Events in the History of Mass Spectrometry

1897 Determination of mass of the electron by J.J. Thomson (Nobel Prize in 1906)
1912 First mass spectrometer (J.J. Thomson)
1934 First double-focusing magnetic analyzer
1946 Discovery of the principle of analysis by measuring the time of flight of ions
1953 Patented quadrupole analyzer and ion trap
1956 Gas chromatography and mass spectrometry (GC-MS) coupling
1966 Chemical ionization discovery; first sequencing of peptides
1968 First description of ionization method by spraying in electric field (electrospray)
1968 Atmospheric pressure ionization (API)
1969 Field desorption (FD)
1974 First Fourier transform ion cyclotron resonance (FT-ICR) mass spectrometry
1980 First thermospray description
1981 Fast atom bombardment (FAB) description
1985 Matrix-assisted laser desorption/ionization (MALDI) discovery
1989 Application of electrospray ionization (ESI) for analysis of biomolecules
1999 No matrix laser desorption (desorption/ionization on porous silicon [DIOS])
1999 Quantitative analysis of proteins with isotope-coded affinity tags (ICAT)
2002 Nobel Prize for Fenn and Tanaka for development of soft ionization techniques for analysis of biomolecules
2002 Application of ion mobility mass spectrometry (IMMS)
2004 Desorption electrospray ionization (DESI)
2004 Application of electron transfer dissociation (ETD) as fragmentation method
2005 Introduction of Orbitrap
2005 Direct analysis in real time (DART)
2013 Demonstration of surgery knife with MS detection (intelligent knife [iKnife])
2014 Human proteome description

3

Basic Definitions

Marek Smoluch and Kinga Piechura

Department of Biochemistry and Neurobiology, Faculty of Materials Science and Ceramics, AGH University of Science and Technology, Kraków, Poland

More definitions can be found in Murray, K.K., Boyd, R.K., Eberlin, M.N. et al. (2013). Definitions of terms relating to mass spectrometry (IUPAC Recommendations 2013). Pure Appl Chem 85 (7): 1515–1609.

Accurate mass an experimentally determined mass of an ion that is used to determine an elemental formula. For ions containing combinations of the elements C, H, N, O, P, S, and the halogens, with mass less than 200 Da, a measurement with 5 ppm uncertainty is sufficient to uniquely determine the elemental composition.

Analyzer part of the mass spectrometer that separates ionized species according to their mass-to-charge ratio (m/z).

Atomic mass unit [AMU; u] a non-SI mass unit that is defined to be 1/12 of the mass of carbon isotope ^{12}C, $1\,u \approx 1.66 \cdot 10^{-27}$ kg.

Average mass the mass of an ion, atom, or molecule calculated using the masses of all isotopes of each element weighted for their natural isotopic abundance.

Base peak the most intense (tallest) peak in a mass spectrum.

Chromatogram a graph showing the result of the dependence of the analytical signal from the chromatograph from the retention time of individual components of the sample. The chromatogram obtained by using a UV detector provides information on retention time, absorbance intensity, area under the peak, and peak symmetry.

Dalton [Da] a non-SI mass unit equal to the atomic mass unit. Often used in biochemistry and molecular biology, especially for proteins. For large molecular compounds, it is useful to use multiples of unit, especially kilodalton, 1 kDa = 1000 Da.

Mass Spectrometry: An Applied Approach, Second Edition. Edited by Marek Smoluch, Giuseppe Grasso, Piotr Suder, and Jerzy Silberring.
© 2019 John Wiley & Sons, Inc. Published 2019 by John Wiley & Sons, Inc.

Daughter ion See Fragment ion.

Deconvolution a mathematical process involving the transformation of the mass spectrum of a substance that has been multiply ionized (charges +2, +3 etc., or −2, −3 etc.) to a mass spectral derivative on which all of the observed ions have a unit charge (charge +1 or −1).

Detector part of the mass spectrometer where ion current is quantitatively transformed in electrical current. The ions are coming to detector after their separation in analyzer.

Dimeric ion an ion formed by ionization of a dimer or by association of an ion with its neutral counterpart such as $[M_2]^{+\bullet}$ or $[M–H–M]^+$.

Electron volt (eV) a non-SI unit of energy defined as the energy acquired by a particle containing one unit of charge through a potential difference of one volt, $1\,eV \approx 1.6 \cdot 10^{-19}\,J$.

Extracted ion chromatogram (EIC) a chromatogram created by plotting the intensity of the signal observed at chosen m/z value or series of values in a series of mass spectra recorded as a function of retention time. EIC allows the calculation of the content of the selected mixture component based on the value of the area under the peak.

Field-free region any region of a mass spectrometer where the ions are not dispersed by magnetic or electric field.

Fragment ion (daughter ion, product ion) an electrically charged product of reaction of a particular parent ion (precursor ion).

Ionization efficiency ratio of the number of ions formed to the number of atoms or molecules consumed in the ion source.

Mass accuracy difference between measured and actual mass. Can be expressed either in absolute or relative terms.

Mass number the sum of the protons and neutrons in an atom, molecule, or ion.

Mass precision root-mean-square (RMS) deviation in a large number of repeated measurements.

Mass range the range of m/z over which a mass spectrometer can detect ions or is operated to record mass spectrum.

Mass resolution the smallest mass difference Δm (Δm in Da or $\Delta m/m$ in, e.g. ppm) between two equal magnitude peaks such that the valley between them is a specified fraction of the peak height.

Mass resolving power ($m/\Delta m$) in a mass spectrum, the observed mass divided by the difference between two masses that can be separated, $m/\Delta m$. The method by which Δm was obtained and the mass at which the measurement was made should be reported.

Mass spectrometer an instrument that measures the m/z values and relative abundances of ions.

Mass spectrometry branch of science that deals with all aspects of mass spectrometers and the results obtained with these instruments.

Mass spectrum a plot of the detected intensities of ions as a function of their *m/z* values.

Metastable ion an ion that is formed with internal energy higher than the threshold for dissociation but with a lifetime long enough to allow it to exit the ion source and enter the mass spectrometer where it dissociates before detection.

Mobile phase the liquid, gas, or fluid in supercritical phase that flows through a chromatography system, moving the materials to be separated over the stationary phase. It should enable detection and keep high purity and stability during the chromatographic run, but it cannot react with components of the sample and stationary phase. The eluent composition during the run can be fixed (isocratic elution) or changeable (gradient elution).

Molar mass mass of one mole of a compound.

Molecular ion an ion formed by the removal of one or more electrons to form a positive ion or the addition of one or more electrons to form a negative ion.

Monoisotopic mass exact mass of an ion or molecule calculated using the mass of the lightest isotope of each element.

MS/MS (tandem mass spectrometry) the acquisition and study of the spectra of the electrically charged products of precursors of *m/z* selected ion or ions or of precursor ions of a selected neutral mass loss.

Multiple reaction monitoring (MRM) See Selected reaction monitoring (SRM).

Neutral loss loss of an uncharged species from an ion during either a rearrangement process or direct dissociation.

Nominal mass mass of an ion or molecule calculated using the mass of the lightest isotope of each element rounded to the nearest integer value and equivalent to the sum of the mass numbers of all constituent atoms.

Precursor ion (parent ion) ion that reacts to form particular fragment ions. The reaction can be unimolecular dissociation, ion/molecule reaction, isomerization, or change in charge state.

Proteome the entire set of proteins that is, or can be, expressed by a genome, cell, tissue, or organism at a certain time.

Parent ion See Precursor ion.

Product ion See Fragment ion.

Protonated molecule an ion formed by interaction of a molecule with a proton and represented by the symbol $[M+H]^+$. The term protonated molecular ion is deprecated; this would correspond to a species carrying two charges. Also the term pseudomolecular ion is deprecated.

Radical ion a radical that carries an electric charge. A positively charged radical is called a radical cation.

Retention time the total time from the injection of the sample to the time of the maximum of the peak of compound elution.

Selected ion monitoring (SIM) operation of a mass spectrometer in which the abundances of one or several ions of specific m/z values are recorded rather than the entire mass spectrum.

Selected reaction monitoring (SRM) data acquired from specific fragment ions corresponding to m/z selected precursor ions recorded via two or more stages of mass spectrometry. SRM can be performed as tandem mass spectrometry in time or in space.

Space charge effect result of mutual repulsion of particles of like charge that limits the current in a charged particle beam or packet and causes some ion motion in addition to that caused by external fields.

Stationary phase the chromatographic column filling present in solid or liquid phase. Interaction with filling allows separation of the mixture components.

Tandem mass spectrometry See MS/MS.

Torr non-SI unit for pressure, 1 torr = 1 mmHg = 1.333 mbar = 133.3 Pa.

Total ion current (TIC) sum of all the separate ion currents carried by the different ions contributing to a mass spectrum.

Total ion current chromatogram chromatogram obtained by plotting the total ion current detected in each of a series of mass spectra recorded as a function of retention time.

Transmission the ratio of the number of ions leaving a region of a mass spectrometer to the number entering that region.

4

Instrumentation

4.1 Ionization Methods

4.1.1 Electron Ionization (EI)

Claudio Iacobucci

Institute of Pharmacy, Martin Luther University Halle-Wittenberg, Halle (Saale), Germany

The electron ionization (EI) was pioneered by Dempster [1, 2] since 1916 and subsequently improved by Bleakney and Nier [3, 4]. After one century from its invention, EI is currently the most common interface between gas chromatography and mass spectrometry and is widely used in organic chemistry for structural determination of small molecules.

The EI source (see Figure 4.1) consists of a heated chamber maintained under high vacuum, while the gaseous analyte is introduced through a sample hole. Additionally, there is a heated filament used for the thermionic emission of electrons, which are subsequently accelerated to 70 eV in proximity of the sample hole. The collision between an accelerated electron and the gaseous molecule of the analyte M causes its ionization to the radical ion $[M]^{+\bullet}$, the so-called molecular ion. During this process, a part of the kinetic energy of the accelerated electron, which is higher than the ionization energy of M, is transferred to the nascent molecular ion. Thus, the resulting high internal energy of the open-shell $[M]^{+\bullet}$ ion induces its extensive fragmentation (Figure 4.2 upper panel). The so-formed product ions as well as the eventually survived $[M]^{+\bullet}$ ion are finally extracted through an exit hole to the MS analyzer. The mass spectra obtained by EI are highly reproducible, allowing the generation of large libraries of spectra for an automated and reliable identification of compounds. Moreover, the efficiency of the EI process provides a high sensitivity without being selective for specific classes of molecules. However, the analyte molecules need to be transferred to the gas phase before EI analysis. Therefore, the molecular weight and the polarity of analyzable compounds are limited. In cases of small polar compounds, the volatility can be increased by specific derivatizations, i.e. chemical reactions increasing their ionization. The range of EI applications has been recently extended by the direct coupling of EI with liquid chromatography (LC) [6].

The fragmentation pathways of open-shell molecular cations depend on the functional groups present in the analytes and have been extensively studied [5, 7].

Mass Spectrometry: An Applied Approach, Second Edition. Edited by Marek Smoluch, Giuseppe Grasso, Piotr Suder, and Jerzy Silberring.

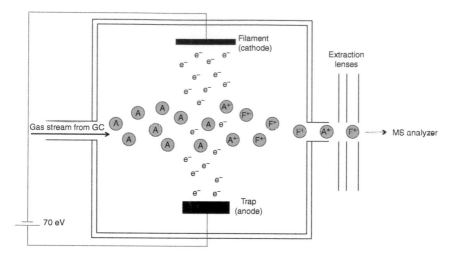

Figure 4.1 Schematic layout of an electron ionization ion source. The accelerated electrons ionize the gaseous analyte molecules (A). The high internal energy of the molecular ions (A$^{+\cdot}$) determines extensive fragmentation processes and the formation of product ions (F$^{+\cdot}$).

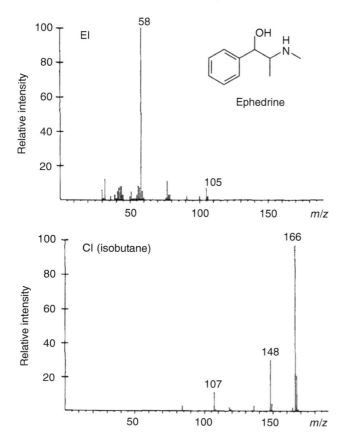

Figure 4.2 Comparison of mass spectra of ephedrine ionized by electron ionization (EI) and chemical ionization (CI) using isobutane as reagent gas. The pseudomolecular ion of ephedrine is clearly visible when CI is used at m/z 166 as [M + H]$^+$ ion. *Source*: Reprinted with permission from Ref. [5].

Dissociation reactions are initiated by either the positive charge or the radical of the molecular ion. While the positive charge induces heterolytic cleavages by attracting an electron pair, the free radical has a strong tendency for electron pairing, thus inducing homolytic cleavages. On top of that, rearrangements, followed by dissociation, are also frequently observed.

The molecular weight determination plays a central role for MS-based structure elucidation of analytes. Since the majority of elements have more than one stable isotope, the elemental composition of the analyte generates specific signal patterns. Beyond the m/z ratio of the molecular ion, the specific isotopic pattern can confirm its identity. Although EI is the most convenient tool for ionizing relatively apolar molecules up to c. 900 u, it does not provide information about the mass and the elemental composition of the intact analyte in a number of cases. The extensive molecular ion fragmentation under EI conditions is caused by the high energy (c. 1400 kJ/mol at 70 eV electron kinetic energy) transferred from an accelerated electron to the neutral analyte. Reducing the acceleration potential of the EI electron beam partially mitigates the over-fragmentation issue but dramatically decreases the ionization efficiency. This, in turn, lowers the sensitivity and negatively affects the reproducibility of product ion mass spectra.

4.1.2 Chemical Ionization (CI)

Claudio Iacobucci

Institute of Pharmacy, Martin Luther University Halle-Wittenberg, Halle (Saale), Germany

In the effort to overcome the limitations of EI, "soft" ionization techniques have been developed since the 1950s of last century. In particular, chemical ionization (CI), pioneered by Talrose [8], has been definitely established as an analytical tool by Munson and Field in the late 1960s [9–12] redefining the boundaries of MS. The CI process involves the transfer of a charge carrier from an ionized gas to the analyte of interest by mild collisions. Thus, the resulting low internal energy of the pseudomolecular ion allows for measuring its intact mass and isotopic pattern (Figure. 4.2 bottom panel).

CI was initially a major breakthrough in gas chromatography–mass spectrometry (GC-MS) and subsequently evolved to atmospheric pressure chemical ionization (APCI), which can be coupled with LC for the analysis of more polar compounds. CI and APCI characteristics provide ideal experimental conditions for a number of analytical tasks, and they have been applied to the large variety of systems in the last 50 years [13, 14].

4.1.2.1 Principle of Operation: Positive and Negative Ion Modes

The CI process is based on the gas-phase ion/molecule reactions between ions, generated by EI of a reagent gas, and the analyte of interest resulting in

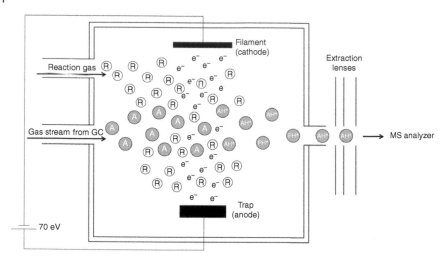

Figure 4.3 Schematic layout of a chemical ionization ion source. The accelerated electrons ionize the molecules of a reagent gas (R). Subsequently, the gaseous analyte molecules are ionized by ion/molecule reactions. The internal energy of pseudomolecular ions (AH$^+$) is sufficient to induce the formation of product ions (FH$^+$).

formation of the secondary ions. Therefore, the CI (see Figure 4.3) source can be seen as an EI source filled with a reagent gas, such as CH$_4$ [8, 11, 15], NH$_3$ [16–19], or isobutane [16, 20–22], at a pressure of c. 10 Pa. Modern EI ion sources can be operated in EI and CI modes. The pressure of the reagent gas is four to six orders of magnitude higher than in the EI source, and the free electron, generated by the heated filament, ionizes almost exclusively the molecules of the reagent gas. The large excess of the reagent gas compared to the analyte effectively shields the latter from ionizing electrons, thus suppressing the EI process. The primary ionization is followed by a cascade of ion/molecule reactions, which finally generate the analyte ions by either *proton transfer, electrophilic addition, anion abstraction*, or *charge exchange* [23]. The ratio of competitive reactions occurring in the plasma strongly depends on the reagent gas employed and on its pressure in the CI source [11, 24].

The plasma generated by methane, the most commonly used reagent gas, contains the highly acidic [CH$_5$]$^+$ ion [15, 25, 26], which is mainly responsible for the secondary ionization of the analyte. In particular, the [CH$_5$]$^+$ ion can protonate the analyte molecule (M). Thus, [M + H]$^+$ ions can be detected in a positive ionization mode. The capability of the analyte to accept the proton from [CH$_5$]$^+$ depends on its proton affinity. Additionally, the [CH$_5$]$^+$ ion generates other reactive species, such as [C$_2$H$_5$]$^+$, upon collision with other methane molecules, which can, in turn, ionize the analyte to $[M + C_2H_5]^+$ by *electrophilic addition*. The formation and reactivity of the $[CH_5]^+$ ion in the plasma can be explained, as a first approximation, by Eqs. (1)–(5) [11, 13, 27]:

$$CH_4 + e^- \rightarrow [CH_4]^{+\bullet}, [CH_3]^+ \tag{1}$$

$$CH_4 + [CH_4]^{+\bullet} \rightarrow [CH_5]^+ + CH_3^\bullet \tag{2}$$

$$CH_4 + [CH_3]^+ \rightarrow [C_2H_7]^+ \rightarrow [C_2H_5]^+ + H_2 \tag{3}$$

$$M + [CH_5]^+ \rightarrow [M+H]^+ + CH_4 \tag{4}$$

$$M + [C_2H_5]^+ \rightarrow [M+C_2H_5]^+ \tag{5}$$

On the contrary, a common ionization process by *anion abstraction* is the hydride abstraction, which yields $[M – H]^+$ ions and is frequently observed for aliphatic alcohols [23]. Ionization by *charge exchange* is caused by the presence of open-shell species in the reagent plasma. The resulting analyte radical cations resemble the ones occurring in EI and follow different dissociation rules compared to even-electron cations formed by CI [23].

Different types of anions coexist alongside the abovementioned cations in the gas phase and can be extracted with a negative potential by the MS anodic orifice. In negative CI, an analyte M can be ionized by two types of ion/molecule reactions, namely, *proton transfer* ($[M – H]^-$ ions) and *anion attachment* ($[M + X]^-$ ions). While the deprotonation is predominant for analytes, which comprise acidic groups, the *anion attachment*, usually Cl^-, requires the use of halogenated reagent gases [28–31].

Additionally, the analyte can collide with one thermal electron in the plasma and can directly be ionized by *electron capture* (EC) [29, 31–34]. In this case, three different types of molecular ions can be formed, depending on the stability of the $[M]^{-\bullet}$ open-shell anion (Eqs. 6–8) [23]:

$$M + e^- \rightarrow [M]^{-\bullet} \tag{6}$$

$$M + e^- \rightarrow [M-A]^- + A^\bullet \tag{7}$$

$$M + e^- \rightarrow [M-B]^- + [B]^+ + e^- \tag{8}$$

The efficiency of EC depends on the electron affinity of the target compound. The high electronegativity of halogen atoms (Figure 4.4) enables to analyze polychlorinated biphenyls (PCBs) and fire retardants with high sensitivity and selectivity [33].

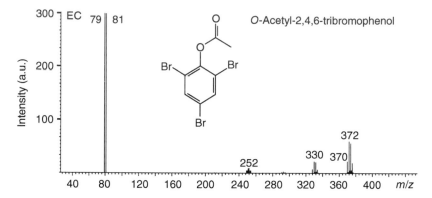

Figure 4.4 Spectrum of *O*-Acetyl-2,4,6-tribromophenol ionized by electron capture (EC) using methane as reagent gas. The isotopic pattern of [M]⁻· at 370 m/z testifies the presence of three bromine atoms in the molecular ion. 2,4,6-Tribromophenol is present in Lagavulin 16-year-old whisky. *Source*: Adapted from Ref. [35].

Questions

- What is the method of ionization in the EI source?
- Discuss the main limitations and advantages of EI.
- What is the method of ionization in the CI source?
- What is the meaning of the term "soft ionization method?" Which ionization method is "softer," EI or CI?
- What is the main feature of the target compound to be considered in order to evaluate the efficiency of the EC process?
- What is APCI and how it works?

References

1 Dempster, A. (1916). LII. The ionization and dissociation of hydrogen molecules and the formation of H3. *The London, Edinburgh, and Dublin Philosophical Magazine and Journal of Science 31*: 438–443.

2 Dempster, A. (1918). A new method of positive ray analysis. *Physical Review 11*: 316–325.

3 Bleakney, W. (1929). A new method of positive ray analysis and its application to the measurement of ionization potentials in mercury vapor. *Physical Review 34*: 157–160.

4 Nier, A. (1947). Electron impact mass spectrometry. *The Review of Scientific Instruments 18*: 415.

5 Fales, H.M., Milne, G.W.A., Winkler, H.U. et al. (1975). Comparison of mass spectra of some biologically important compounds as obtained by various ionization techniques. *Analytical Chemistry 47*: 207–219.

6 Cappiello, A., Famiglini, G., Pierini, E. et al. (2007). Advanced liquid chromatography– mass spectrometry interface based on electron ionization. *Analytical Chemistry 79*: 5364–5372.

7 Märk, T.D. and Dunn, G.H. (2013). *Electron Impact Ionization*. Vienna: Springer.

8 Talrose, V.L. and Lubimova, A.K. (1998). Secondary process in the ion source of the mass spectrometer (Reprint From 1952). *Journal of Mass Spectrometry 33*: 502–504.

9 Munson, M.S. and Field, F. (1965). Reactions of gaseous ions. XV. Methane+ 1% ethane and methane+ 1% propane. *Journal of the American Chemical Society 87*: 3294–3299.

10 Munson, M. (1965). Proton affinities and the methyl inductive effect. *Journal of the American Chemical Society 87*: 2332–2336.

11 Munson, M.S. and Field, F.-H. (1966). Chemical ionization mass spectrometry. I. General introduction. *Journal of the American Chemical Society 88*: 2621–2630.

12 Munson, B. (2000). Development of chemical ionization mass spectrometry. *International Journal of Mass Spectrometry 200*: 243–251.

13 Richter, W. and Schwarz, H. (1978). Chemical Ionization: a mass spectrometric analytical procedure of rapidly increasing importance. *Angewandte Chemie 90*: 449–469.

14 Harrison, A.G. (1992). *Chemical Ionization Mass Spectrometry*. Boca Raton, FL: CRC Press.

15 Heck, A.J., de Koning, L.J., and Nibbering, N.M. (1991). On the structure of protonated methane. *Journal of the American Society for Mass Spectrometry 2*: 453–458.

16 Takeda, N., Harada, K.I., Suzuki, M. et al. (1982). Application of emitter chemical ionization mass spectrometry to structural characterization of aminoglycoside antibiotics—2. *Organic Mass Spectrometry 17*: 247–252.

17 Hancock, R. and Hodges, M. (1983). A simple kinetic method for determining ion-source pressures for ammonia CIMS. *International Journal of Mass Spectrometry and Ion Physics 46*: 329–332.

18 Keough, T. and DeStefano, A. (1981). Factors affecting reactivity in ammonia chemical ionization mass spectrometry. *Journal of Mass Spectrometry 16*: 527–533.

19 Hunt, D.F., McEwen, C.N., and Upham, R.A. (1971). Chemical ionization mass spectrometry II. Differentiation of primary, secondary, and tertiary amines. *Tetrahedron Letters 12*: 4539–4542.

20 Milne, G., Fales, H., and Axenrod, T. (1971). Identification of dangerous drugs by isobutane chemical ionization mass spectrometry. *Analytical Chemistry 43*: 1815–1820.

21 McGuire, J.M. and Munson, B. (1985). Comparison of isopentane and isobutane as chemical ionization reagent gases. *Analytical Chemistry 57*: 680–683.

22 Maeder, H. and Gunzelmann, K. (1988). Straight-chain alkanes as reference compounds for accurate mass determination in isobutane chemical ionization mass spectrometry. *Rapid Communications in Mass Spectrometry 2*: 199–200.

23 Gross, J.H. (2011). *Mass Spectrometry: A Textbook*. Berlin, Heidelberg: Springer-Verlag.

24 Drabner, G., Poppe, A., and Budzikiewicz, H. (1990). The composition of the CH4 plasma. *International Journal of Mass Spectrometry and Ion Processes 97*: 1–33.

25 Mackay, G., Schiff, H., and Bohme, D. (1981). A room-temperature study of the kinetics and energetics for the protonation of ethane. *Canadian Journal of Chemistry 59*: 1771–1778.

26 Fisher, J., Koyanagi, G., and McMahon, T. (2000). The C2H7+ potential energy surface: a Fourier transform ion cyclotron resonance investigation of the reaction of methyl cation with methane1. *International Journal of Mass Spectrometry 195*: 491–505.

27 Field, F. and Munson, M. (1965). Reactions of gaseous ions. XIV. Mass spectrometric studies of methane at pressures to 2 torr. *Journal of the American Chemical Society 87*: 3289–3294.

28 Dougherty, R.C. and Weisenberger, C. (1968). Negative ion mass spectra of benzene, naphthalene, and anthracene. A new technique for obtaining relatively intense and reproducible negative ion mass spectra. *Journal of the American Chemical Society 90*: 6570–6571.

29 Dillard, J.G. (1973). Negative ion mass spectrometry. *Chemical Reviews 73*: 589–643.

30 Bouma, W.J. and Jennings, K.R. CIMS of explosives. *Organic Mass Spectrometry 16*: 330–335.

31 Budzikiewicz, H. (1986). Negative chemical ionization (NCI) of organic compounds. *Mass Spectrometry Reviews 5*: 345–380.

32 Ardenne, M., Steinfelder, K., and Tümmler, R. (2013). *Elektronenanlagerungs-Massenspektrographie organischer Substanzen*. Berlin, Heidelberg, New York: Springer-Verlag.

33 Dougherty, R.C. (1981). Negative chemical ionization mass spectrometry. *Analytical Chemistry 53*: 625–636.

34 Hunt, D.F. and Crow, F.W. (1978). Electron capture negative ion chemical ionization mass spectrometry. *Analytical Chemistry 50*: 1781–1784.

35 Bendig, P., Lehnert, K., and Vetter, W. (2014). Quantification of bromophenols in Islay whiskies. *Journal of Agricultural and Food Chemistry 62*: 2767–2771.

4.1.3 Atmospheric Pressure Ionization (API)

4.1.3.1 Atmospheric Pressure Chemical Ionization (APCI)

Claudio Iacobucci

Institute of Pharmacy, Martin Luther University Halle-Wittenberg, Halle, Germany

Atmospheric pressure chemical ionization (APCI) [36–38], which derives from merging CI and electrospray ionization (ESI), relies on the ion/molecule reactions at atmospheric pressure to generate ions of apolar or slightly polar species, usually up to c. 1500 u [39]. APCI can serve as an interface, complementary to ESI, between liquid chromatography (LC) and mass spectrometry (MS) and has a number of applications, such as steroid analysis [40].

The sample solution is introduced through a capillary in the APCI chamber and nebulized by a coaxial jet of nitrogen. The so-formed aerosol is heated to further vaporize the initial droplets and push the analyte in the direction of a metal pin by the gas stream. The metal pin, also known as corona pin, is maintained at a few kV potential relative to the exit counter electrode, causing an electric discharge of several μA. In APCI sources (Figure 4.5), the corona discharge plays the same role as heated filament in CI sources, which in this case cannot be employed under atmospheric pressure conditions. Free electrons ionize N_2 molecules, thus inducing a cascade of charge transfer reactions (Eqs. 9–13) and the formation of a

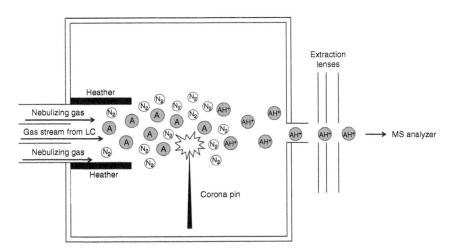

Figure 4.5 Schematic layout of an atmospheric pressure chemical ionization ion source. The electrons generated by the corona discharge ionize the nebulizing nitrogen gas. Subsequently, the vaporized analyte molecules are ionized by ion/molecule reactions.

plasma. Finally, the analyte is ionized upon collision with a protonated molecule of the polar solvent employed, e.g. water:

$$N_2 + e \rightarrow N_2^+ + 2e \tag{9}$$

$$N_2^{+\bullet} + N_2 \rightarrow N_4^{+\bullet} \tag{10}$$

$$N_4^+ + H_2O \rightarrow H_2O^+ + 2N_2 \tag{11}$$

$$H_2O^+ + H_2O \rightarrow H_3O^+ + OH^\bullet \tag{12}$$

$$H_3O^+ + M \rightarrow MH^+ + H_2O \tag{13}$$

APCI is a soft ionization technique whereby the protonated molecular ions have to be subsequently fragmented by tandem MS.

4.1.3.2 Electrospray Ionization (ESI)

Piotr Suder

Department of Biochemistry and Neurobiology, Faculty of Materials Science and Ceramics, AGH University of Science and Technology, Kraków, Poland

4.1.3.2.1 Introduction

Electrospray is presently (along with matrix-assisted laser desorption/ionization [MALDI]) the most popular ion source used in biological sciences. This source can be relatively easily connected to virtually any analyzer used in modern instruments. Simple design, and therefore, small cost of production, reliability, and ease of use also contribute to its popularity. The main advantages of ESI based on its analytical capabilities are listed below:

1) *Atmospheric pressure ionization*
   Electrospray ionizes a sample under atmospheric pressure. This is one of the most important features, as it enables analysis of substances in solution, without significant preliminary pretreatment (e.g. evaporation, crystallization, precipitation). Thus, it is possible to easily link the mass spectrometer to high pressure liquid chromatography (HPLC) system, capillary electrophoresis, and related techniques.
2) *Multiply charged ions*
   During ionization by ESI, more than one proton can be attached or detached from the molecule. This phenomenon is most common for larger molecules (usually bigger than 200–300 Da) and depends on the number of polar groups in the protein/peptide sequence. It should be noted that the spectrometer measures the mass-to-charge ratio (m/z). Thus, an increase in "z" in m/z ratio allows for observation of high molecular mass molecules in the low m/z range along the mass spectra. This makes possible to apply almost any analyzer and identify high molecular weight (MW) substance in a relatively low range of m/z. For the multiply charged ions, it is also possible to

determine their masses with very high precision. As the ions appear at the low m/z range, also resolution is significantly improved (Section 4.1.3.2.5.2).

3) *Soft ionization technique*

Electrospray is the so-called "soft" ionization method. It means that during MS analysis, unwanted fragmentation is usually not observed (in contrast to electron impact [EI]). By applying appropriate analytical conditions (voltages and temperature in the ion source), we may also measure MW of non-covalent complexes [41].

4.1.3.2.2 ESI Principle of Operation

Analyte in the solvent is introduced into the ionization chamber at a controlled flow rate (usually from 1 to 100 µl min^{-1}) through the inlet capillary (Figure 4.6). Due to the high voltage generated between the ESI tip and MS inlet, the solvent reaching the capillary end forms the Taylor cone (Figure 4.6). Spraying of the solvent is supported by the sheath gas (e.g. nitrogen or air), at a flow rate of 5–15 l min^{-1} and pressure between 0.5 and 2.0 bar. Gas flow rate depends on the ion source design and varies significantly between constructions.

Generation of Taylor cone allows for the massive release of tiny droplets of solvent, containing analyte molecules, thus leading to the formation of a jet. As surface tension of the liquid counteracts droplets formation, a relatively high voltage (up to 5 kV) is necessary for proper operation of the ion source. Such influence of electric potential on liquids was described by Sir Geoffrey Taylor in 1964. Formation of the Taylor cone is useful in various branches of sciences (electrospinning of nanofibers, electrospraying in analytical sciences, space exploration, etc.). To achieve effective ionization, it is necessary to dry droplets before they enter the analyzer. As the distance is short (millimeters to centimeters), evaporation process must be very effective. Therefore, sheath gas is heated (100–350 °C) to provide additional thermal energy for evaporation. When the flow rate is very low, sheath gas in ESI may not be necessary, but at higher flow rates, the gas stabilizes Taylor cone and supports efficient drying of the droplets traveling

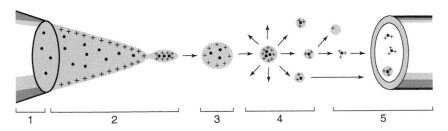

Figure 4.6 Electrospray source: (1) capillary delivering analyte, (2) Taylor cone, (3) aerosol droplet, (4) droplet disintegration ("Coulomb explosion"), and (5) ions' desolvation and heated capillary inlet (black dots = analyte ions).

between capillaries. Droplets migrating through the ion source are depleted from solvent molecules, which causes concentration of the charge at the droplet's surface. This leads to droplet instability, when Coulomb repulsion forces overcome surface tension. Following the next phase, called "Coulomb explosion," the arising smaller droplets recover stability until the moment when the remaining part of solvent evaporates. Coulomb explosions last until complete evaporation of solvent is achieved and dry ions are released [42]. Only such ions can be analyzed in the mass spectrometer (see Figure 4.6). The whole process lasts no longer than micro- or milliseconds and occurs in the ion source and partially in the heated capillary (if this is a part of source construction). It is worth to mention that ions are guided into the analyzer by two forces: electric potential and higher vacuum in the internal part of the ion source.

Heated capillary, independently from its main function (drying of the ions), is the main connecting device between external atmospheric pressure and internal parts of the mass spectrometer where high vacuum is necessary to avoid unwanted collisions of the ions with air components. Due to a high difference of the pressure values between the capillary and the instrument, the beam of ions is directed to the analyzer at the rate close to the sonic speed.

Formation of the ions in ESI sources can be affected by a set of factors, such as solvent properties, nature of the ionized molecule, physical conditions of the ionization process, etc. Often, the knowledge on such settings can be beneficial. The most important factors are described in the following subchapters.

4.1.3.2.3   Electrospray: Working Principles

The analyte dissolved in the appropriate solution is sprayed in the ionization chamber, forming a cloud of tiny droplets (Figure 4.6). Capillary axis is usually tilted at 60–90 °C toward heated capillary axis (orthogonal setup). It seems to be unfavorable setup, as majority of the spray should not enter the instrument, but in fact, almost independently on the angle between capillaries, the spray is guided by the potential difference reaching values 2–5 kV.

Spraying capillary is usually grounded, having potential 0, but heated capillary (or its cover) has high negative potential in case of the positively charged ions (positive ion mode) or high positive potential for the negatively charged ions (negative ion mode). Due to potential difference, spray droplets containing charged ions are guided directly to the heated capillary inlet. Of course, it is possible to ground this capillary and generate an appropriate potential on the spraying capillary – it mainly depends on the manufacturer's setup.

Orthogonal (or close to orthogonal) arrangement causes less non-ionized impurities to reach internal parts of the ion source in comparison to the coaxial approach. All impurities, which do not carry the charge, have enough kinetic energy to travel straight, without deflection, toward inlet capillary. This is the reason why they stuck in the ionization chamber (which must be regularly cleaned) without entering the mass spectrometer in the orthogonal setup. Similar cleaning process is used inside the instrument for remaining, small amounts of dirt, driven into MS by the gas stream. In the instrument construction, so-called skimmers, funnels, beam guides, or square quadrupoles with devices removing non-ionized impurities (neutral blockers) are applied to prevent the system from contamination.

An interesting phenomenon, already mentioned earlier in this chapter, is size reduction of the solvent droplets during their flight between capillaries in the ion source. This effect is strongly connected with fast evaporation of the pure solvent present in each droplet. In the positive ion mode, droplet surface predominantly concentrates positively charged analyte molecules. Evaporation of the solvent causes concentration of the positive charges, which increases repulsion of the charges, by magnifying Coulomb forces (Eq. 1) between them. When diameter of the droplet constantly decreases, Coulomb forces, upon critical conditions, can overcome surface tension forces, thus tearing the droplet into the smaller ones. This is characterized by Eq. (1):

$$F_C = k \left(\frac{Q_1 \times Q_2}{r^2} \right) \tag{1}$$

where k is the constant coefficient, Q_1, Q_2 are ion charges, and r is the distance between charges.

The process is repeated several times and is limited by the amount of solvent and analyte molecules. Boundary diameter of the droplets can be calculated using the Rayleigh equation (Eq. 2). After reaching the smallest possible diameter, droplets release the so-called "dry" ions (see Figure 4.6):

$$q^2 = 8\pi^2 \varepsilon_0 \gamma d^3 \tag{2}$$

where q is the electric charge, ε_0 is the electrical permeability, γ is surface tension, and d is the droplet diameter.

It seems that in some setups, additional heating of drying gas (nitrogen or air), as well as high temperature of the heated capillary (up to 360 °C), should trigger dissociation of covalent bonds in the analytes, but in practice such phenomenon is not observed. This additional thermal energy is consumed to evaporate solvent during its very short travel between capillaries. It is important to select an appropriate temperature in relation to other parameters, such as the type of solvent (more volatile = lower temperature), flow rate of the solution (higher flow rate = higher temperature), or analyte stability (e.g. analysis of non-covalent complexes requires lower temperature). Usually, temperatures of the heated capillary (if present) and drying gas should be chosen empirically.

Dry ions (depleted of solvent) are traveling inside the inlet capillary almost at the sonic speed. At least two factors regulate their high velocity:

1) Fast drying gas flow between external (ambient pressure) and internal (low-pressure) parts of the ion source. Typical difference between pressures in the

ambient and low-pressure parts of the ion source reaches three levels of magnitude (externally c. 101.3 kPa, internally c. 100 Pa). Pressure difference regulates the speed of traveling ions in this part of instrument. Under normal working conditions, this parameter does not significantly influence the overall spectra quality, but too high pressure in the ion source can deteriorate spectrum due to collisions of the ions with air components. MS manufacturers usually do not allow to change default pressure settings in their constructions.

2) Potential difference between capillaries and skimmers, which directs and accelerates ions toward the mass spectrometer inlet. This parameter may be of additional use as an alternative way of fragmentation (in source decay). For such purpose, only pure substance can be subjected for fragmentation, as all ions present in the source will be decomposed.

It should also be noted that pressure in the remaining compartments of the ion source (just behind inlet capillary) is still too high and does not allow for proper work of MS (analyzers and detectors). Mass spectrometer equipped with the electrospray ion source has very efficient pumping system providing vacuum to achieve parameters necessary for the proper operation. Usually, it is a group of two or more rotary pumps and high performance turbomolecular pumps providing pressure gradient in the consecutive parts of the instrument, reaching 1×10^{-5} mbar in the ion traps or even 1×10^{-8} mbar in time-of-flight (TOF) analyzers. Such low pressure is beneficial for the sensitivity of MS analysis as it minimizes undesirable collisions between analyte ions and gas molecules (air components), which may lead to unwanted fragmentation. Also, residual solvents are pumped out along with the remaining non-ionized impurities. Under such conditions, the ions can be easily guided by electric or magnetic fields generated inside MS to the analyzer (or analyzers) and then detected in the last part of the MS – detector.

4.1.3.2.4 Properties of the Solvent

One of the most important factors of ESI solvents is their capability to attract or remove protons. If it is thermodynamically possible for the solvent to accept or release proton(s), ionization may occur without additional supplementation with acids or bases. When protic solvents (containing labile protons) are used, physical parameters in the ion source, like potential difference between capillaries or elevated temperature, will be responsible for ion formation. In opposition to protic solvents, aprotic ones prevent from spontaneous ions formation. In other words, if the analyte dissolves in nonpolar aprotic solvents only, it will be necessary to introduce additional constituents, which effectively support ionization process, for example, small quantities of protic solvents to supplement solution used to dissolve the sample.

The most common solvents (mobile phase) applied in ESI are water, alcohols (methanol, ethanol, *n*-propanol, isopropanol), nitriles (acetonitrile), and trichloromethane (usually mixed with methanol). Depending on the positive or negative ion modes, frequently addition of preferred acids or bases is considered. Concentration of such additives is usually low (0.01–0.1% v/v). Working in the positive ion mode, function of proton donor is provided by acids (formic or acetic). Trifluoroacetic acid (TFA) should be avoided as it may cause ion suppression. In case of the negative ion mode, small addition of various bases is applied (the most popular used is NH_{3aq}).

Solvents listed above, like alcohols, water, and nitriles, are predominantly applied during ESI-MS analyses. They are selected due to their physicochemical properties described below:

1) *Low boiling point temperature*

   As the solvent must be effectively evaporated from the sample during ionization process, it is common to avoid solvents having high boiling point temperatures (Table 4.1) or difficult to evaporate (low values of partial pressures of their vapors). Solvents with high boiling point temperatures usually do not evaporate completely in the ion source. They are transferred in significant amounts through the heated capillary into the low-pressure section of the mass spectrometer. Too high quantities of liquid in the low-pressure area or even in the analyzer effectively influence pressure conditions inside the instrument, which in turn significantly impacts spectra quality or may even destroy internal parts of an instrument due to eventual shortcuts.

2) *Low surface tension*

   This parameter has also big impact on the final spectra quality in ESI-MS. At a decreased solvent surface tension, it is easier to reach instability

Table 4.1 Properties of the selected solvents applied during electrospray ionization.

Solvent	Boiling point temperature (°C) (at ambient pressure)	Surface tension $(10^3 \, J \, m^{-2})$ (for +25 °C)
Acetonitrile	81.6	28.45
Dimethylformamide (DMF)[a]	152.3	36.3
Dimethyl sulfoxide (DMSO)[a]	189.0	43.0
Ethanol	78.3	22.0
Methanol	64.7	22.1
Propanol	97.5	23.4
Tetrahydrofuran (THF)	65.4	26.4
Trichloromethane	61.2	26.6
Water[a]	100.0	71.98

[a] Usually mixed with other solvents.

of droplets during ionization process and initiate their decay upon Coulomb explosion (see also Eq. (2): Rayleigh model). In ESI-MS analyses pure water is unflavored due to its relatively high surface tension (Table 4.1). To omit this problem, water is usually mixed with other solvents (e.g. methanol, acetonitrile) to decrease surface tension and make the analysis more effective.

Solution containing the analyte should be free from molecules having low ionization energy or completely dissociated, detergents at high concentrations, and insoluble materials (low MW involatile salts, metal ions), except for acids/bases being proton donors/acceptors promoting ionization process. These requirements are important for few reasons:

1) Additional ions in solution (especially salts) can disturb spraying process in the ionization chamber.
2) Some metal ions, interacting with the analyte, can form clusters of unpredictable charge, which makes spectra interpretation difficult or impossible. Moreover, formation of adducts generates several ions of the same substance, which leads to decreased sensitivity (the signal is "divided" among all adducts).
3) Substances fully dissociated in solution, having their own charge, can block ionization or compete for free charges in solution, what can affect ionization process.

The above phenomena are together referred to as an ion suppression effect, which unfavorably influences ionization process. As a consequence, it is strongly recommended to avoid the presence of these substances during sample preparation; otherwise sample purification might be necessary before analysis to receive satisfactory results.

It should be stated here that there are some exceptions from the rules given above. Addition of some salts like silver trifluoroacetate (AgTFA) might be necessary during analysis of substances, which cannot be ionized under typical conditions. For example, when the analyte does not have nucleophilic center capable of accepting free proton (like aliphatic hydrocarbons), formation of a non-covalent metal ion adduct (with silver) allows for detection of such molecules. Similar procedures can be applied during analyses of nonpolar polymers, like polystyrene, polyethylene, etc.

4.1.3.2.5   *Principles of ESI-MS Spectra Interpretation*

Mass spectrometers equipped with electrospray ion sources can provide a huge quantity of data generated during analysis. Below, basics of data interpretation are given. It is important to keep in mind that it is possible to collect much more information about the analytes than presented below.

4.1.3.2.5.1 Low MolecularWeight Molecules Figures 4.7–4.9 show mass spectra of a small MW substance obtained by ESI-MS.

Figure 4.7 presents spectrum of the peptide of MW = 691.4 Da, analyzed in the positive ion mode. Peptide concentration was $0.1\,\mu mol\,ml^{-1}$ in the acidified water/methanol solution (69.9% H_2O + 30% CH_3OH + 0.1% HCOOH v/v/v). Peptide sequence was RLWAF (Arg-Leu-Trp-Ala-Phe). Independently on the 692.4^{1+} pseudomolecular ion, there is also a small quantity of impurities in the lower m/z range, and a peak at m/z = 675.4 probably comes from the loss of $-NH_3^+$ group. Insert in this figure shows isotopic distribution of the main peak: distances between isotopic peaks are equal to 1 Da, which indicates that charge per molecule is equal to +1. Doubly and triply charged species and higher are not visible. It can be thus assumed that the molecule is a ligand accepting only single proton. Additional signal at m/z 545.2 is a result of spontaneous fragmentation in the ion source (removal of C-terminal phenylalanine), or the substance was contaminated by peptide at the same sequence without C-terminal Phe (which is quite possible during peptide synthesis).

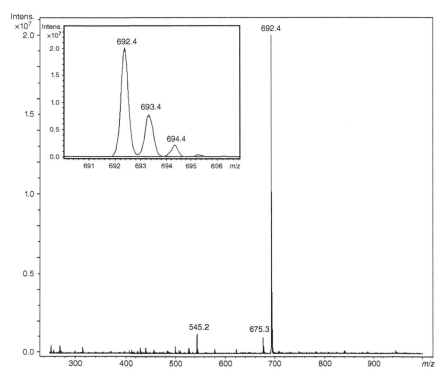

Figure 4.7 MS spectrum of a small MW peptide (RLWAF, positive ions). Insert shows the clearly visible isotopic pattern of the pseudomolecular ion of the most abundant peak on the spectrum.

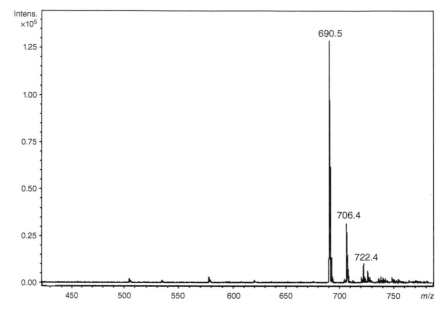

Figure 4.8 Mass spectrum of the peptide RLWAF (see also Figure 4.7) in the negative ion mode.

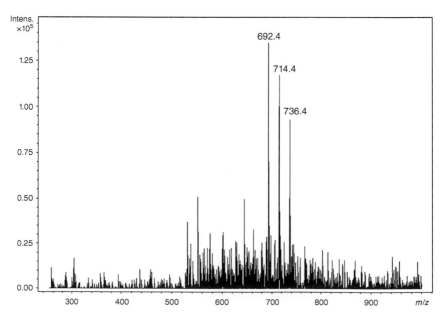

Figure 4.9 Ion suppression effect in the ESI ion source.

Mass spectrum shown in Figure 4.8 was generated at similar conditions to those from Figure 4.7 with few exceptions: the same peptide was dissolved in alkaline solvent (69.9% H_2O + 30% CH_3OH + 0.125% NH_{3aq}; v/v/v), and the spectrum was acquired in the negative ion mode. The pseudomolecular ion arises after removal of a single proton from the peptide molecule. Also visible are $-NH_2$ adducts at m/z 706.4 and 722.4. This also indicates that spectra taken in different ion modes may not be identical, which should be taken into consideration during complete analysis of a given molecule.

Figure 4.9 presents spectrum of RLWAF analyzed in the positive ion mode, similarly to data from Figure 4.7. Instead of pure solvents, the peptide was dissolved in the phosphate buffered saline (PBS) consisting of 50 mM NaH_2PO_4, 50 mM Na_2HPO_4, 1.4 mM KCl, and 70 mM NaCl (analyte solution contained 50% PBS and 50% acidified water/methanol like in the case of analysis illustrated in Figure 4.7).

Very low intensity of the ions can be observed (as compared with Figures 4.7 and 4.8). Additionally, sodium adducts (+22 Da) are clearly visible: a peak at m/z = 714.4 can be defined as $[M + Na]^+$, while the signal at m/z 736.4 belongs to $[M + 2Na - H]^+$. In both cases M is the molecular weight of the peptide.

4.1.3.2.5.2 Substances of Higher Molecular Masses ESI is a technique allowing for determination of higher MW substances in a low m/z range, due to their capability of attracting/detaching more than one protons to/from the molecule. m/z of the ions can be described by equations

$$m/z = \left[M + nH\right]^{n+} \text{ for positive ions}$$

and

$$m/z = \left[M - nH\right]^{n-} \text{ for negative ions.}$$

Figure 4.10 presents mass spectrum of the neuropeptide dynorphin A (YGGFL RRIRP KLKWD NQ, monoisotopic mass 2146.19 Da) at a concentration of 5 nmol ml^{-1}, taken in a positive ion mode. At least three most intense signals are visible at m/z 537.8, 716.6, and 1074.5. Isotopic pattern of the ion at 1074.5 (insert) allows the charge of this ion to be easily calculated. As a distance between adjacent isotopic peaks is 0.5, this ion carries +2 charge (charge = 1/distance between peaks, thus z = +2). To calculate MW of the substance represented by a peak at a known m/z ratio, the equation below can be applied:

$m/z = 1074.0$ (monoisotopic ion : the first, not the most intense ion in an envelope)

$M = [(m/z)*z] - z = [1074.0*2] - 2 = 2146 \text{ Da}$

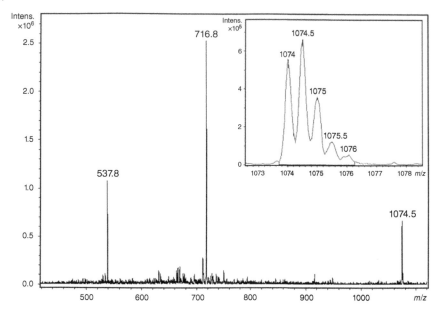

Figure 4.10 Mass spectrum of dynorphin A.

To confirm the calculated MW, similar calculations should be performed for all other peaks (m/z 537.8 and 716.8). It can be deduced that with a decrease of m/z ratio, the charge z increases by a value of +1. Thus, for the indicated peaks, the corresponding charges will be for 716.8: 3+ and for 537.8: 4+. Going into detailed analysis of isotopic pattern of peak 716.8, it can be found out that m/z of the monoisotopic peak (first from the group) is equal to 716.3. Calculations for those peaks were as follows:

$$\text{for } m/z = 716.3\left[\left(716.3*3\right)-3\right]M = 2145.9\,\text{Da};$$
$$\text{for } m/z = 537.8\left[\left(537.8*4\right)-4\right]M = 2147.2\,\text{Da}.$$

After averaging of all three values, MW of dynorphin A was calculated as MW = 2146.37 Da. The real monoisotopic weight of this peptide is M = 2146.19 Da. For this spectrum, inaccuracy in MW measurement was 0.18 Da. It is worth to mention that for the appropriately calibrated, modern high-resolution mass spectrometers, such inaccuracy can be a few levels of magnitude lower. The above procedure can be fully automated and is termed spectrum deconvolution.

4.1.3.2.5.3 High MW Molecules Mass spectrum observed in Figure 4.11 shows horse heart myoglobin in an active form (UniProtKB: MYG_HORSE, lub ID P68082, MW = 16951.3 Da). Concentration of the analyte was $2\,\text{nmol}\,\text{ml}^{-1}$

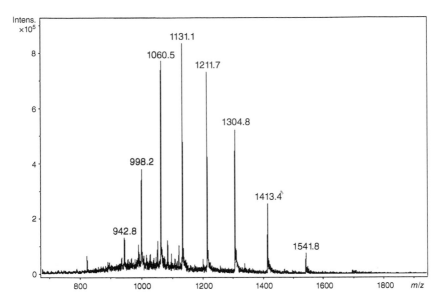

Figure 4.11 Mass spectrum of the horse heart myoglobin.

(positive ion mode, solvent: water/methanol/formic acid; 69.9 : 30 : 0.1; v/v/v). A series of ions at various charges is visible on the spectrum. The resolution of the mass spectrometer was limited to c. 5000; therefore it was impossible to estimate charge of the ions directly from their isotopic patterns. However, it is still possible to calculate the charge of every ion of myoglobin using simple equation:

$$n = \frac{m_{n+1} - 1.008}{m_n - m_{n+1}}$$

where n is charge (as a number of protons) per ion m_n, m_n is m/z of the ion having n charge, m_{n+1} is m/z of the ion having $n+1$ charge, and 1.008 is the proton mass.

Using data from the mass spectrum (Figure 4.11), it is easy to calculate charges of detected ions arising from myoglobin ionization. Exemplarily, using the equation presented above and two neighboring signals (m/z 1304.8 and 1413.4), we can calculate charge (z) of the ion of $m/z = 1413.4$ Da as follows:

$$n = \frac{1304.8 - 1.008}{1413.4 - 1304.8} = \frac{1303.79}{108.6} = 12.005 \approx 12$$

Keeping in mind that adjacent peak at lower m/z ratio is $n+1$ charged, we can assign charges for the detected ions. Calculated data were collected in Table 4.2.

Table 4.2 Myoglobin MW based on the peaks *m/z* from MS spectrum (Figure 4.11).

m/z	z	MW calculated	Deviation
942.8	18	16 952.4	+1.76
998.2	17	16 952.4	+1.76
1060.5	16	16 952	+1.36
1131.1	15	16 951.5	+0.86
1211.7	14	16 949.8	−0.84
1304.8	13	16 949.4	−1.24
1413.4	12	16 948.8	−1.64
1541.8	11	16 948.8	−1.64
Average MW		16 950.6	SD = 1.59

Calculated average MW of the protein, M = 16950.6 Da (standard deviation [SD] = 1.59 Da), is very close to theoretically calculated MW of the myoglobin sequence ($MW_{theor} = 16951.3$ Da), and the estimated imprecision of our measurement is 0.7 Da and perfectly fits within the range designated by SD value. Currently, calculations of protein MW are performed by the software for deconvolution, dedicated for mass spectra interpretation, which should be provided by the vendor, together with the MS system.

4.1.3.2.5.4 Mixtures Mass spectrum shown in Figure 4.12 arises from tryptic fragments of peptide MW = 6975.5 Da. The visible peaks mainly represent short tryptic peptides of the cleaved protein sequence. Table 4.3 summarizes the results of this complex spectrum analysis.

The remaining peaks visible on the spectrum, not included in the table (e.g. 527.4; 797.9; 1063.5), are peptidergic impurities of the original sample taken for identification.

4.1.3.2.5.5 Polymers At a first glance, mass spectrum of the polymer is very similar to the mass spectrum of a high MW substance (compare Figures 4.11 and 4.13). A closer view allows for finding basic differences, such as distances between main peaks in the case of high MW substance (ascending with *m/z*) in contrast to equal distances between peaks for the polymer ions. It is caused by the properties of polymers themselves. It should be remembered that polymer sample is usually a mixture of chains of various lengths built of monomeric moieties. The charge of every chain is usually the same and equals +1, so distances between peaks on the mass spectrum would reflect the mass of the monomer building polymer molecules. Interpretation is more complicated for heteropolymers or end groups, but some useful

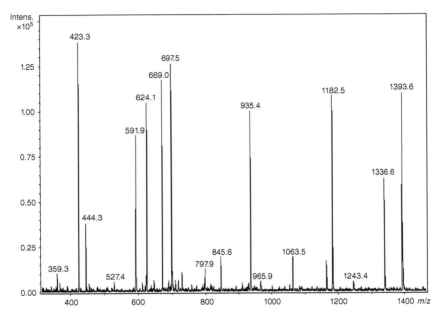

Figure 4.12 Mass spectrum of the peptide map derived after enzymatic cleavage of the high MW peptide derived from wax moth (*Galleria mellonella* L.).

Table 4.3 MW and sequences of peptides identified in the peptide map.

Peptide monoisotopic mass (Da) (sequences in brackets)	Ion 1 (charge 1+)	Ion 2 (charge 2+)	Ion 3 (charge 3+)
1181.5 (1–10: ETESTPDYLK)	1182.5	591.9	—
1392.7 (11–21: NIQQQLEEYTK)	1393.6	697.5	—
1867.9 (22–37: NFNTQVQNAFDSDKIK)	1869.0	935.5	624.1
1335.7 (38–49: SEVNNFIESLGK)	1336.6	669.0	—
844.5 (50–56: ILNTEKK)	845.1	423.3	—
443.2 (57–60: EAPK)	444.3	—	—

information may be received from distances between ions, even without fragmentation procedure [43].

Mass spectrum shown in Figure 4.13 was generated using ESI in the positive ion mode. It shows a mixture of two polymers dissolved in chloroform/methanol 1 : 1 (v/v), acidified with formic acid (0.1%, v). It is easy to find MW

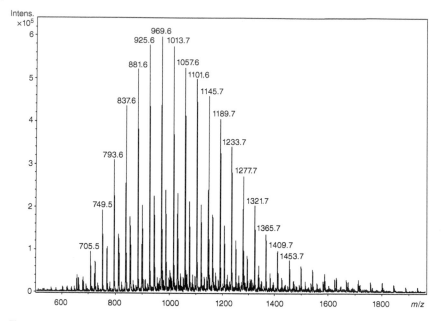

Figure 4.13 Mass spectrum of two homopolymers.

of monomers building these polymers by subtracting *m/z* ratios of adjacent peaks (e.g. 1101.6 – 1057.6 = 44 Da). As polymeric chains from the same polymer have similar properties, it is possible to estimate which chain length dominates in the analyzed polymer (the highest peak comes from the most abundant chain). In the described case, the highest peak has $m/z = 969.9^{1+}$. By dividing its MW by monomer weight, we can anticipate that the most abundant chain length in the polymer mixture is 22 monomers per chain (968.9/44 = 22).

Questions

- What are the main advantages of ESI over other ionization techniques?
- Why is it advantageous to have multiply charged ions in ESI?
- What does "soft ionization" mean?
- What is the "Taylor cone?"
- What are the skimmers?
- Why most ESI sources use an orthogonal arrangement?
- On what bases the temperature of the heated capillary and the drying gas flow are chosen?
- Which factors regulate the velocity of the dry ions inside the inlet capillary?

- What are the most common solvents used in ESI, and on what bases they have been selected?
- What is the "ion suppression" effect and by what it is caused?
- In what cases silver trifluoroacetate can be used?
- How the experimental conditions can affect the ESI spectrum generated by a chemical species? Illustrate with an example.

References

36 Horning, E., Horning, M., Carroll, D. et al. (1973). New picogram detection system based on a mass spectrometer with an external ionization source at atmospheric pressure. *Analytical Chemistry 45*: 936–943.

37 Carroll, D., Dzidic, I., Stillwell, R. et al. (1975). Atmospheric pressure ionization mass spectrometry. Corona discharge ion source for use in a liquid chromatograph-mass spectrometer-computer analytical system. *Analytical Chemistry 47*: 2369–2373.

38 Byrdwell, W.C. (2001). Atmospheric pressure chemical ionization mass spectrometry for analysis of lipids. *Lipids 36*: 327–346.

39 Dass, C. (2007). *Fundamentals of Contemporary Mass Spectrometry*. Hoboken, NJ: Wiley.

40 Bruins, A. (1991). Mass spectrometry with ion sources operating at atmospheric pressure. *Mass Spectrometry Reviews 10*: 53–77.

41 Silberring, J. and Ekman, R. (eds.) (2002). *Mass Spectrometry and Hyphenated Techniques in Neuropeptide Research*. New York: Wiley.

42 Johnstone, R. and Rose, M. (1996). *Mass Spectrometry for Chemists and Biochemists*. Cambridge: Cambridge University Press.

43 Kowalczuk, M. and Adamus, G. (2016). *Mass spectrometry for the elucidation of the subtle molecular structure of biodegradable polymers and their degradation products*. Mass Spectrometry Reviews 35 (1): 188–198.

4.1.3.3 Nanoelectrospray

Piotr Suder

Department of Biochemistry and Neurobiology, Faculty of Materials Science and Ceramics, AGH University of Science and Technology, Kraków, Poland

4.1.3.3.1 *Introduction*
In contrast to the electrospray ion source, nanoelectrospray (or shorter: nano-spray or nanoelectrospray ionization [nanoESI]) does not use sheath gas for sample/mobile phase nebulization. In this technique, the flow rate is significantly lower. Low flow rates are mainly achieved by the reduction of capillary diameter in comparison with the typical electrospray ion source. Capillary diameters can go down to 20 μm for an *online* nanospray or even reach 0.5 μm for the *offline* nanoESI. Nebulization is initiated when the voltage is applied, and, at the capillary tip, solvent containing analyte molecules spontaneously forms Taylor cone (Section 4.1.3.2.2). As an initial diameter of the droplets is lower, the ions undergo drying much faster, due to the subtler spraying. The first, and the most important, consequence is a possibility to receive higher ionization efficiency in comparison with a typical electrospray source. The second feature concerns much lower sample consumption (even two or three levels of magnitude), which is especially important in life sciences where only tiny amounts of precious biological material are available. An additional benefit of the nanoESI source related to the significantly decreased flow rate is a much longer signal duration in the instrument. Hence, the sample can be analyzed longer, allowing for multiple fragmentation.

Basic differences between ESI and nanoESI are collected in Table 4.4.

4.1.3.3.2 *Nanospray Sensitivity*
Initial droplet volume, formed under the influence of electric field, has fundamental importance for the sensitivity of analysis. Generally, a decrease in droplet diameter raises sensitivity. As the flow rates for nanoelectrospray are at least 1–2 levels of magnitude lower, solvent evaporation is usually complete, allowing for ionization efficiency at least 10%, which is roughly 10–100× more effective than in the case of conventional ESI ion sources.

4.1.3.3.3 *Ionization in Nanoelectrospray and Electrospray Ion Sources*
Ionization in both types of ion sources occurs as the effect of conditions promoting attachment or detachment of proton(s) to/from the analyte molecules. The basic parameters are as follows:

a) High voltage between both capillaries in the ion source.
b) Appropriate temperature in the source, achieved by heating inlet capillary and, sometimes sheath gas, which promotes solvent evaporation.
c) Solvent pH, providing deficit or excess of protons.

Table 4.4 Nanoelectrospray versus electrospray.

Parameter	Nanospray	Electrospray
Capillary diameter	0.5–20 µm	>50 µm
Analyte flow rate	1–1000 nl min^{-1}	0.5–100 (1000) µl min^{-1}
Sample consumption	Very low (c. 1 µl analysis^{-1})	Low (5–50 µl analysis^{-1})
Spraying	Forced by potential	Forced by sheath gas and potential
Source voltage	1.5–2.5 kV	4–5 kV
Sensitivity	100 amol µl^{-1} to 10 fmol µl^{-1}	10 fmol µl^{-1} to 1 pmol µl^{-1}
Ionization efficiency	Above 10%	Below 1%
Influence of low molecular weight (MW) salts	Acceptable	Sensitive
Interference with external factors (salt ions, sample impurities)	Small	High

In particular, the latter parameter listed above needs further elucidation; in both approaches the ionization is effective when pH value of the solvent is possibly far from pI of the analyte. Good examples are proteins and peptides. For these molecules, pI is usually in a range between 5 and 7. Dissolution of peptide in a solvent at pH < 4 (e.g. water/methanol acidified with formic acid) causes protonation of amino acid side chains (Arg and Lys residues), as well as N-terminal amino groups. Even if pH of the solution is not significantly different from pI of the analyte, fast solvent evaporation in the ion source results in close interactions between protons generated in residual solution containing sample molecules susceptible to protonation, which finally results in ions' formation.

4.1.3.3.4 Offline *Nanospray and Practical Advices*

As mentioned earlier, nanoelectrospray may be applied in two variants, either *offline* or *online*. The *offline* version is usually used for the analysis of a single sample. The best suited for such analyses are commercially available fused silica capillaries having internal diameter (ID) in a usual range of 0.1–30 µm, additionally tapered at the end. It is possible to manufacture such capillaries using fused silica tubings or graphite-based pipette tips, but some experience and personal protection equipment (fused silica is brittle) are required. Efficient fabrication can be done using a capillary puller, but it is worth to remember that fused silica (quartz) has fairly higher melting point than glass; hence additional heating (laser-based pullers) is necessary.

It is important to prepare such capillaries from electroconductive material (conductive pipette tips) or cover already tapered tips with a thin film of inert metal (e.g. gold, platinum, rhodium) instead of applying liquid junction. Nanospray formation still requires a difference of potentials sufficient to form Taylor cone. For the analysis, the capillary should be filled with at least few nanoliters of the sample and positioned with the aid of micromanipulator to stabilize the outlet in front of mass spectrometer (MS). By turning high voltage on, spraying and ion formation should start spontaneously. It is important to raise the voltage slowly to form Taylor cone without generation of an electric arch between capillaries. In most cases, 1.5 kV is sufficient, but final settings (up to 2.0–2.5 kV) may also depend on solvent conductivity, capillary material, and its ID. Additional fine positioning between capillaries should improve and stabilize the signal. Approximately one hundred nanoliters introduced into the capillary is usually enough to acquire satisfactory spectra for 30 minutes or longer, which is sufficiently long time to complete MS and MS^n analyses. During manipulation and voltage settings, it is substantial to follow safety rules due to a high risk of electric shock.

4.1.3.3.5   Online *Nanospray and Practical Advices*

This type of nanospraying allows for connection between mass spectrometer equipped with the ESI-type ion source and any suitable liquid delivering analytical system like nanoLC or capillary electrophoresis. As a general rule in the *online* nanospray, liquid traveling at a very low flow rate is pumped through the capillary. The mechanism of ions' formation is similar to that observed for an *offline* nanoESI setup, and Taylor cone is formed by electric potential without the sheath gas support. Typically, for chromatographic separations connected *online* with MS, the flow rate is usually set between 100 and 500 nl min^{-1}. Capillaries used here have ID in a range of 5–25 μm to avoid increased backpressure. Materials used for fabrication of capillaries do not need to be electroconductive, as potential is delivered by using so-called liquid junction. Solvent with the analyte is delivered through a metal zero-dead-volume tee, where potential is applied. It is also worth to note that the setup of the mass spectrometer itself is important for the operation. It is easier to connect both systems together when MS potential is applied to the heated/inlet capillary and delivery capillary is grounded than in constructions where the heated/inlet capillary is grounded and high voltage is applied to the sprayer. For the latter variant, it is necessary to consult high pressure liquid chromatography (HPLC) or capillary electrophoresis manufacturer, as such connection may damage chromatographic or capillary systems.

4.1.3.3.6   Summary

Advantages arising from direct connection between nanoLC system and nanospray are obvious: the properly working nanoLC-MS/MS instrument has at least two levels of magnitude higher sensitivity than analytical HPLC linked to ESI-MS. Sample consumption is very low; typically, 100–200 fmol (or less) of a tryptic digest of protein is enough for its complete identification, and solvent consumption is very low: the 75 minutes gradient is performed with a volume of 20–25 µl. Even precolumn and analytical column equilibration does not significantly increase solvent consumption. The entire analysis (sample-to-sample injection) consumes not more than 50–60 µl of solvents.

Disadvantages of such systems are as follows: susceptibility to faults like clogging of capillaries, connectors, or capillary columns and difficult diagnostics of problems like leakages that are invisible at such low flow rates.

Questions

• What are the main differences between ESI and nanoESI?
• What are the basic parameters to be controlled in order to allow ionization in ESI and nanoESI?
• How is it possible to obtain capillaries suitable for the offline nanoESI analysis?
• What are the main advantages and disadvantages of nanoESI?

4.1.3.4 Desorption Electrospray Ionization (DESI)

Anna Bodzon-Kulakowska and Anna Antolak

Department of Biochemistry and Neurobiology, Faculty of Materials Science and Ceramics, AGH University of Science and Technology, Kraków, Poland

4.1.3.4.1 *Introduction*

The desorption electrospray ionization (DESI) ion source was designed and constructed by Graham R. Cooks et al. in 2004. It originates from electrospray ionization (ESI) ion source. The difference is that the stream of charged liquid droplets, which are detached from the Taylor cone, is directed straight toward the surface of the analyzed sample. It means that the sample does not need to be dissolved in the liquid and pumped into the system. Thus, DESI ion source enables analysis of solid samples, such as tablets, tissue sections, and many others [1].

The main advantage of the DESI ion source is its ability to work under ambient conditions (atmospheric pressure). Additionally, in comparison with other methods used for surface analysis (see Section 4.1.7 and Chapter 6), DESI does not require any form of sample preparation. This allows for analysis of different types of samples, even somehow unusual ones, such as the surface of the skin, leaves, fruits, tablets, etc. Tandem mass spectrometry (MS/MS) analysis of the registered ions additionally allows for unambiguous identification of molecular constituents of the analyzed sample. Addition of an internal standard makes quantitative measurements possible. An interesting advantage of this technique is the ability to apply "*reactive* DESI" where the hardly ionizable substances are derivatized by the specific reagent applied directly to the solution, which is sprayed on the sample surface (see below) [2].

4.1.3.4.2 *Construction of an Ion Source*

DESI source is composed of two capillaries: the spray capillary and mass spectrometry (MS) inlet capillary (Figure 4.14). Solvent and gas are delivered to the spray capillary, which then sprays the solvent over the analyzed surface. At the opposite position, an MS inlet capillary is located that collects the resulting ions and transfers them to the mass spectrometer. Below the capillaries, an x–y table with deposited sample is located. This table is controlled by a software and can be moved in the x–y directions, which provides access to any point of the measured surface. Additionally, the spray capillary is equipped with a camera, and a light source directed at the analyzed surface, allowing for tracking the entire analysis.

The position of the spray capillary can be regulated (see Figure 4.15). The adjustable parameters are the height of the capillary above the table, its distance, and the left–right positions in regard to the inlet capillary. Additionally, the angle at which the stream of solution is sprayed toward the surface may also be controlled. The distance between the inlet capillary and a sample can be

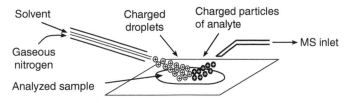

Figure 4.14 Desorption electrospray ionization (DESI) ion source.

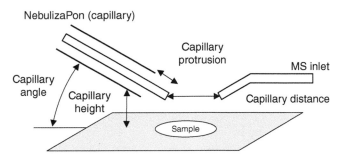

Figure 4.15 Ion source geometry.

modified to a certain extent. The distance between the nebulizing capillary and the inlet capillary, as well as an angle of the nebulizing capillary in relation to the sample table, influences the ionization process. The distance between the inlet capillary and the sample affects the sensitivity of the analysis by influencing the efficiency of collecting the resulting ions. All three-dimensional settings of both capillaries and the table are referred to as the ion source geometry.

Other parameters, such as solution flow rate inside the spray capillary, gas pressure, and applied voltage that are responsible for aerosol generation, must be individually selected for each type of sample. These parameters influence the diameter and velocity of the aerosol droplets directed toward the sample. Any new surface to be analyzed may require slightly different geometry of the ion source and other parameters described above. Hence, a series of measurements may be necessary prior to the analysis to optimize settings for the best results [3].

4.1.3.4.3 *Principle of Operation of DESI Source*
As mentioned above, in DESI ion source, the appropriate solvent is delivered to the spray capillary under a suitable pressure. At the tip of this capillary, the aerosol is formed thanks to the three factors. The first is the pressure, under which the solution is supplied to the capillary. The second is related to the presence of gas (usually nitrogen), which maintains nebulization (see Figure 4.14). The third factor is high voltage applied to the spray capillary, resulting in the formation of a Taylor cone and the aerosol of the charged

solvent droplets. The presence of the high voltage also supports the process of droplets' desolvation, leading to a gradual reduction of their size. Finally, the droplets of a diameter between 0.5 and 10 μm (average 3–4 μm) hit the surface of the sample at a speed of 80–180 m s^{-1} (average speed: 120 m s^{-1}).

4.1.3.4.4   *Mechanism of Ion Formation*

During the first stage of the ion formation, the so-called primary droplets are formed. The flow of the solution in the spray capillary and the suitable gas pressure with the electric charge allow for the formation of electrospray plume with the charged droplets of appropriate size. Considering the same liquid flow in the capillary, the higher the gas pressure, the faster the resulting droplets move, and the smaller size they have. Velocity of the primary droplets directly affects the number of secondary droplets produced – hence the sensitivity of the analysis. On the other hand, smaller droplets resulted from the high gas pressure (considering the same liquid flow) may evaporate before reaching the surface. Therefore, excessive gas pressure may result in a poor spectra quality. What is more, the strong gas stream may damage the material being analyzed. Therefore, the value of gas pressure has to be carefully chosen. Liquid flow also influences the size of the droplets. Having the same gas pressure, if the flow of liquid is too low, it results in droplets of small diameters. In the opposite situation, at the same gas pressure, the high liquid flow may cause the formation of larger droplets and intense soaking of the sample surface.

The next step is formation of secondary droplets (see Figure 4.16) (*droplet pickup mechanism*). The primary beam of droplets strikes the analyzed surface at a speed of several hundred meters per second and spills over the area 3–10 times larger than the diameter of the single droplet. This results in the formation of a liquid film, in which the substances, present in the analyzed sample, dissolve. At the same time, another series of primary droplets hits the surface and produces secondary droplets containing analyte. At first, small droplets are being formed, which move quickly in a direction almost parallel to the sample surface. Next, larger but slower droplets appear, bouncing from the surface at a bigger angle. It seems that only the first small droplets get inside the spectrometer. Hence, the optimum setting for the capillary that delivers ions to the spectrometer (MS inlet) is usually at a low height from the sample table. The transport of these droplets to the spectrometer is caused by the pressure difference arising at the inlet of the spectrometer and by the flow of the nebulizing gas, which is being reflected from the surface. Further formation of the ions in a gas phase occurs as in the case of a typical electrospray by subsequent evaporation of the solvent and Coulomb fission [4].

Figure 4.16 *Droplet pickup* ionization.

4.1.3.4.5   The Role of the Surface, on Which the Sample Is Deposited

For DESI measurements, the sample must be placed on the appropriate surface, as its nature may affect the quality of the analysis. In the case of DESI, conductive surfaces, such as metals or graphite, should be avoided. It is to prevent the loss of charge by droplets striking the surface and by the resulting ions. Depending on the ionization mode, in which the sample is measured, better results can be achieved by using a suitable surface for sample deposition. For measurements in the negative ion mode, Teflon (polytetrafluoroethylene [PTFE]) is a very good candidate as it is a highly electronegative polymer. In the case of a positive ion mode, the use of acrylic glass (poly(methyl methacrylate) [PMMA]) improves quality of the obtained results. Of course, it is also important to assure that the surface does not interact with the analyzed substance, which will reduce the sensitivity of the analysis. The most commonly used surfaces during DESI analysis are soda glass, Teflon, acrylic glass, and paper.

4.1.3.4.6   Reactive DESI

Reactive DESI is the way of analysis that allows for rapid detection of substances that are not ionized under normal conditions and do not generate a clear signal on the spectrum. During this procedure, specific substrate is introduced into the solution to be sprayed onto the surface of the sample. This substrate either quickly reacts with the analyzed substance, creating an easily ionized product, or forms an adduct with individual sample component, which enables its detection. This increases the selectivity and sensitivity of the analysis. To name some examples of such procedure, we may mention the use of hydroxylamine to detect steroids in urine [5]; dinitrophenylhydrazine (DNPH) for the detection of malondialdehyde (MDA), an oxidative stress marker [6]; or betaine aldehyde chloride for cholesterol detection [7] (Figure 4.17).

Other analyses with the use of reactive DESI include addition of HCl or TFA to the solution to detect explosive substances RDX (hexahydro-1,3,5-trinitro-1,3,5-triazine) and HMX (1,3,5 7-tetranitro-1,3,5,7-tetraazacyclooctane) as chloride or TFA anion adducts [8].

DESI ion source can be used in many different types of assays, such as analysis of alkaloids in plant tissues [9], investigation of medications, or as a detector for thin layer chromatography (TLC). DESI is also used in forensic sciences to detect the presence of explosives, such as trinitrotoluene (TNT), RDX, HMX, pentaerythritol tetranitrate (PETN) on various surfaces (plastic, paper, leather, glass), as fingerprints or for ink analysis. This method is widely used in biochemical analysis of drug metabolites or steroids in urine [10, 11]. Use of DESI in MS imaging techniques is discussed in Chapter 6. An exemplary spectrum derived from lipid analysis in cell culture by the use of DESI ion source is presented in Figure 4.18. The analysis was performed in both positive and negative ion modes. Signals visible on the spectra correspond to the lipids found in the fibroblast cell membrane.

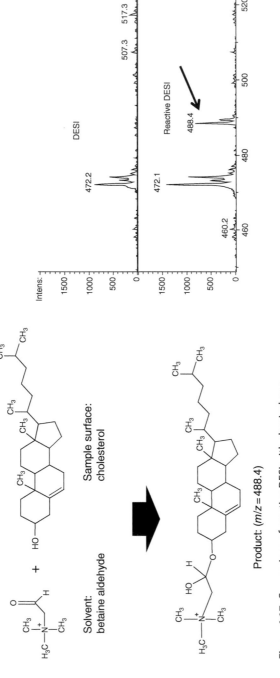

Figure 4.17 Comparison of reactive DESI with classical one.

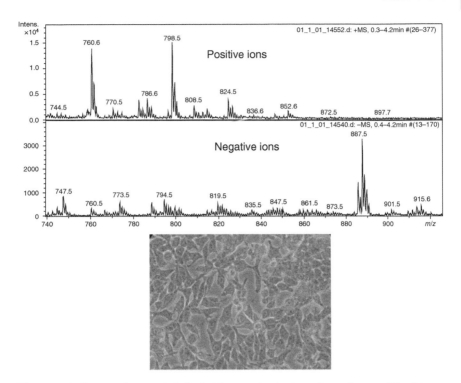

Figure 4.18 Spectra characteristic for lipids occurring in the cell membrane of fibroblasts. Analysis of the positive and negative ions. Insert shows the photo of fibroblast culture at 400 times magnification.

References

1 Takats, Z., Wiseman, J.M., and Cooks, R.G. (2005). Ambient mass spectrometry using desorption electrospray ionization (DESI): instrumentation, mechanisms and applications in forensics, chemistry, and biology. *Journal of Mass Spectrometry 40* (10): 1261–1275.
2 Alberici, R.M., Simas, R.C., Sanvido, G.B. et al. (2010). Ambient mass spectrometry: bringing MS into the 'real world'. *Analytical and Bioanalytical Chemistry 398* (1): 265–294.
3 Bodzon-Kulakowska, A., Drabik, A., Ner, J. et al. (2014). Desorption electrospray ionisation (DESI) for beginners: how to adjust settings for tissue imaging. *Rapid Communications in Mass Spectrometry 28* (1): 1–9.
4 Costa, B. and Cooks, R.G. (2008). Simulated splashes: elucidating the mechanism of desorption electrospray ionization mass spectrometry. *Chemical Physics Letters 464* (1): 1–8.

5 Huang, G., Chen, H., Zhang, X. et al. (2007). Rapid screening of anabolic steroids in urine by reactive desorption electrospray ionization. *Analytical Chemistry 79* (21): 8327–8332.

6 Girod, M., Shi, Y., Cheng, J.-X., and Cooks, R.G. (2011). Mapping lipid alterations in traumatically injured rat spinal cord by desorption electrospray ionization imaging mass spectrometry. *Analytical Chemistry 83* (1): 207–215.

7 Wu, C., Ifa, D.R., Manicke, N.E., and Cooks, R.G. (2009). Rapid, direct analysis of cholesterol by charge labeling in reactive desorption electrospray ionization. *Analytical Chemistry 81* (18): 7618–7624.

8 Cotte-Rodriguez, Z., Takats, N., Talaty, H.C., and Cooks, R.G. (2005). Desorption electrospray ionization of explosives on surfaces: sensitivity and selectivity enhancement by reactive desorption electrospray ionization. *Analytical Chemistry 77* (21): 6755–6764.

9 Talaty, N., Takáts, Z., and Cooks, R.G. (2005). Rapid in situ detection of alkaloids in plant tissue under ambient conditions using desorption electrospray ionization. *Analyst 130* (12): 1624.

10 Chen, H., Talaty, N.N., Takats, Z., and Cooks, R.G. (2005). Desorption electrospray ionization mass spectrometry for high-throughput analysis of pharmaceutical samples in the ambient environment. *Analytical Chemistry 77* (21): 6915–6927.

11 Kauppila, T.J., Talaty, N., Kuuranne, T. et al. (2007). Rapid analysis of metabolites and drugs of abuse from urine samples by desorption electrospray ionization mass spectrometry. *Analyst 132* (9): 868.

4.1.3.5   Laser Ablation Electrospray Ionization (LAESI)

Anna Bodzon-Kulakowska and Anna Antolak

Department of Biochemistry and Neurobiology, Faculty of Materials Science and Ceramics, AGH University of Science and Technology, Kraków, Poland

Laser ablation electrospray ionization (LAESI) ion source is a combination of laser desorption (average wavelength for infrared radiation is 3–8 μm) and ionization with charged droplets generated in the electrospray source. This technique was introduced by Peter Nemes and Akos Vertes in 2007. Its main feature is the possibility of analysis of hydrated samples and operation under atmospheric pressure.

During LAESI analysis, a laser of the wavelength of 2940 nm is directed toward the sample, which absorbs laser energy (Figure 4.19). The laser pulse lasts about 5 ns. The wavelength is correlated with the oscillation frequency of the O—H bond in the water molecule; thus, LAESI uses the water present in the sample as a matrix. Energy absorption causes ablation, i.e. evaporation of the sample and desorption of the molecules into the gas phase, which are then ionized by interaction with charged droplets arising from an ESI-type ion source located above the sample. The commonly used solution in the ESI ion source is 50% methanol with 0.1% acetic acid. Interestingly, the distance of ESI source from the surface of analyzed sample is quite large – at the level of 10–30 mm. This relates to the mechanism of ion formation. After the initial laser pulse, the first batch of ions and neutral molecules is desorbed until the atmospheric pressure stops its spreading. After that, the second batch of particles is desorbed as a result of recoil. This effect lasts about 300 μs and contains larger molecules that can move further. These particles are ionized by the charged droplets produced in the ESI ion source.

The resulting ions enter the spectrometer where they are analyzed. Separation of both processes, i.e. (i) desorption of the sample initiated by the laser pulse and (ii) ionization of molecules by ESI-charged droplets, provides high-quality spectra that are easy to interpret. The obtained spectra resemble those generated by ESI, with a detection range of up to 100 kDa.

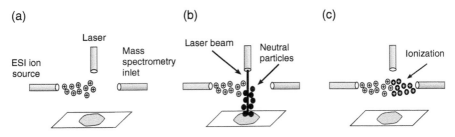

Figure 4.19 Laser ablation electrospray ionization (LAESI) ion source. Subsequent steps of sample ionization.

During one laser pulse, 20–40 μm of material from the sample surface is removed, allowing in-depth profiling of substance distribution. The depth of the resulting crater depends on the nature of the substance being analyzed, including its mechanical, optical, and thermal properties and power characteristics of the laser. The diameter of the laser beam is c. 200 μm; therefore it has lower resolution when compared with matrix-assisted laser desorption/ionization (MALDI) or desorption electrospray ionization (DESI). Reduction of the beam diameter leads to a decrease in the sensitivity of the analysis, which is due to the smaller amount of desorbed material [1].

As mentioned above, LAESI does not require any special sample preparation, as it is in the case of MALDI technique. Additionally, the analysis can be carried out under the atmospheric pressure. Moreover, hydration of the sample, which is usually an obstacle in other types of ionization, such as MALDI or SIMS, is an advantage and an essential element of LAESI analysis. In addition to standard analyses, LAESI method allows for measurements of atypical samples, e.g. gel-embedded samples, such as bacterial colonies in agarose gel, which cannot be analyzed under vacuum, or *in vivo* bacterial strains, for bacteria or fungi profiling, or in search for substances produced by them [2]. Additionally, analysis may be performed directly in the droplets of aqueous solutions. Ionization caused by charged droplets generated in the ESI source assures multiple ionization of the analyzed substances. This is of great importance as it allows for the *top-down* analysis of proteins [3]. This means that once a protein has been detected on the surface, it can be identified and its posttranslational modifications can be evaluated through its fragmentation (Section 7.5). This type of analysis is unattainable in other types of ion sources used in mass spectrometry, e.g. MALDI, where samples usually become ionized only one- or twofold.

By the use of this technique, metabolites, peptides, lipids, proteins, and other molecules derived from cells, tissues, and other hydrated samples can be analyzed. In addition, this examination can be performed very fast due to the lack of specific sample preparation procedures. The possibility of using internal standards in the case of complex samples, such as plasma, allows quantitative analysis.

References

1 Nemes, P. and Vertes, A. (2007). Laser ablation electrospray ionization for atmospheric pressure, in vivo, and imaging mass spectrometry. *Analytical Chemistry 79* (21): 8098–8106.
2 Kiss, A., Smith, D.F., Reschke, B.R. et al. (2012). Evaluation of a novel laser ablation electrospray ionization source for the imaging of bacteria from high salt content liquid medium. Poster presentation at Ourense Conference on Imaging Mass Spetrometry, University of Vigo at Ourense, Ourense, Spain.
3 Kiss, A., Smith, D.F., Reschke, B.R. et al. (2014). Top-down mass spectrometry imaging of intact proteins by laser ablation ESI FT-ICR MS. *Proteomics 14*: 1283–1289.

4.1.3.6 Photoionization

Jerzy Silberring[1,2]

[1] Department of Biochemistry and Neurobiology, Faculty of Materials Science and Ceramics, AGH University of Science and Technology, Kraków, Poland
[2] Centre of Polymer and Carbon Materials, Polish Academy of Sciences, Zabrze, Poland

Photoionization (PI) process has been applied in the mid-1970s of the twentieth century and was linked to gas chromatography and later on to liquid chromatography. It is now a popular method, in particular PI under ambient conditions for the analysis of explosives and in forensic and environmental sciences.

The method utilizes UV light photons for ionization of molecules, predominantly low molecular mass hydrophobic compounds that are difficult to analyze using other sources. Three phases of the entire process can be listed as follows:

- Vaporization of the sample
- Generation of photoions
- Detection

The most frequently used source of UV photons is a krypton lamp, with photon energy of 10 eV. This energy is lower than the one needed to ionize the majority of air components and solvents but is sufficient to ionize/excite analyte molecules. Thus, spectra do not contain too high background and are easier to interpret. For further reading of the development of PI, we recommend an excellent review by *Rafaelli and Saba* [1]. PI under high vacuum was applied to enhance matrix-assisted laser desorption/ionization (MALDI) ionization (vacuum UV single-photon post-ionization [VUV SP]), which significantly improved sensitivity of the obtained spectra [2], and the technique was successfully applied by R. Kostiainen and his group [3] for identification of compounds in sage leaves and chili pepper.

PI may involve many processes, including photoexcitation, photodissociation, radiative decay, recombination, and collisional quenching [1]. Generally, ionization leads to the formation of protonated species (pseudomolecular ion; $[M+H]^+$) and is shown in Eq. (1) [1, 4]:

In the presence of protic solvent or charge carrier (ambient conditions),

$$M + R^+ \rightarrow MH^+ + R^- \tag{1}$$

A minor reaction may also occur during direct ionization of the molecule by UV irradiation:

$$M + h\nu \Rightarrow M^{+\bullet} + e^- \tag{2}$$

From the relationship (1), it is clear that the sample can be first mixed with a solvent, which acts as a donor of protons. Such conditions apply to atmospheric pressure ion sources. Depending on the type used, the solvent could

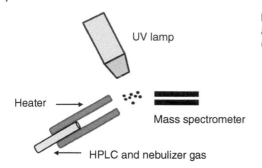

Figure 4.20 Construction of the atmospheric pressure photoionization (APPI) ion source.

UV lamp

Heater →

Mass spectrometer

← HPLC and nebulizer gas

increase the number of ions that are formed. Moreover, other substances may facilitate enhanced ionization and are called dopants. Typically, toluene, anisole, and acetone are applied for such purpose; however the latter may degrade tubings, frequently used for liquid pumping.

Such constructions comprise an interesting and promising variant of PI, i.e. atmospheric pressure photoionization (APPI). This method resembles electrospray or atmospheric pressure chemical ionization (APCI) to such extent that it was possible to replace APCI source with APPI where corona discharge needle has been replaced by a UV source to link liquid chromatography with the mass spectrometer [5]. Such solutions are now commercialized for fast replacement between various ion sources operating under ambient pressure (Figure 4.20). There are also vendors offering PI detectors that can easily be mounted to the existing gas chromatographs.

Another option is called desorption atmospheric pressure photoionization (DAPPI), which utilizes hot solvent vapors for desorption followed by PI [3] and resembles desorption electrospray ionization (DESI) technique (Chapter 4.1.3.4).

The liquid solution containing solvent and sample is then vaporized with the help of a nebulizing gas, such as nitrogen, and enters the ion source.

Summarizing, PI is a constantly developing soft ionization technique involving simple construction and does not require advanced operation skills. It can function as a complementary method to other sources, such as electrospray ionization (ESI), MALDI, and APCI for the analysis of a broad number of volatile organic compounds (VOCs) but also inorganic species.

Questions

- What kind of samples is most suitable to be ionized by PI?
- What is the most frequently used source of UV photons in PI?
- What are the "dopants" in PI?
- What is DAPPI?

References

1 Rafaelli, A. and Saba, A. (2003). Atmospheric pressure photoionization mass spectrometry. *Mass Spectrometry Reviews* 22: 318–331.

2 Hsu, H.C. and Ni, C.-K. (2018). Vacuum ultraviolet single-photon postionization of amino acids. *Applied Sciences* 8: 699.

3 Haapala, M., Pól, J., Saarela, V. et al. (2007). Desorption atmospheric pressure photoionization. *Analytical Chemistry* 79 (20): 7867–7872.

4 Syage, J.A. (2004). Mechanism of [M + H]$^+$ formation in photoionization mass spectrometry. *Journal of the American Society for Mass Spectrometry* 15: 1521–1533.

5 Robb, D.B., Covey, T.R., and Bruins, A.P. (2000). Atmospheric pressure photoionization: An ionization method for liquid chromatography–mass spectrometry. *Analytical Chemistry* 72: 3653–3659.

4.1.4 Ambient Plasma-Based Ionization Techniques

Marek Smoluch

Department of Biochemistry and Neurobiology, Faculty of Materials Science and Ceramics, AGH University of Science and Technology, Kraków, Poland

4.1.4.1 Introduction

Ambient desorption/ionization mass spectrometry (ADI-MS) allows for rapid and direct analysis of substances in their native state at atmospheric pressure without or with minor sample preparation. In recent years, many techniques were developed to allow desorption and ionization of the sample under such conditions. Some of those methods found wide applicability, but some of them appeared to be not attractive enough to gain attention. Table 4.5 lists the ambient desorption/ionization (ADI) plasma ionization methods, and in the following subsections, some of the most widely used ones are discussed in more details. It is worth to note that the ADI methods differ in several ways from the atmospheric pressure ionization (API) methods. ADI requires no sample preparation, which is positioned directly between the ion source and the inlet to the mass spectrometer. Furthermore, for ADI methods, the sample can be directly ionized practically in each physical state. ADI sources do not need a specialized connection system, but only the standard source (e.g. electrospray ionization [ESI]) needs to be removed. In general, the plasma-based ADI methods differ in the plasma-forming mechanism and the reactions involved in the ionization/desorption processes.

4.1.4.2 Direct Analysis in Real Time (DART)

4.1.4.2.1 *Introduction*

Direct analysis in real time (DART) belongs to the group of ionization/desorption techniques held under the atmospheric pressure (ADI), and next to DESI described in Section 5.1.3.4, it is the second oldest method in this group. The DART ion source is usually used to analyze substances placed or adsorbed on the surface. Analysis can be performed directly on clothes, banknotes, the surface of fruits or vegetables, etc. Such capabilities make DART an excellent technique for identifying, among others, illegal substances, traces of explosives, pesticide monitoring, drugs, etc. Other applications of DART are, for example, quality control or monitoring the progress of chemical synthesis. DART even allows for direct analysis of the surface of living organisms.

4.1.4.2.2 *Construction of the Ion Source*

The description of DART construction is shown in Figure 4.21. The helium (or argon) is introduced into the ion source and next passes through the corona discharge region and then through electrodes generating ions, electrons, and

Table 4.5 Ambient plasma-based ionization techniques.

Plasma formation	Abbreviations/name	Power supply	Principle of operation	State of the sample
Corona discharge	DART (direct analysis in real time)	DC (direct current)	Sample surface exposed to plasma stream	Solid, liquid, gas
	DAPCI (direct atmospheric pressure chemical ionization)	DC	Sample surface exposed to the reactant molecules produced by APCI	Solid, liquid
	ASAP (atmospheric solids analysis probe)	DC	Sample is evaporated and then ionized by corona discharge	Solid, liquid
	DCBI (desorption corona beam ionization)	DC	Plasma stream directly interacts with the sample	Solid, liquid
Glow discharge	FAPA (flowing atmospheric pressure afterglow)	DC	Sample introduced outside the glow discharge zone	Solid, liquid, gas
	PADI (plasma-assisted desorption ionization)	RF (radio frequency)	Sample surface is in constant contact with low temperature plasma	Solid, liquid
	Mikroplazma – MHCD (microhollow cathode discharge – microplasmas)	DC	Sample surface introduced into the plasma stream	Solid, liquid, gas
	MFGDP (micro-fabricated glow discharge plasma)	DC	Sample surface introduced into the plasma stream	Solid, liquid, gas
Dielectric barrier discharge	DBDI (dielectric barrier discharge ionization)	AC (alternating current)	A sample ionized by plasma ion bombardment generated by DBD	Solid, liquid, gas
	LTP (low temperature plasma)	DC	Sample ionized by a plasma stream generated in a DBD source	Solid, liquid, gas
	MIPDI (microwave-induced plasma desorption ionization) MPT (microwave plasma torch)	Microwave induced discharge	Sample ionized directly by a high-energy plasma stream	Solid, liquid, gas

Source: Adopted from Ref. [1].

(a)

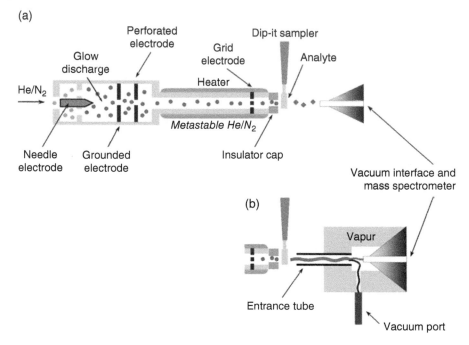

(b)

Figure 4.21 Schematic diagram of direct analysis in real time (DART) ion source. *Source*: Reprinted with permission from Ref. [2].

excited atoms/molecules (metastable particles). Electrical discharge in the DART source is classified as corona discharge-to-head (C-G), which is transition between the corona discharge and glow discharge. The temperature of the helium introduced into the discharge region is usually in the range of 50–60 °C and then grows up to 350 °C at the outlet. The gas leaving the ion source contains only excited, neutral molecules. A typical flow rate of the gas in the DART source is about $3\,l\,min^{-1}$. In rare cases, instead of helium or argon, nitrogen is used, mainly because of lower costs, but the ionization efficiency of most compounds is lower. The use of argon is more common due to better selectivity and weaker fragmentation in the ion source, but the sensitivity of the analysis is worse than that of helium. The excited atoms of helium leaving the source ionize constituents of atmospheric air and the analyte. In most cases, the ionization of the analyte is secondary and proceeds using ionized air components. The mechanism of ionization is described later.

4.1.4.2.3 Mechanism of Ion Generation
In the case of DART, the sample ionization takes place in the gas phase, and its mechanism is similar to the atmospheric pressure chemical ionization (APCI) described in Section 4.1.3.1. Ionization takes place in several stages. In the first

phase, metastable helium (He*), after being excited at the source, can interact with nitrogen, oxygen, or water present in the air, causing so-called Penning ionization. The above process can be described using Eq. (1):

$$He^* + N_2 \rightarrow He + N_2^{+\bullet} + e^- \tag{1}$$

Metastable helium can also react with other molecules present in the gas phase, including molecules of the analyzed substance (M). The ionization process can then be written as

$$He^* + M \rightarrow He + M^{+\bullet} + e^- \tag{2}$$

Due to the large amount of nitrogen in the air, the process (Eq. 2) is relatively inefficient, and the ionization of the analyte occurs mainly in an alternative way. This mechanism can be represented by a series of reactions. After the formation of $N_2^{+\bullet}$ (Eq. 1), dimerization of nitrogen occurs (Eq. 3), as well as ionization of water as a result of charge transfer (Eq. 4):

$$N_2^{+\bullet} + N_2 + N_2 \rightarrow N_4^{+\bullet} + N_2 \tag{3}$$

$$N_4^{+\bullet} + H_2O \rightarrow 2N_2 + H_2O^{+\bullet} \tag{4}$$

At atmospheric pressure, radical cations $H_2O^{+\bullet}$ reacts with water by reactions of Eqs. (5) and (6), in which the protonated water clusters are generated:

$$H_2O^{+\bullet} + H_2O \rightarrow H_3O^+ + OH^\bullet \tag{5}$$

$$H_3O^+ + nH_2O \rightarrow \left[(H_2O)_n + H \right]^+ . \tag{6}$$

Protonated water clusters can react with an analyte, ionizing it by generation of pseudo-molecular ions (Reaction 7):

$$M + \left[(H_2O)_n + H \right]^+ \rightarrow [M + H]^+ + nH_2O. \tag{7}$$

Ionization of the analyte may also occur as a result of the charge transfer from radical cations $N_4^{+\bullet}$, $O_2^{+\bullet}$, or NO^+ ion. In this case, we are dealing with Penning ionization, where molecular ions are formed:

$$N_4^{+\bullet} + M \rightarrow 2N_2 + M^{+\bullet} \tag{8}$$

$$O_2^{+\bullet} + M \rightarrow O_2 + M^{+\bullet} \tag{9}$$

$$NO^+ + M \rightarrow NO + M^{+\bullet}. \tag{10}$$

In rare cases, the ionization mechanism is mixed, which leads to the simultaneous formation of M^+ and $[M + H]^+$ ions. This phenomenon is disadvantageous because it distorts the typical isotope pattern and thus the interpretation of the spectrum.

DART can also be used to analyze substances in negative ion mode. In this case, the Reaction (11), which for positive ions was the source of nitrogen radical cations, provides electrons in negative ion mode. Electrons are captured by the oxygen atoms present in the air, forming anion radicals:

$$O_2 + e^- \rightarrow O_2^{-\bullet} \tag{11}$$

Oxygen radical anions participate in the formation of analyte adducts and then dissociate into the analyte radical anions (Reactions 12 and 13):

$$O_2^{-\bullet} + M \rightarrow [M+O_2]^{-\bullet} \tag{12}$$

$$[M+O_2]^{-\bullet} \rightarrow M^{-\bullet} + O_2. \tag{13}$$

Ionization of the analyte can also occur directly by electron capture (Eqs. 14 and 15), deprotonation (Eq. 16), or anion attachment (Eq. 17):

$$M + e^- \rightarrow M^{-\bullet} \tag{14}$$

$$MX + e^- \rightarrow M^- + X^\bullet \tag{15}$$

$$MH \rightarrow [M-H]^- + H^+ \tag{16}$$

$$M + X^- \rightarrow [M+X]^- \tag{17}$$

where X is an organic fragment (usually a halogen).

The presence of individual ions is closely related to the type of solvent used in the analysis. Typical ions generated in the DART technique depending on the polarity of the analyte are listed in Table 4.6.

4.1.4.2.4 *Applications of DART*

Because of the great utility of the DART method for analyzing both solid and liquid samples under atmospheric pressure, the method has many interesting applications. The most popular are detection of illegal substances on banknotes or clothes, monitoring of the presence of pesticides in vegetables, and identification of traces of explosives. In addition, DART was used to control the quality of food products to determine the authenticity of the oil, to analyze coffee beans from different manufacturers, and to identify mycotoxins in beer. In forensic science, the technique was used to classify inks of all types, in the

Table 4.6 Types of ions, generated in DART.

Analyte polarity	Positive ions	Negative ions
Unpolar	$M^{+\bullet}$, $[M+H]^+$, $[M+O+H]^+$	$M^{-\bullet}$, $[M-H]^-$, $[M-H+O_2]^-$, $[M+O_2]^{-\bullet}$, $[M+Cl]^-$
Polar	$[M+H]^+$, $[M+O+H]^+$, $[M+NH_4]^+$ (or other adducts)	$[M-H]^-$, $[M+OH]^-$, $[M+CN]^-$, $[M+Cl]^-$ (other adducts possible)

study of the authenticity of printed documents, to identify drug mixtures so-called bath salts, and for analysis of synthetic cannabinoids. For the quality control, DART was used to analyze cosmetic ingredients and to detect stabilizers in polypropylene. DART has also found use in clinical diagnostics for the detection of phenylketonuria in infants and for urine and plasma analysis. In environmental analysis UV filters and the presence of parabens in cosmetic products and soil for contamination were examined. Other biological applications include the study of flavonoids in plants and analysis of hydrocarbons on the surface of living flies, before and after mating dance.

4.1.4.2.5 Coupling with Chromatographic Techniques

DART can be combined with thin layer chromatography (TLC). In this case, the chromatographic plate or its interesting fragment can be directly analyzed in the ion source. The sensitivity of such analysis reaches the nanograms of the test substance. DART properties, which allow for direct ionization of gases, were used to combine gas chromatography (GC). In this case, the outlet of the chromatography column is located directly in the ionization zone of the DART source. An online combination with capillary electrophoresis was also described. The capillary outlet is positioned just as in GC case. A successful attempt was also made to combine the high performance liquid chromatography (HPLC) with DART technique, where the outlet capillary was made of fused silica, the end of which was located in the ionization zone of the DART source.

4.1.4.3 Flowing Atmospheric Pressure Afterglow (FAPA)

4.1.4.3.1 Introduction

The flowing atmospheric pressure afterglow (FAPA) belongs to a group of APCI methods. It is derived from the atmospheric pressure glow discharge (APGD) technique and differs in terms of the ionization place of the sample. In the case of APGD ionization, the sample is introduced into the glow discharge zone, while in the FAPA method the sample is introduced outside the zone. In both cases, the ionization occurs as a result of an electrical discharge in the gas, which is called a glow discharge. The plasma temperature is about 500 °C, but outside the plasma zone (for the FAPA method), the temperature drops radically to 40–50 °C.

4.1.4.3.2 The Construction of the Ion Source

In the case of FAPA method, two types of ion source geometry are used. One is *pin-to-capillary* geometry (Figure 4.22), and the other *pin-to-plate* geometry. The difference lies in the different shape of the anode; for one method, it is a small diameter tube (on the order of 1 mm), and the other is an anode plate. In both cases the discharge occurs in a chamber, to which from one side the capillary in form of small rod is inserted. The opposite side of the chamber is the anode described above. Near the cathode, the discharge gas, usually helium, is delivered. The gas ensures the flow of generated excited particles. This stream

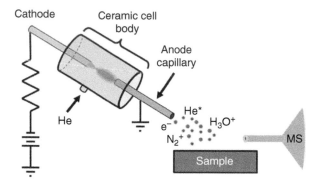

Figure 4.22 Schematic diagram of flowing atmospheric pressure afterglow (FAPA) ion source. *Source*: Reprinted with permission from Ref. [3].

will then pass outside the chamber toward the inlet of the mass spectrometer. It is also possible to operate the source in reverse polarity; in this case cathode and anode location are swapped. In the area between the outlet from the FAPA source and the inlet to the mass spectrometer, a sample is introduced in the form of an aerosol, gas, or solid phase.

4.1.4.3.3 Mechanism of Ionization

Ionization is caused directly, or more frequently indirectly, by metastable helium atoms (He*), which are generated by electrical discharges. In the first case the so-called Penning ionization occurs with the formation of molecular ions. The reaction describing this process is described as follows:

$$He^* + M \rightarrow He + M^{\bullet+} + e^- \tag{18}$$

In the indirect variant, however, first the ionization of gaseous components of the atmosphere such as nitrogen or water vapor occurs:

$$He^* + N_2 \rightarrow He + N_2^+ + e^- \tag{19}$$

$$He^* + H_2O \rightarrow He + H_2O^+ + e^- \tag{20}$$

In the background of FAPA mass spectra, under appropriate conditions of analysis, large clusters of water $(H_2O)_nH^+$ are visible, resulting from the following reaction sequence:

$$N_2^+ + 2N_2 \rightarrow N_4^+ + N_2 \tag{21}$$

$$N_4^+ + H_2O \rightarrow H_2O^+ + 2N_2 \tag{22}$$

$$H_2O^+ + H_2O \rightarrow H_3O^+ + OH \tag{23}$$

$$H_3O^+ + (n-1)H \tag{24}$$

Analytes that have a higher affinity for protons than water clusters can be ionized by reaction with them, with the formation of pseudomolecular ions:

$$(H_2O)H^+ + M \rightarrow nH_2O + MH^+ \tag{25}$$

A typical FAPA mass spectrum is almost identical to ESI technique, where we mainly observe the signal corresponding to a protonated molecule, generally without the presence of fragmentation ions.

4.1.4.3.4 *Applications of FAPA*

The FAPA method as relatively new slowly begins to gain wider use. So far, its utility was demonstrated in pesticide analysis where a detection limit of at least 10 fmol was reached. Monitoring of pesticides in fruits has also potential in the food industry. In this case the presence of pesticides in lettuce leaves, or apple peel, was monitored, and the filtered pulp from various fruit juices was examined. One of the most interesting applications of the FAPA source is the analysis of explosives. Compounds such as trotyl (TNT), hexogen (RDX), or pentrite (PENT) were analyzed. For the latter, a detection limit of 500 amol was obtained. The synthetic and natural homo- and copolymers were also analyzed by FAPA such as starch, cellulose, poly(ethylene terephthalate) (PET), Nylon 6/6, and acrylonitrile butadiene styrene (ABS). Samples were tested directly allowing for identification of the polymer at a very short time of a single measurement of 30 seconds. The method is also well suited for analyzing the products of the synthesis of illegal psychoactive compounds. Methcathinone produced from easily available medication was analyzed by this technique.

4.1.4.3.5 *Couplings with Other Techniques*

The miniaturized FAPA source was used with capillary electrophoresis. In this case, the capillary outlet of the electrophoretic system was integrated into the outlet of the nebulizing gas (nitrogen). The FAPA source was set at an optimal angle to the nebulizer stream. Miniaturization of the source is the first step toward the construction of a small portable instrument that would be able to measure not only in the lab but also outside it, e.g. in public (stations, airports, border crossings, supermarkets). Possibility to use the FAPA source for analyzing gaseous substances was used to couple this technique with GC. This construction was used to analyze the mixture of 13 herbicides and mixtures of 6 warfare agents extracted from the soil. The FAPA source was also linked to a flow-through electrochemical cell (EC-FAPA). This combination was used to simulate the metabolism of psychoactive compounds.

4.1.4.4 Dielectric Barrier Discharge Ionization (DBDI)
4.1.4.4.1 *Introduction*

The dielectric barrier discharge ionization (DBDI) method also belongs to a group of ionization/desorption techniques performed under atmospheric

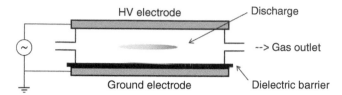

Figure 4.23 The schematic diagram of dielectric barrier discharge ionization (DBDI) ion source.

pressure (*atmospheric pressure desorption/ionization* [ADI]). The method was first used in 2007 as an alternative to other ADI methods already present on the market [4].

4.1.4.4.2 The Construction of the Ion Source
The schematic diagram of the ion source is shown in Figure 4.23. It is based on the principle of electric discharge between two electrodes separated by a dielectric barrier. This process typically employs high voltage with an alternating current. The source provides a stable low temperature plasma, rich in high-energy electrons.

4.1.4.4.3 Mechanism of Ionization
The ionization process is dependent on several factors such as discharge gas polarity, proton affinity, and analyte type. The process is quite complex in that it involves several mechanisms, such as fast atom bombardment, electron ionization, and electron transfer. It can also lead to direct Penning ionization or what is usually dominant mechanism, indirect with the use of water vapor present in the air. The DBDI method belongs to the so-called soft ionization methods; however, fragmentation in the source can occur due to collisions with electrons or metastable particles generated in the plasma.

4.1.4.4.4 Applications of DBDI
Despite the fact that DBDI technique is quite shortly present on the market, it found a relatively wide application range. A mixture of 43 pesticides in complex matrix (apple, orange, tomato), medicaments, personal hygiene, psychoactive substances, explosives, amino acids, water-soluble vitamins, and polycyclic aromatic hydrocarbons were analyzed. Most of the above listed applications were made in combination with liquid chromatography, which proves that these methods are fully compatible and have a high application potential with relatively simple handling. The DBDI source was also used to build a portable mass spectrometer, which can significantly increase the importance of the method in the future.

Questions

- What are the main differences between ADI and API methods?
- What are the typical ions generated by the DART technique and how the polarity of the analyte affects their generation?
- What are the main applications of DART?
- Is it possible to combine DART with chromatographic techniques?
- What is the main difference between FAPA and APGD?
- What are the main applications of FAPA?
- What is DBDI and which are its main applications?

References

1 Gross, F.H. (2014). Direct analysis in real time: a critical review on DART-MS. *Analytical and Bioanalytical Chemistry 406*: 63–80.
2 Hajslova, J., Cajka, T., and Vaclavik, L. (2011). Challenging applications offered by direct analysis in real time (DART) in food-quality and safety analysis. *TrAC Trends in Analytical Chemistry 30*: 204–218.
3 Shelley, J.T., Wiley, J.S., and Hieftje, G.M. (2011). Ultrasensitive ambient mass spectrometric analysis with a pin-to-capillary flowing atmospheric-pressure afterglow source. *Analytical Chemistry 83* (14): 5741–5748.
4 Na, N., Zhao, M., Zhang, S. et al. (2007). Development of a dielectric barrier discharge ion source for ambient mass spectrometry. *Journal of the American Society for Mass Spectrometry 18*: 1859–1862.

4.1.5 Matrix-Assisted Laser Desorption/Ionization (MALDI)

4.1.5.1 Introduction

Przemyslaw Mielczarek[1] and Jerzy Silberring[1,2]

[1] *Department of Biochemistry and Neurobiology, Faculty of Materials Science and Ceramics, AGH University of Science and Technology, Kraków, Poland*
[2] *Centre of Polymer and Carbon Materials, Polish Academy of Sciences, Zabrze, Poland*

Matrix-assisted laser desorption/ionization (MALDI), desorption/ionization on porous silicon (DIOS), desorption/ionization using modified surface-enhanced laser desorption/ionization (SELDI), and nanostructure-assisted laser desorption/ionization (NALDI) are currently and commonly applied in laser-assisted ionization mass spectrometry techniques. Their utilization is extensive because of wide applicability, ranging from high molecular weight (MALDI) analysis, through the possibility of selectively analyzing individual components of the sample (SELDI), to the identification of low molecular weight substances (DIOS and NALDI).

MALDI is an ionization technique used in mass spectrometry that allows analysis of labile macromolecules in the gas phase (in form of pseudomolecular ions) and is classified as the so-called "soft" ionization technique (does not provide spontaneous fragmentation).

The ionization in MALDI ion source is a result of laser radiation, usually using a wavelength of the UV range, of the co-crystallized mixture of sample and matrix. The role of the matrix is to absorb laser energy and transfer it to sample molecules that do not absorb laser light of this range. As a result, the desorption and ionization of the analyte takes place after laser beam shot. A principle of this process is shown in Figure 4.24.

The most commonly observed in the mass spectra are signals coming from protonated molecules, but also sodium and potassium adducts may be observed. The formation of adducts may be affected by many factors, for example, the presence of sodium or potassium ions that are present in solution, synthesis in sodium glass vessels, and presence of other molecules, such as ammonia (e.g. ammonium adducts).

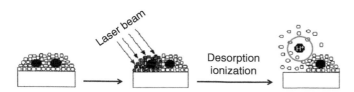

Figure 4.24 Ionization process in the MALDI ion source.

MALDI source works under high vacuum (below 10^{-6} Tr), although atmospheric pressure MALDI (AP-MALDI) is also available (see Section 4.1.5.3). AP-MALDI also offers the ability to perform MS/MS experiments using ion trap and other analyzers, depending on the type of a mass spectrometer. In the case of AP-MALDI, lasers with UV or infrared wavelengths are used, which additionally eliminates the need for special matrices (the matrix may be, for example, an aqueous solution). The limitation of the AP-MALDI technique is the m/z range, limited by the analyzer's parameters (ion trap, quadrupole).

MALDI technique is most often associated with a time-of-flight (TOF) analyzer, which allows rapid identification of molecular weights of peptides, proteins, carbohydrates, nucleic acids, synthetic polymers, or other high molecular weight substances (even above 100 kDa). MALDI is also useful in proteome analysis (see Section 8.1). An additional advantage over the electrospray ionization (ESI) (see Section 4.1.3.2) is better tolerance to impurities, such as salts or detergents. As a result, in MALDI, we mainly obtain signals corresponding to the singly charged pseudomolecular ions, which facilitates interpretation of more complex spectra and does not require deconvolution (see Section 4.1.3.2.5.2) in protein analysis.

The first step in MALDI analysis is to apply the sample and matrix mixture to the target plate. The plate, commonly made of stainless steel, usually contains 384 or 1536 predefined spots. Preparation of the mixture containing sample with matrix should be is crucial to obtain homogeneous crystalline layer after evaporation of the solvent. The most popular technique is the dried droplet method, which involves mixing equal volumes of the sample and the matrix solution and then applying it on the plate with a pipette. One of the most advanced ways is to apply an evenly distributed matrix layer onto tissue sections. The technique of atom sputtering is used in the case of imaging mass spectrometry, when the distribution of substances in tissues is analyzed. In this case, electrospray can be utilized to apply the matrix, which allows the tissue to be covered with a more homogeneous layer, as compared with other methods. After crystallization of the mixture, an excess of the salts present in the sample can also be removed by brief washing of the crystals with water, just prior to the analysis. The differences in solubility of salts and crystals of the sample with the matrix may be used here.

In addition to the standard stainless steel plates, samples can be concentrated on the prepared surfaces, AnchorChip® (Bruker Daltonics). The spot on the plate is covered with a hydrophilic layer, while the surroundings are hydrophobic. As a result, the sample is concentrated on a very small (hydrophilic) surface with a diameter of 0.1–0.5 mm, which results in a significant increase in sensitivity, even up to 1000 times.

Currently, there are disposable plastic plates available, also with prespotted matrices, which further simplifies preparation of the sample prior to analysis and minimizes contamination from other samples previously analyzed on the same target. It is worth noting that application of a too high laser beam energy can desorb polymers from the plastic that effectively interferes with the mass spectra.

4.1.5.2 The Role of Matrix

Przemyslaw Mielczarek[1] and Jerzy Silberring[1,2]

[1] Department of Biochemistry and Neurobiology, Faculty of Materials Science and Ceramics, AGH University of Science and Technology, Kraków, Poland
[2] Centre of Polymer and Carbon Materials, Polish Academy of Sciences, Zabrze, Poland

The matrix plays a key role in the process of desorption and ionization of a sample in MALDI. Its main task is to absorb the energy from laser radiation and ionize molecules of the analyte. The consequence of this process is transfer of the proton from excited matrix molecule to analyte molecules that facilitates ionization. Matrix molecule is usually a weak organic acid that absorbs radiation at the wavelength of the applied laser. The most commonly used matrices and their structures are shown in Table 4.7. Ionization of substances that are not readily protonated (hydrophobic compounds, e.g. polymers) can be assisted by the addition of salts containing suitable ions like Ag^+ (the most popular is AgTFA). This type of treatment is often used in the analysis of synthetic polymers (e.g. polystyrene). On the other hand, the acidic nature of matrices often makes the analysis of non-covalent complexes difficult, due to spontaneous dissociation of these complexes at low pH [1].

Table 4.7 The most commonly used MALDI ionization matrices.

Sinapinic acid (SA)	α-Cyano-4-hydroxycinnamic acid (CHCA)
2,5-Dihydroxybenzoic acid (DHB)	1,5-Diaminonaphthalene (1,5-DAN)

1,5-Diaminonaphthalene has an interesting property, as it promotes spontaneous fragmentation of peptides and proteins. Thus, in the case of pure sample, this matrix can be used for at least partial sequence identification.

4.1.5.3 Atmospheric Pressure MALDI

Giuseppe Grasso

Dipartimento di Scienze Chimiche, Università di Catania, Catania, Italy

Atmospheric pressure (AP)-MALDI was described for the first time in 2000 in two consecutive papers by Laiko et al. [2, 3]. The latter are highly cited papers, demonstrating the large impact that the introduction of this new analytical technique had on the scientific community. Indeed, AP-MALDI differs from traditional MALDI because it does not need vacuum to analyze samples, which can be ionized in air at ambient pressure, thus adding AP-MALDI to the realm of soft ionization techniques working at ambient pressure already in use at that time (ESI and APCI).

Ionization of the sample at ambient pressure rather than in vacuum presents several pros and cons, which should be clearly understood before deciding which experimental approach is the most convenient for a particular sample. In Table 4.8, advantages and disadvantages of vacuum MALDI and AP-MALDI

Table 4.8 Comparison of vacuum MALDI and AP-MALDI.

Type of ionization	Vacuum MALDI	AP-MALDI	References
Advantages	Low limits of detection; low laser fluency; low chemical noise; high mass range (up to 500 kDa in combination with a TOF analyzer)	Interchangeable with other ambient pressure ion sources; preserves intact precursor ions, low PSD fragments; possibility of liquid matrices; use of ion traps for MS/MS experiments; high salt tolerance	[4, 5]
Disadvantages	High post-source decay (PSD); poor pulse-to-pulse; reproducibility	Excessive appearance of analyte salt and analyte matrix adducts; high in source decay fragments; poor pulse-to-pulse reproducibility; low ion yields; low mass range (up to 6 kDa)	[4, 6]
Examples of applications	Biomarkers discovery; food chemistry and natural product analyses; clinical applications	Works of art; proteolytic enzymes; solid-state arrays; metabolites, lipids, and small peptides	[7–16]

are reported together with some examples of application for both ionization techniques given in recent literature.

For a generic proteomic study where tryptic digests of proteins have to be analyzed, the use of AP-MALDI or traditional MALDI produces significantly different results that have to be considered when public accessible search algorithms and databases are used to identify the proteins. Indeed, the two approaches can produce quite different fragmentation patterns, which include the number of generated fragment ions, the main types of fragment ions, and the resolution of the fragment ions [17]. Particularly, most databases commonly used do not consider fragment ion types derived from mono-protonated precursor ions, and this constitutes a major limitation using AP-MALDI low-energy CID for database search derived identification of proteins. Although this problem can easily be circumvented by manually identifying peptide fragments and their possible modifications, in the case of a mixture of several different unknown proteins, the limitation of database automatic search dictated by the use of AP-MALDI must be taken into serious consideration.

On the contrary, the AP-MALDI source is particularly advantageous in the case of solid-state arrays, which require a high-throughput screening analysis. Application of AP-MALDI to enzyme arrays on solid-state supports for activity screening has been widely reported. Indeed, unlike in the case of vacuum MALDI-TOF configuration, in AP-MALDI, the samples to be analyzed are located outside the vacuum and alongside the associated robotics and process lines, giving the possibility to work on all sorts of solid-state arrays [12, 13]. Another advantage of AP-MALDI in this case is the possibility to carry out a multiplexed approach for the sample analysis, being able to recycle the biomolecules for further studies. Indeed, while the vacuum can be a problem for some biological molecules, the use of AP preserves the state as well as the activity of most biomolecules. Before or even after the MS investigation, areas of the sample not contaminated with matrix molecules can easily be used for further solid-state analyses such as AFM, SPR, etc. As an example, Figure 4.25 shows a scheme of the procedure applied for the *in situ* tryptic digestion and AP-MALDI-MS characterization of immobilized matrix metalloprotease catalytic domains (MMPcds). The high salt tolerance of AP-MALDI, necessary to avoid cdMMP degradation and deactivation, represents a key feature in this case.

Analogously, the possibility of placing objects of any sort and shape in front of the AP source without the need to put them in vacuum makes the use of AP-MALDI particularly suitable for the analytical investigation of the artwork. Indeed, this technique has been applied to obtain qualitative and semiquantitative data about organic and inorganic components of inks of ancient documents in a manner that is minimally destructive and does not suffer from significant interference from the substrate (see Figure 4.26) [19].

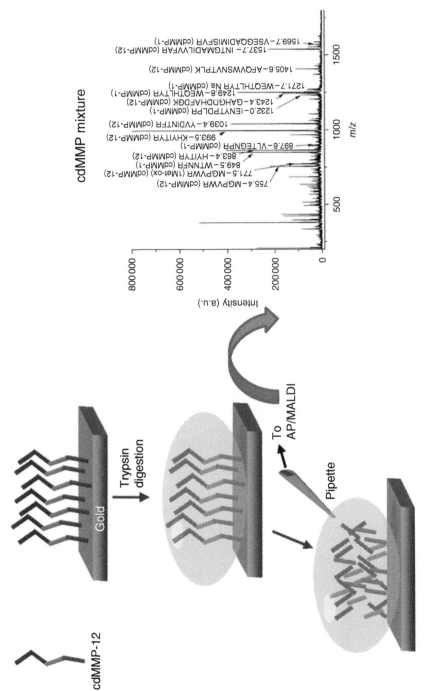

Figure 4.25 Scheme of the procedure applied for in situ tryptic digestion and AP-/MALDI-MS characterization. Enzyme molecules, previously immobilized on the gold surface (a mixture of different cdMMPs in this particular case), are put in contact with the trypsin solution in a humid chamber at 37°C overnight. The supernatant solution (0.5 µl) is sampled after 6 hours for AP-/MALDI-MS analysis (resultant spectrum on the right side). The same experimental procedure can also be applied using a different peptidase, such as endoproteinase Glu-C. *Source:* Adapted with permission from Ref. [12], Copyright 2006, Wiley.

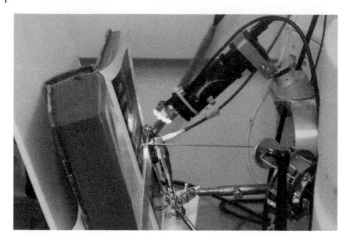

Figure 4.26 Home-modified AP-/MALDI- MS apparatus used for investigating the chemical composition of inks used in ancient books. *Source*: Reprinted with permission from Ref. [18], Copyright 2009, Springer.

It has also been reported that AP-MALDI generates pseudomolecular ions with lower internal kinetic energies than those produced by traditional MALDI [20], preserving the ions from PSD fragmentation. However, such a low internal kinetic energy of the ionized molecule also causes the formation of cooler molecular ions. The latter can result in clustering of molecular species with matrix molecules, which are largely detected in most AP-MALDI analyses, hindering the analysis of samples containing very low abundance and/or not well ionizable species. In addition, an increase in the laser energy threshold in AP-MALDI compared with vacuum MALDI has also been observed, and it can be explained by considering that protonated gas-phase ions derive from fragmentation of hot charged clusters. Indeed, a rapid drop in the cluster temperature in AP is observed due to a higher collisional cooling rate of the clusters with air, resulting in cluster fragmentation and escape of the protonated ions. Therefore, in order to support an efficient ion production, relatively higher laser power than in vacuum MALDI must be used in the AP system [4].

Particular attention has to be paid when using AP-MALDI for the identification of the cleavage sites of an enzyme along a peptide substrate, as an appropriate use of specific matrix/sample ratios must be taken into account. Indeed, it has been reported that as in the case of vacuum MALDI, a low matrix/sample ratio can be responsible for peptide fragment detection not due to the action of the enzyme but ascribed to the ionization process, possibly leading to wrong enzyme cleavage site assignment [21].

Finally, one important advantage of AP-MALDI over vacuum MALDI is its better suitability for the use of liquid matrices. It is well known that MALDI sample preparations have a finite lifespan, the latter depending on the high matrix consumption observed in the traditional ionization process, especially when higher laser fluency has to be used to obtain a detectable signal. Enhanced ionization efficiency and signal durability have been reported using 2,5-DHB liquid matrix, which provided up to 100% sequence coverage for peptides [22].

4.1.5.4 MALDI Mass Spectra Interpretation

Przemyslaw Mielczarek[1] and Jerzy Silberring[1,2]

[1] *Department of Biochemistry and Neurobiology, Faculty of Materials Science and Ceramics, AGH University of Science and Technology, Kraków, Poland*
[2] *Centre of Polymer and Carbon Materials, Polish Academy of Sciences, Zabrze, Poland*

MALDI allows for the observation of pseudomolecular ions produced by the addition of proton to the analyte, but also adducts with sodium or potassium ions can be formed. Due to the presence of a matrix, the interpretation of mass spectra in the low *m/z* range is difficult. Matrix ions cause the formation of many abundant signals visible below 500 *m/z*. This significantly impedes analysis of small molecules using this technique. Exemplary mass spectra obtained by MALDI technique are shown in Figure 4.27 (protein) and Figure 4.28 (polymer).

MALDI is a mild ionization technique (does not cause spontaneous fragmentation of the sample during ionization process). It enables analysis of multicomponent samples without prior chromatographic separation and also has high sensitivity at femtomolar or even attomolar level [23].

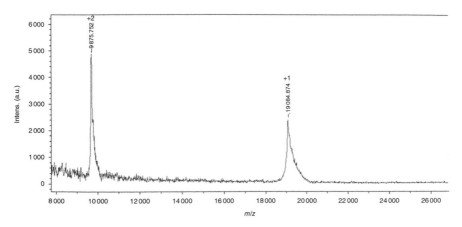

Figure 4.27 Mass spectra of staphopain C obtained by MALDI-TOF technique. Ion at *m/z* 19 084.874 (singly charged) is a pseudomolecular ion [M+H]⁺. Another ion (*m/z* 9675.752) is a doubly charged ion of staphopain C.

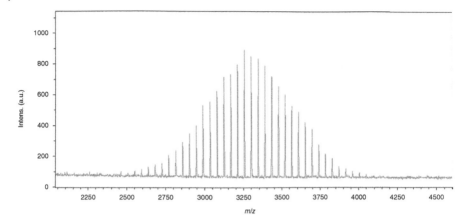

Figure 4.28 Spectrum of polyethylene glycol (PEG) obtained by MALDI-TOF showing equidistant peaks 44.026 Da apart.

4.1.5.5 Desorption/Ionization on Porous Silicon (DIOS)

Przemyslaw Mielczarek[1] and Jerzy Silberring[1,2]

[1] *Department of Biochemistry and Neurobiology, Faculty of Materials Science and Ceramics, AGH University of Science and Technology, Kraków, Poland*
[2] *Centre of Polymer and Carbon Materials, Polish Academy of Sciences, Zabrze, Poland*

DIOS is a technique that enables analysis of low molecular weight substances using MALDI-TOF instrument. DIOS or porous silicon dioxide (DIOSD) plates can be used for direct ionization of the deposited sample without any use of a matrix, enabling low molecular weight substances to be analyzed that are not detected by MALDI. The presence of the matrix is the reason why MALDI is not used for the analysis of low molecular weight compounds due to the abundant matrix signals in the range below 500 m/z [24].

Porous silicon is a material that can be used for ionization of many molecules and has been invented by G. Siuzdak's research team for a wide range of compounds, such as sugars, peptides, glycolipids, or low molecular weight medicines, ranging from 500 to 12 000 Da. For these compounds, mainly pseudomolecular ions were observed. For peptides, a clear signal was obtained even for 700 attomoles of the substance deposited on the target plate. In DIOS or DIOSD, the function of the matrix is taken by porous silicon, onto which the analyzed sample is deposited.

The plate made of porous silicon, before its introduction into the mass spectrometer, is attached to a standard target plate used in commercial mass spectrometers. Care should be taken to avoid oversized dimensions as this may irreversibly damage the ion source. The appropriate adapter can easily be manufactured by the university workshop. For both MALDI and DIOS, the most commonly used analyzer is TOF. DIOS allows analysis of low molecular weight compounds, such as neurotransmitters, prostaglandins or steroids, psychoactive substances, and other compounds. This technique has also been used in forensic science, e.g. to detect traces of low molecular weight polymers (nonoxynol-9 and octoxynol-9) in the suspect and victim. Because the use of DIOS does not require additional preparation of the sample, it does not require the use of advanced separation methods like chromatography. As a result, the identification of the substance is fast and easy.

DIOS is also used to monitor enzymatic reactions directly on the surface of porous silicon, for example, the formation of choline from acetylcholine. It is also possible to analyze peptide maps without matrix, and the protein digestion can be performed directly on the DIOS target plate.

4.1.5.6 Surface-Enhanced Laser Desorption/Ionization (SELDI)

Przemyslaw Mielczarek[1] and Jerzy Silberring[1,2]

[1] Department of Biochemistry and Neurobiology, Faculty of Materials Science and Ceramics, AGH University of Science and Technology, Kraków, Poland
[2] Centre of Polymer and Carbon Materials, Polish Academy of Sciences, Zabrze, Poland

SELDI is a relatively new technique utilizing chromatography and MALDI-TOF mass spectrometry. Surface-modified SELDI chips are available in various arrangements as hydrophobic (like C18 stationary phase), hydrophilic, or ion exchangers that interact with proteins, nucleic acids, or other molecules with well-defined physicochemical properties [25].

Sample can be directly deposited on the SELDI chips (target plates), without any cleaning or other preparation, regardless of salt content and other undesirable substances. After a short period of incubation, contaminants should be removed by simply washing the surface of the plate with a suitable solvent. Samples prepared in such a way can be analyzed in a MALDI-TOF instrument. Surface-modified SELDI plates adsorb only a specific group of substances characterized by similar physicochemical properties, e.g. hydrophilic or hydrophobic. Other components of the sample are not adsorbed on the surface and are removed during the plate washing procedure. By applying different surface modifications, it is possible to obtain a predefined biological sample and to significantly improve the quality of the information obtained by this method. The sequential binding and rinsing procedure is similar to a typical liquid chromatography, and is often referred to as retentate chromatography, by analogy to the interaction with the stationary phase in a liquid chromatography column.

SELDI is a technique that allows rapid identification of biomolecules in complex matrices, such as blood, cerebrospinal fluid, urine, etc. with potential clinical applications for fast screening for biomarkers of pathophysiology. This was the main idea to design this system with strong expectations to speed up proteomic research. This method is applicable in many areas of science. It was used to identify protein markers in early stages of cancer, such as prostate cancer, pancreas, or skin. Protein profiles were tested in urine and blood. Alzheimer's disease was also investigated using SELDI to monitor changes in C-terminally digested ß-amyloid peptides. For this purpose, antibodies specific for the N-terminal ß-amyloid peptides have been prepared on the SELDI chip. In this case, the advantage of SELDI over ELISA is the simultaneous identification of multiple peptides with mass spectrometry during one measurement rather than a series of separate tests for each peptide and complicated data interpretation. After initial excitement with this technique, which was carrying a lot of expectations, critical feedback on poor reproducibility appears, thus limiting its potential applications in clinical diagnosis.

4.1.5.7 Nanostructure-Enhanced Laser Desorption/Ionization (NALDI)

Przemyslaw Mielczarek[1] and Jerzy Silberring[1,2]

[1] Department of Biochemistry and Neurobiology, Faculty of Materials Science and Ceramics, AGH University of Science and Technology, Kraków, Poland
[2] Centre of Polymer and Carbon Materials, Polish Academy of Sciences, Zabrze, Poland

NALDI has been introduced as a new alternative ionization technique in mass spectrometry. It has been shown that nanostructures, such as carbon nanotubes, act in the same way as a matrix used in MALDI. Additionally, the use of silicon substrate (DIOS) makes desorption/ionization process efficient without matrix. The combination of these two concepts, utilizing nanostructure-coated plates, results in the new ionization technique being called NALDI. The target plate is coated with a layer of semiconducting nanotubes or nanofibers, made of zinc oxide (ZnO), tin(IV) oxide (SnO_2), gallium nitride (GaN), or silicon carbide (SiC). Such nanostructure layers are applied or "grown" on a silicon substrate. Currently, mass spectrometry target plates with various types of nanostructures are commercially available for NALDI analyses [26].

Measurement with the aid of NALDI is very simple. It is based on the application of sample directly on the target plate (usually 0.5–1.0 μl of the sample solution). The mass limit may vary depending on the plate type and nature of the sample. After solvent evaporation, the plate is ready to be measured. For comparison with the MALDI-type analysis, NALDI has much higher sensitivity for the low molecular weight compounds that are not well ionized in MALDI, and the ions are visible in the range of a mass spectrum where ions from matrix might be visible as well.

The surface of the NALDI target plate is strongly hydrophobic. This property can be used to purge the sample or selectively bind to the plate, similarly to SELDI analyses. After depositing the sample to the surface of the target plate, the hydrophobic compounds are adsorbed on the surface. After washing the surface, interfering substances (salts and other contaminants) can be removed before measurement. This technique has been used to analyze biological fluids, such as urine, blood, serum, cell culture media, and more. SELDI also simplifies analysis of drugs and their metabolites in the urine, thus highly reducing preparation procedure.

4.1.5.8 Summary

Przemyslaw Mielczarek[1] and Jerzy Silberring[1,2]

[1] *Department of Biochemistry and Neurobiology, Faculty of Materials Science and Ceramics, AGH University of Science and Technology, Kraków, Poland*
[2] *Centre of Polymer and Carbon Materials, Polish Academy of Sciences , Zabrze, Poland*

MALDI and DIOS are two complementary ionization methods. They allow observation of ions formed by the addition of protons to the molecules (pseudomolecular ions) in a wide range of molecular weights. For DIOS, surface modifications can help to achieve better sensitivity. SELDI is used for the identification of unknown substances and can be applied to analyze protein and peptide profiles. The advantage of this technique is an increased selectivity, which allows to "pick up" selected types of molecules and identify them even in complex mixtures, such as body fluids (e.g. blood, urine, and cerebrospinal fluid). However, the disadvantage is low reproducibility of the results and high cost of the apparatus. The latest laser ionization technique, NALDI, combines the advantages of DIOS (no matrix) and SELDI (high selectivity). This approach has been widely used in biological analyses and will undoubtedly be further developed due to its large analytical potential.

Questions

- What is the role of the matrix in MALDI?
- What are the advantages of MALDI over ESI?
- What problems can be encountered in the analysis of non-covalent complexes by MALDI?
- What are the main advantages and disadvantages of AP-MALDI versus MALDI?
- Give some examples of analysis where the use of AP-MALDI is particularly useful.

- Discuss the main applications of DIOS and in which case represents a better alternative to MALDI.
- What is the main drawback of SELDI?
- What are the main advantages and applications of NALDI?

References

1 Silberring, J. and Ekman, R. (2002). *Mass Spectrometry and Hyphenated Techniques in Neuropeptide Research*, 259–275. New York: Wiley.
2 Laiko, V.V., Baldwin, M.A., and Burlingame, A.L. (2000). Atmospheric pressure matrix assisted laser desorption/ionization mass spectrometry. *Analytical Chemistry 72*: 652–657.
3 Laiko, V.V., Moyer, S.C., and Cotter, R.J. (2000). Atmospheric pressure MALDI/ion trap mass spectrometry. *Analytical Chemistry 72*: 5239–5243.
4 Moskovets, E., Misharin, A., Laiko, V., and Doroshenko, V. (2016). A comparative study on the analytical utility of atmospheric and low-pressure MALDI sources for the mass spectrometric characterization of peptides. *Methods 104*: 21–32.
5 Moyer, S.G. and Cotter, R.J. (2002). Atmospheric pressure MALDI. *Analytical Chemistry 74*: 468A.
6 Bierstedt, A., Stindt, A., Warschat, C. et al. (2014). High repetition rate atmospheric pressure matrix-assisted laser desorption/ionization in combination with liquid matrices. *European Journal of Mass Spectrometry 20*: 367–374.
7 Hajduk, J., Matysiak, J., and Kokot, Z.J. (2016). Challenges in biomarker discovery with MALDI-TOF MS. *Clinica Chimica Acta 458*: 84–98.
8 Yoshimura, Y., Goto-Inoue, N., Moriyama, T., and Zaima, N. (2016). Significant advancement of mass spectrometry imaging for food chemistry. *Food Chemistry 210*: 200–211.
9 Silva, R., Lopes, N.P., and Silva, D.B. (2016). Application of MALDI mass spectrometry in natural products analysis. *Planta Medica 82*: 671–689.
10 Ucal, Y., Durer, Z.A., Atak, H. et al. (2017). Clinical applications of MALDI imaging technologies in cancer and neurodegenerative diseases. *Biochimica et Biophysica Acta, Proteins and Proteomics 1865* (7): 795–816.
11 Kiss, A. and Hopfgartner, G. (2016). Laser-based methods for the analysis of low molecular weight compounds in biological matrices. *Methods 104*: 142–153.
12 Grasso, G., Fragai, M., Rizzarelli, E. et al. (2006). In Situ AP-MALDI characterization of anchored MMPs. *Journal of Mass Spectrometry 41*: 1561–1569.
13 Grasso, G., Fragai, M., Rizzarelli, E. et al. (2007). A new methodology for monitoring the activity of cdMMP-12 anchored and freeze-dried on Au (111). *Journal of the American Society for Mass Spectrometry 18*: 961–969.

14 Grasso, G., Rizzarelli, E., and Spoto, G. (2007). AP-MALDI/MS complete characterization of insulin fragments produced by the interaction of IDE with bovine insulin. *Journal of Mass Spectrometry 42*: 1590–1598.

15 D'Agata, R., Grasso, G., Parlato, S. et al. (2007). The use of atmospheric pressure laser desorption mass spectrometry for the study of iron-gall ink. *Applied Physics A: Materials Science and processing 89*: 91–95.

16 Kompauer, M., Heiles, S., and Spengler, B. (2017). Atmospheric pressure MALDI mass spectrometry imaging of tissues and cells at 1.4-μm lateral resolution. *Nature Methods 14*: 90–96.

17 Mayrhofer, C., Krieger, S., Raptakis, E., and Allmaier, G. (2006). Comparison of vacuum matrix-assisted laser desorption/ionization (MALDI) and atmospheric pressure MALDI (AP-MALDI) tandem mass spectrometry of 2-dimensional separated and trypsin-digested glomerular proteins for database search derived identification. *Journal of Proteome Research 5*: 1967–1978.

18 Giurato, L., Candura, A., Grasso, G., and Spoto, G. (2009). In situ identification of organic components of ink used in books from the 1900s by atmospheric pressure matrix assisted laser desorption ionization mass spectrometry. *Applied Physics A 97*: 263–269.

19 Grasso, G., Calcagno, M., Rapisarda, A. et al. (2017). Atmospheric pressure MALDI for the non-invasive characterization of carbonaceous ink from Renaissance documents. *Analytical and Bioanalytical Chemistry 409*: 3943–3950.

20 Wolfender, J.-L., Chu, F., Ball, H. et al. (1999). Identification of tyrosine sulfation in Conus pennaceus conotoxins alpha-PnIA and alpha-PnIB: further investigation of labile sulfo- and phosphopeptides by electrospray, matrix-assisted laser desorption/ionization (MALDI) and atmospheric pressure MALDI mass spectrometry. *Journal of Mass Spectrometry 34*: 447–454.

21 Grasso, G., Mineo, P., Rizzarelli, E., and Spoto, G.M.A.L.D.I. (2009). AP/MALDI and ESI techniques for the MS detection of amyloid-β peptides. *International Journal of Mass Spectrometry 282*: 50–55.

22 Ait-Belkacem, R., Dilillo, M., Pellegrini, D. et al. (2016). In-source decay and pseudo-MS3 of peptide and protein ions using liquid AP-MALDI. *Journal of the American Society for Mass Spectrometry 27*: 2075–2079.

23 Kraj, A. and Silberring, J. (2008). *Proteomics: Introduction to Methods and Applications*, 89–99. Hoboken: Wiley.

24 Peterson, D.S. (2007). Matrix-free methods for Laser desorption/ionization mass spectrometry. *Mass Spectrometry Reviews 26*: 19–34.

25 Tang, N., Tornatore, P., and Weinberger, S.R. (2004). Current development in SELDI affinity technology. *Mass Spectrometry Reviews 23*: 34–44.

26 Daniels, R.H., Dikler, S., Li, E., and Stacey, C. (2008). Break free of the matrix: sensitive and rapid analysis of small molecules using nanostructured surfaces and LDI-TOF mass spectrometry. *Journal of Laboratory Automation 13*: 314–321.

4.1.6 Inductively Coupled Plasma Ionization (ICP)

Aleksandra Pawlaczyk and Małgorzata Iwona Szynkowska

Faculty of Chemistry, Institute of General and Ecological Chemistry, Lodz University of Technology, Łódź, Poland

4.1.6.1 Introduction

Analytical methods using inductively coupled plasma (ICP) as the source of excitation and ionization include the ICP-OES (optical emission spectrometry) and ICP-MS (mass spectrometry) techniques. Both marriages of ICP with optical and mass spectrometers resulted in creation of instruments capable of performing multielemental determinations. Since their birth, they evolved rapidly, and nowadays, they are among the most useful methods of trace element analysis and find many applications in the analysis of biological, clinical, geological, forensic, and environmental materials, as they allow liquid, solid, gaseous, organic, and inorganic samples to be introduced into ICP system [1–9].

Inductively coupled plasma mass spectrometry (ICP-MS) can be recognized as a universal technique for almost every material elemental characterization. Since its commercial introduction into the market in 1983, it developed during the next two decades into a sophisticated and irreplaceable in many laboratories instrument of the first choice. New millennium marked important directions of its future evolution. One of them is the possibility of effectively coupling this instrument with other techniques, devices, or new approaches to sample introduction systems that has opened new doors for a wide range of new applications, including elemental bioimaging or speciation analysis, which made the ICP-MS even more powerful [1–3]. Along with these developments, some difficulties with matrix interferences were solved by introduction of the collision cell technology (CCT), high-resolution mass spectrometers (HR-ICP-MS), and tandem technology. All those implemented innovations led to ICP-MS technique ready to be customized for a specific sample type or form of the studied analyte. The main features, which render the ICP-MS instrument so unique and perfectly suited to solve a broad range of many analytical problems, are ability to precisely identify and measure (in semiquantitative or quantitate way) most of the elements in the periodic table in a multielement analysis mode, measure individual signals from isotopes of the analyte elements and isotope ratios, detect and determine the concentrations of analyte elements at trace levels, and perform the analysis with high accuracy and precision and at a large linear dynamic working range [1, 4–8].

4.1.6.2 ICP as a Technique of Elemental Analysis and ICP Principle

Among the spectroscopic methods, using different physical and chemical phenomena based on the interaction of a light with matter, four groups of techniques can be distinguished, namely, absorption, emission, luminescence, and light scattering, leading to redirection of the radiation and/or transitions between the energy levels of the atoms or molecules. In spectroscopy, these four phenomena can be used to measure atoms or molecules in qualitative or quantitative way or to study physical processes. In the case of light scattering, some of the electromagnetic radiation is transferred through the matter and scattered in other directions, instead of traveling in the original direction (redirection of light). Scattering may or may not be accompanied by energy transfer, depending on whether the light is shifted or not, and to which degree different types of scattering can be distinguished. Adsorption occurs when there is a transfer of energy from the radiation field to atom or molecule, and, as a result, a transition from a lower level to a higher level takes place. Consequently, if the reverse process called emission appears, there is a decay from a higher energy to a lower level with energy transfer or with no emission of radiation (nonradiative decay). In emission techniques, atoms, molecules, or ions are excited to the appropriate energy levels by means of a specific external factor that supplies this energy. This factor may be, for example, a flame, electric spark, spark discharge, or plasma. If the excitation is caused by a high temperature energy source, then those techniques are called optical emission techniques; if by light they are named fluorescence techniques. Excited atoms, particles, or ions reluctantly stay in the excited state, and they return to their basic states, emitting extra energy in the form of radiation of a specific wavelength [9]. The composition of the emitted radiation at specific wavelengths is characteristic for a given atom or molecule, and the emission intensity itself is linearly proportional to the analyte concentration and can be used for quantification in ICP-OES technique. Both emission spectrometry and mass spectrometry refer to elements because high energy of the excitation source destroys the molecular structure of the sample and the spectra obtained by ICP-OES technique are closely related to the electronic structure of the atom. For ICP-MS, the ionization part in the high-energy source is crucial from the point of view of gathered signal, whereas for ICP-OES technique the emission of radiation of a specific wavelength plays the key role [3,10–11].

Plasma can be treated as a coexistence, in a confined space, of the positive ions, electrons, and neutral species of an inert gas, which most frequently is argon or helium. The most common plasma source in atomic spectroscopy is radio-frequency (RF) inductively coupled plasma. The power, typically in the range of 0.5–1.5 kW at a frequency of 27 or 40 MHz, is supplied by a coil made of copper tubing with recirculating water inside. The plasma is initiated by a spark from a Tesla coil. The spark generates so-called seed electrons, leading to ionization of the carrier gas. This process is sustained by constant delivery of the power to the plasma [3, 4, 11–15].

Other plasma sources are also used, such as microwave-induced plasma (MIP), direct current plasma (DCP), and glow discharge (GD). GD phenomena are applied as a source of light in many devices, such as plasma screen TV, and are typically low-pressure plasmas based on the cathodic sputtering to atomize solid samples. The process is initiated by the flow of electric current through the glow discharge. As soon as the negatively charged electrons are attracted by the anode and moved toward it, they may collide with the fill-gas atoms and generate gas (e.g. argon) ions. Thus, after the sample is atomized as a result of a collision with electrons and metastable fill-gas atoms, it would be excited and ionized. In the case of MIP, plasma carrier gas such as argon, nitrogen, or helium is directed through the capillary, and ignition starts with a spark from a Tesla coil. The power of 50–200 W at 2.45 GHz is delivered to the plasma from the microwave generator. This plasma source is characterized by quite high excitation temperatures reaching 7000–9000 K with a relatively low gas temperature of 1000 K, which seems to be suitable for the emission of nonmetals. In the DCP, an electrical discharge is struck between two anodes, through where argon gas flows and cathode situated above the central apex. As a consequence, an inverted V-shaped plasma is created. In the section below, we only focus on the RF-ICP plasma [4].

The basic advantages of ICP-MS technique include short analysis time, good precision and accuracy, low detection limits, a wide range of linearity of the calibration curve, and the possibility of performing simultaneous multielement analysis (Table 4.9). By using this technique, most elements of the periodic table can be examined, while noble gases, elements with a short half-life, and H, C, O, N, and F atoms are not determined. ICP-MS technique also offers the possibility of identifying the isotopic composition. In terms of the analytical range of ICP-OES and ICP-MS detection, they complement each other. ICP-MS measures concentrations from $pg\,g^{-1}$ (ppt) to several tens of $ng\,g^{-1}$ (ppb), while the ICP-OES covers the concentrations from $ng\,g^{-1}$ (ppb) to several tens of $mg\,g^{-1}$ [1–6].

It should also be highlighted that for some elements, like B, Br, Ca, I, K, P, S, Se, and Si, very high detection limits are achieved. For Br and I, small amounts of positive ions are generated in the ICP plasma. For example, the ionization potential of elements like As ($E_0 = 9.81$ eV), Se ($E_0 = 9.75$ eV), or P ($E_0 = 10.49$ eV) still enables to perform quantitative analysis since their ionization efficiencies reach 52, 33, and 33%, respectively. In the case of elements, such as S ($E_0 = 10.36$ eV) or F ($E_0 = 17.42$ eV), their ionization efficiencies, not exceeding 14% for S and 0.001% for F, make their determination almost impossible. Moreover, for elements like K, P, S, Se, and Ca, many isobaric and polyatomic interferences resulting from sample matrix or plasma species may be created, which will overlap with the signal of the primary isotope. One of the solutions can be the choice of less abundant isotopes with minor or no interferences for their quantification, which will highly influence detection capabilities for these elements [12–18].

Table 4.9 Selected analytical parameters of RF ICP-MS technique, where RSD is a relative standard deviation.

Parameter	ICP-MS
Information gathered	Elemental (not molecular), detection and/or quantification
Mass range	5–260 AMU
Detection limits	Even below 10 ppt
Range of the linearity of the calibration curve	10^7
Short-term precision[a] (RSD)	0.1–3%
Long-term precision[b] (RSD)	<5%
Acceptable concentration of salts in the solution	0.1–0.4%
Analytical rage	From ppt to several dozen ppb
Time of analysis	2–3 minutes
Sample consumption	3–4 ml
RF power	0.5–1.5 kW
RF frequency	27 MHz (40 MHz – frequency commonly used in ICP-OES)
Vacuum	High $\sim 10^{-7}$ to 10^{-8} mbar

[a] Short-term precision understood as multiple measurement of the same value in a short time interval.
[b] Long-term precision determined on the basis of measuring the same sample over a longer period of time, e.g. several hours.

4.1.6.3 Ionization of Elements and Ionization Efficiency

In contrast to the concept of plasma in biology, plasma in physics is defined as the state of matter characterized by a considerably high ionization of particles, similar to the total ionization. In analytical chemistry, plasma is an ionized gas (usually argon), which is a mixture of free positive ions and electrons, with a positive and negative total charge being approximately the same. The temperature distribution in the plasma ranges from 5000 K outside to even 10000 K in the center of the plasma channel. Such high temperatures ensure high excitation efficiency and ionization and small chemical interferences [13]. At a temperature of approximately 6000 K, even about 90% of the elements are ionized. Based on the calculations made by Niu and Houk [16], the relation between the efficiency of ionization depending on the plasma temperature can be studied. In general, with higher plasma temperature, the better conditions are assured for the ionization of most elements. A drop in plasma temperature reduces significantly ICP-MS applicability. Elements with ionization energies not exceeding 6 eV are ionized almost in 100% at plasma temperature around 6500 K. The increase of plasma temperature to 500 K causes the elements with ionization energy of 7.5 eV or less to be ionized in nearly 100%. Only in some specific situations, lower plasma temperatures are preferable, e.g. cool plasma conditions are implemented to determine the concentration of elements such as K, Ca, or Fe, which have low ionization energies [16–18].

4.1.6.4 Mechanism of ICP Formation

The basic scheme of the ICP-MS system is presented in Figure 4.29. The ICP-MS spectrometer consists of the sample introduction system, ICP ion source, interface region, mass analyzer (e.g. quadrupole, time of flight [TOF]), detector (electron multiplier), and signal processor (Figure 4.29). Ions excited in the plasma pass through the interface to the mass analyzer. Plasma is generated at atmospheric pressure, while the analyzer is in a vacuum system. Therefore, an interface consisting of two types of cones, sample cone and a skimmer, is a bridge between the part of instrument working under normal low pressure and high vacuum. Ions from the plasma are sucked into the zone of gradually decreasing pressure. Diameters of holes in cones do not exceed 1 mm, and their size affects the number of ions reaching the analyzer and the vacuum. Behind the cones there are electrostatic lenses (focusing lens) that focus, form, and direct the ion beam into the analyzer axis, in which they are separated according to their mass-to-charge ratio [18–21].

Plasma ICP (mostly argon) is produced by heating in a plasma torch. A typical torch consists of three concentrically arranged channels. The innermost channel (injector), through which the carrier gas flows, is responsible for the transport of the sample to the plasma. Plasma gas flow is used to create plasma and cooling. An auxiliary gas flows between the carrier gas and the plasma gas, which separates the plasma from the inside of the torch. Highly pure argon gas flows through the quartz tube, which is surrounded, in its upper part, by an induction coil connected to an RF generator (c. 27–40 MHz). As a result of the short-term high-voltage discharges from the Tesla coil, the seeding electrons are created and initiate generation of argon ions. High-frequency current flowing in the induction coil generates a time-varying magnetic field (closed ellipses outside the coil), which in turn induces the flow of electron whirl current. This is accompanied by the production of large amounts of Joule's heat. All these processes cause intense ionization of the gas and plasma formation. The conductive plasma is therefore enclosed in a high-frequency electromagnetic field, and a continuous supply of energy from the induction coil is needed [16].

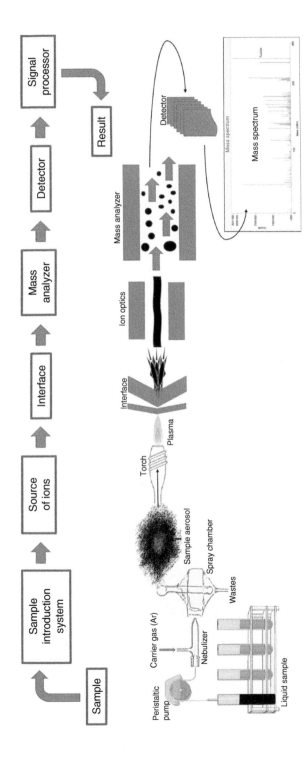

Figure 4.29 Schematic diagram of ICP-MS.

4.1.6.5 Ways of Plasma View and Plasma Generation

The plasma is generated by seeding ions in a stream of argon, which flows through an RF and magnetic fields in the area of the induction coil (Figure 4.30). An initial step is a high-voltage spark, which finally leads to initial argon ionization. Created electrons and ions, after being accelerated by the magnetic field, can impinge with other argon atoms. In consequence, more argon is ionized, and plasma is generated. The energy is delivered to the electrons and ions, and whole process of energy transfer is termed as inductive coupling [19]. The sample aerosol introduced into the innermost region of the plasma known as axial channel passes through the following zones: the preheating zone (PHZ), the initial radiation zone (IRZ), and the normal analytical zone (NAZ). At the end, there is a plasma tail (Figure 4.30) where the ions may further react to form additional neutral atoms and oxides [16–19]. The analytical zone is considered to be an area of 15–25 mm above the inductive coil where the emission is measured. Thus, the optimal sampling area within the plasma is located few mm downstream from the tip of the IRZ.

Before the analytes reach the plasma tail, four main processes take place: desolvation, evaporation (in the PHZ), atomization (in the IRZ), and ionization (in the NAZ). In the PHZ, the solvent is removed by evaporation. Then, the aerosol particles are turned into gases and subsequently undergo atomization and ionization processes, i.e. decomposition into atoms and ions. The outer region of the plasma is cooled by the plasma gas, flowing tangentially at a considerable velocity to prevent the torch from melting. This process supports a compact and stable plasma formation. The upstream part of the induction region may reach temperatures even as high as $10\,000\,K$; however, the typical ionization temperature ranges from 6500 to $7500\,K$ under hot plasma conditions [16–21].

The NAZ constitutes of mostly neutral atoms of Ar, H, O, and N that are ionized to produce a squad of ions and electrons within the plasma region. In general, main components of the sample injected into the plasma are singly charged atomic cations M^+, but depending on the plasma conditions (composition, temperature), sample matrix, type of element, etc., the ionization efficiency within the NAZ may change, and some amounts of neutral atoms (M), polyatomic MO^+ ions, or doubly charged M^{2+} ions may also appear. In the NAZ the contribution of the MO^+ and M^{2+} is much smaller than the population of M^+ ions, which dominate in this plasma region. Concentration of the element in the sample should be determined based on the calibration curve [13–14, 16–18, 21].

4.1.6.6 Sample Introduction

In most cases, the sample is delivered to the plasma as a result of the nebulization (spraying) process. A sample in the form of a solution is transported to a nebulizer with the aid of peristaltic pump. The resulting aerosol disperses the

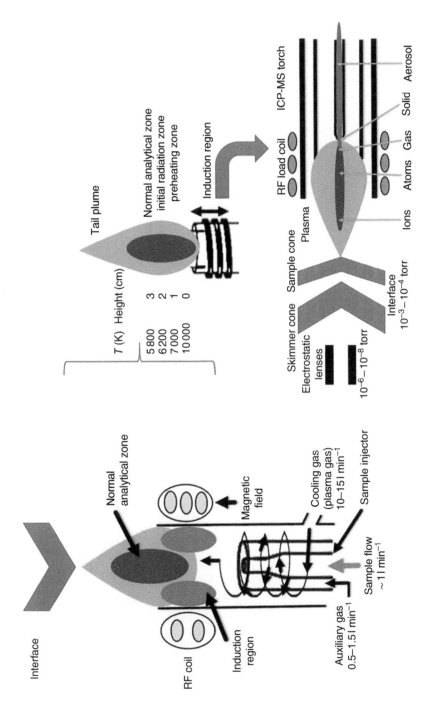

Figure 4.30 The simplified diagram of the plasma torch used in ICP techniques.

liquid sample into the carrier gas in the form of droplets of different sizes. Droplets having less than 10 μm in diameter are favorable and reach the ICP torch. Droplets of larger sizes can significantly reduce plasma energy and thus limit the number of atoms (ions) in the excited state, which in consequence will reduce the intensity of emitted radiation. The most commonly used type of nebulizer is pneumatic concentric glass nebulizer, where argon is introduced in the sidearm and is directed into the nozzle. The physical interaction of a carrier gas and a liquid sample transported through the capillary tube to the nozzle causes generation of an aerosol. In another type of nebulizer called cross-flow, both capillary needles transporting carrier gas and liquid sample, respectively, are positioned at 90° to each other, and the carrier gas is sufficient to form a coarse aerosol at the exit point. In ultrasonic nebulizers the liquid sample is transported onto a vibrating piezoelectric transducer working between 200 kHz and 10 MHz, which transforms the sample into an aerosol. This aerosol flows through the heated tube and reaches a condenser. As a result, dry and desolvated aerosol is delivered into ICP plasma [4, 5, 17].

Other methods of sample introduction that are commonly used include:

- *Hydride generation* (HG): Suitable solution for the determination of elements forming volatile hydrides at ambient temperature, e.g. As, Bi, Ge, Pb, Sb, Se, Sn, and Te. The principle of HG is based on the formation of covalent hydrides (like SbH_3 or H_2Se) in acid conditions in the presence of a reducing agent (metal/acid reductant) such as THB (sodium tetraborohydride, $NaBH_4$), Zn/HCl, $SnCl_2$/HCl-KI, and Mg/HCl-$TiCl_3$. The general chemical reaction leading to the formation of hydrides can be presented as $A^{m+} + (m+n)H^\bullet \rightarrow AH_n + mH^+$, where A denotes an analyte. The most commonly applied is THB due to its versatile reducing and hydride transfer properties. The generated volatile hydrides and other gaseous by-products are transported to the ICP system and ionized in the plasma [3, 20].
- *Electrothermal evaporation* (ETV): Used to evaporate small amounts of liquids or solids. In this method, a tiny amount of sample is deposited on the graphite surface where it is heated at the controlled temperature in such a way that the matrix of the sample will be destroyed or removed (ashing) but the studied analyte will remain intact. In many cases, modifiers are required to support the formation of more stable compounds of analyte to avoid any losses. After separation of the matrix from an analyte, the temperature is increased to vaporize and introduce the sample into the plasma [4].
- *Ablation*: Involves evaporation of a material from the surface of solid samples as a result of arcing or sparking. An efficient technique is a laser ablation, where the evaporation occurs due to the action of a laser beam (electromagnetic radiation beam in the optical range, strictly monochromatic). The ablated sample is then transported into the plasma by a carrier gas. The amount of ablated material is related to the laser used and sample properties. Traditionally,

nanosecond laser pulses have been used for ICP-MS for quantitative and qualitative analysis. However, quantitative analysis is limited by the accessibility of the matrix-matched standards necessary to perform calibration. In many cases, the matrix-matched calibration standards are difficult to obtain or manufacture. Another disadvantage of this method is sample fractionation and a difficulty to analyze low melting point elements. The transport efficiency of the generated aerosol, which is then transported into the ICP, seems to be crucial as well. In the case of femtosecond laser ablation with very short laser pulses (fs-LA), this transport may reach 100%, independently on the ablation cell geometry or type of a carrier gas, because the particle size in the generated aerosol is typically between 5 nm and 3 μm. The size of generated particles also affects the plasma conditions. The particles should be small enough to be totally atomized in the ICP plasma, and, at the same time, the total mass of the aerosol should not impair the ionization temperature [5, 13, 14].

4.1.6.7 Measurement in the ICP-MS Technique

Using a mass spectrometer, the entire mass range can be scanned to identify elements present in the sample. The whole mass spectrum consisting of signals within a range from 6 to 254 AMU can be collected in a very short time, typically in less than 1 second. For quantitation, the instrument should be calibrated using standard solutions and the blank sample. The reliability of the obtained results should be confirmed by the analysis of certified reference materials having the same matrix as the studied samples after being subjected to the same analytical procedures as the sample.

In the case of ICP-OES technique, sample solution is transported by means of a peristaltic pump to the spray chamber, from where it is injected into the plasma center in the form of an aerosol. Then, the excited atoms emit radiation, and the generated beam is split into individual wavelengths using a diffraction grating spectrometer (polychromator or monochromator). The selected monochromatic beams are directed to the detector to measure radiation intensity, which is proportional to the concentration of atoms or ions present in the sample [4].

4.1.6.8 Analyzers in ICP-MS Spectrometers

The most popular current configuration of the ICP-MS instrument is the device with quadrupole analyzer (ICP-QMS). Less common are ICP-MS with a TOF analyzer (ICP-TOF-MS) due to their considerable cost compared to quadrupole-based spectrometers. Below, the most important differences between ICP-QMS and ICP-TOF-MS are presented:

- ICP-TOF-MS spectrometers are lower ion transmission spectrometers and have weaker amplification of the signal by the detector in comparison with quadrupole-based ICP mass spectrometers.

- ICP-TOF-MS spectrometers are characterized by inferior detection limits compared with the ICP-QMS technique.
- ICP-TOF-MS spectrometers can be characterized by a much better mass resolution (normally <2000) when compared with ICP-QMS spectrometers (normally ≪1000). Resolution at mid-mass range for the spectrometer with a TOF analyzer is approximately 1500; however much better resolutions can be achieved for heavy masses like Pb or U (even exceeding 2000).
- TOF analyzer allows the user to separate ions c. 400 times faster than quadrupole analyzer, resulting in a shorter analysis time and lower sample consumption.
- The acquisition of the complete mass spectrum in the range of 5–260 AMU for the ICP-TOF-MS spectrometer is possible in c. 25–30 µs, while using a quadrupole mass filter the time interval would be about 10 ms [22–23].

A typical time-signal profile for ICP-MS analysis is presented in Figure 4.31. After introduction of the sample, the signal is detected and reaches its steady state. Washing is performed between introduction of individual samples to remove all residuals before injection of the next sample. This washout time is typically measured as a time taken for the ICP instrument to detect 1% of the steady-state signal. The washout time is strongly dependent on the matrix, and some elements, such as mercury, generate very high memory effects and need longer washout times [4].

The time-signal profile can be used to measure distribution of selected elements (isotopes) on the sample surface (Figure 4.32). Figure 4.32a shows strongly

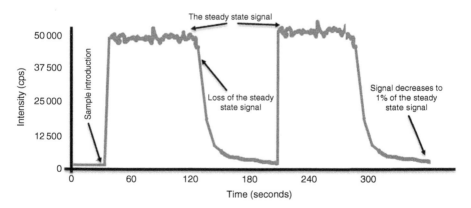

Figure 4.31 A typical time-signal profile for ICP analysis after the sample was introduced via laser ablation system (a), signal reached its steady state (b), signal lost its steady (c), and signal decreased to 1% of the steady state.

(a)

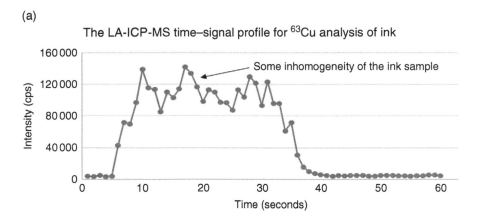

(b)

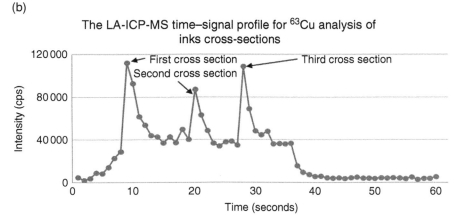

Figure 4.32 The LA-ICP-MS real data time-signal profile for ^{63}Cu signal obtained for unevenly distributed blue ink material on paper (a) and after performing the analysis along the liner of one ink material and three cross sections with different covering material (b).

inhomogeneous distribution of blue ink covering the paper, which was monitored by analysis of a signal from Cu (mostly originating from copper(II) phthalocyanine). Figure 4.32b presents the analysis of three cross sections with two different ink materials on a paper base. This type of research can be very useful in the forensic analysis of inks.

Typical mass spectra collected for NIST standards in glass for the whole mass range (Figure 4.33) and for a selected mass range (Figure 4.34) are presented below. In the spectra all isotopes have been identified, and the achieved resolution for heavy masses has been determined.

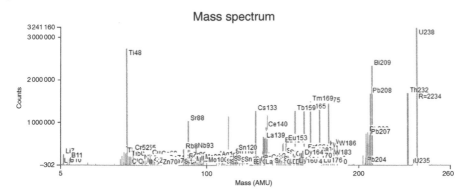

Figure 4.33 Exemplary mass spectrum (from 5 to 260 AMU) collected for 612 glass NIST standard (acquisition time = 1 second) with highlighted resolution for heavy masses reaching 2234.

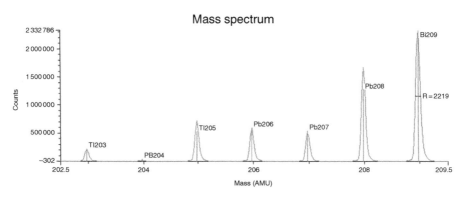

Figure 4.34 Exemplary mass spectrum for chosen mass range (from 203 to 209 AMU) showing the distribution of Pb and Tl isotopes, collected for 612 glass NIST standard (acquisition time = 1 second) with highlighted resolution for heavy masses reaching 2219.

Questions

- What are the main features that render ICP-MS instruments so useful and applicable to a wide range of analytical problems?
- What are the possible factors causing excitation of atoms and/or molecules?
- Describe a plasma.
- How does an MIP work?
- What are the analytical ranges of detection for ICP-OES and ICP-MS?
- How is the detection limit correlated with the ionization potential of the element?
- Describe the mechanism of ICP formation.

- What kind of processes the samples undergo before reaching the plasma tail?
- How can the sample be delivered to the plasma? Provide practical examples.
- For what kind of measurements it is necessary to carry out a calibration before the analysis of the sample?
- What analyzers are commonly used in ICP-MS spectrometers? Discuss advantages and limits of the various options.

References

1 Pröfrock, D. and Prange, A. (2012). Inductively coupled plasma–mass spectrometry (ICP-MS) for quantitative analysis in environmental and life sciences: a review of challenges, solutions, and trends. *Applied Spectroscopy* *66* (8): 843–868.
2 Wolf, R.E. (2005). What is ICP-MS? And more importantly, what can it do? USGS/Central Region/Crustal Imaging & Characterization Team, March 2005. https://crustal.usgs.gov/laboratories/icpms/What_is_ICPMS.pdf (accessed 28 January 2019).
3 Hou, X. and Jones, B.T. (2000). Inductively coupled plasma/optical emission spectrometry. In: *Encyclopedia of Analytical Chemistry* (ed. R.A. Meyers), 9468–9485. Chichester: Wiley.
4 Dean, J.R. (2007). *Practical Inductively Coupled Plasma Spectroscopy*. Chichester: Wiley.
5 Nelms, S.M. (ed.) (2005). *ICP Mass Spectrometry Handbook*. Boca Raton, FL: Blackwell Publishing, CRC Press.
6 Holland, G. and Tanner, S.D. (eds.) (2001). *Plasma Source Mass Spectrometry: The New Millennium*. Cambridge: The Royal Society of Chemistry.
7 Holland, G. and Tanner, S.D. (2005). *Plasma Source Mass Spectrometry: Current Trends and Future Developments*. Cambridge: RSC Publishing.
8 Ramos, J. (2008). Celebrating 25 years of inductively coupled plasma-mass spectrometry. *American Laboratory 40* (16): 30–34.
9 Meyers, R.A. (ed.) (2007). *Encyclopedia of Analytical Chemistry*. Weinheim: Wiley.
10 Holland, G. and Bandura, D.R. (2005). *Plasma Source Mass Spectrometry*. Cambridge: RSC Publishing.
11 Broekaert Jose, A.C. (2002). *Analytical Atomic Spectrometry with Flames and Plasmas*. Waltham: Wiley-VCH Verlag GmbH.
12 Barshick, C.M., Duckworth, D.C., and Smith, D.H. (eds.) (2000). *Inorganic Mass Spectrometry: Fundamentals and Applications*. New York: Marcel Dekker.
13 Thomas, R. (2004). *Practical Guide to ICP-MS*. New York: Marcel Dekker.
14 Douglas, D.J. and Tanner, S.D. (1998). Fundamental considerations in ICPMS. In: *Inductively Coupled Plasma Mass Spectrometry* (ed. A. Montaser), 615–679. New York: Wiley-WCH.

15 Perkin Elmer (2001). *The 30-Minute Guide to ICP-MS*, Technical Note. USA: PerkinElmer.

16 Niu, H. and Houk, R.S. (1996). Fundamental aspects of ion extraction in inductively coupled plasma mass spectrometry. *Spectrochimica Acta Part B: Atomic Spectroscopy 51*: 779–815.

17 Houk, R.S. (1986). Mass spectrometry of inductively coupled plasmas. *Analytical Chemistry 58*: 95A–105A.

18 Jacobs, J.L. and Houk, R.S. (2015). Updated ionization efficiency calculations for inductively coupled plasma mass spectrometry. In: *Diagnostic Studies of Ion Beam Formation in Inductively Coupled Plasma Mass Spectrometry with the Collision Reaction Interface*, Graduate Theses and Dissertations, vol. *14814* (ed. J.L. Jacobs), 16–38. Ames: Iowa State University.

19 Thermo Fisher Scientific (2017). High Performance Radio Frequency Generator Technology for the Thermo Scientific iCAP 7000 Plus Series ICP-OES. *Thermo Scientific Technical Note 43334.*

20 Kumar, A.R. and Riyazuddin, P. (2005). Mechanism of volatile hydride formation and their atomization in hydride generation atomic absorption spectrometry. *Analytical Sciences 21*: 1401–1410.

21 Olesik, J.W. (1991). Elemental analysis using ICP-OES and ICP/MS. *Analytical Chemistry 63* (1): 12A–21A.

22 Balcerzak, M. (2003). An overview of analytical applications of time of flight-mass spectrometric (TOF-MS) analyzers and an inductively coupled plasma-TOF-MS technique. *Analytical Sciences 19* (7): 979–989.

23 GBC Scientific Equipment (1998), The Advantages of Time of Flight Mass Spectrometry for Elemental Analysis. *Technical Note No. 01087600.*

4.1.7   Secondary Ion Mass Spectrometry with Time-of-Flight Analyzer (TOF-SIMS)

Nunzio Tuccitto

Dipartimento di Scienze Chimiche, Università di Catania, Catania, Italy

4.1.7.1   Introduction

Secondary ion mass spectrometry (SIMS) can be defined as the mass spectrometry of ionized particles that are emitted when a solid surface is bombarded by highly energetic primary particles, typically ions. SIMS is one of the major surface characterization techniques available today and is applied in various fields of science and technology. It is used to obtain chemical information from inorganic and molecular solids and even from biological samples. It is quite versatile for the analysis of samples that are compatible with high vacuum conditions; due to these conditions, SIMS can be used to analyze solid or solidified samples (solidified by treatment at low temperatures or freeze-drying). Chemical data in SIMS are determined by the atomic and molecular secondary ions that are emitted from any surface under primary ion bombardment. Mass analysis of these so-called secondary ions supplies direct information on their nature and therefore on the chemical composition of the uppermost monolayer of the bombarded surface area. TOF-SIMS, a type of SIMS where the emitted secondary ions are analyzed by means of a time-of-flight (TOF) analyzer, was considered as a method for the elemental analysis of solid materials because the sputtering process can be exploited as an efficient means for ion formation from solid surfaces. After the instruments were improved, TOF-SIMS was also applied for the characterization of samples of biological interest [1], including cells and diseased tissues. Additionally, TOF-SIMS is very useful for the characterization of microelectronics and corrosion science samples, polymeric materials, and organic samples in general. The technique is used to acquire laterally resolved analysis and the in-depth characterization of structured samples with a resolution down to the nanometer scale, allowing a three-dimensional (3D) rendering of the chemical composition of the materials.

4.1.7.2   TOF-SIMS Principle of Operation

In the TOF-SIMS analysis, particles are emitted from a solid surface (secondary particles) when it is bombarded with high energetic ions (primary particles). Although the main portion of the species emitted from the sample is neutral, in TOF-SIMS, the minimal charged part of those species is detected and analyzed by a TOF mass spectrometer. TOF-SIMS analysis obtains elementary and molecular information by atomic and molecular secondary ions that are emitted from any solid sample under primary ion bombardment. Mass analysis of these so-called secondary ions supplies direct information on their nature and therefore on the chemical composition of the bombarded surface area. A general equation describing the emission of the secondary ions during TOF-SIMS analysis is

$$I_s = I_p Y \alpha \chi \tau$$

where I_s is the secondary ion current, I_p is the primary particle flux, Y is the sputter yield, α is the ionization probability, χ is the concentration of species at the emerging surface layer, and τ represents the instrumental transmission of the analysis apparatus.

As a result of primary ion bombardment, the sample surface is modified not only because of the loss of secondary particles but also by a variety of further radiation effects, such as amorphization, primary ion implantation, and fragmentation of surface molecules or molecular structures. However, under specific conditions, it is possible to maintain the integrity of the surface layer within the timescale of the analytical experiment; within this regime, the operation is called "static" TOF-SIMS. In contrast, during "dynamic" TOF-SIMS, high primary particle flux is used, and the uppermost layers are eroded during analysis where in-depth chemical information is obtained from the sample.

4.1.7.3 The Sputtering of the Sample Surface

Sputtering is the erosion of a solid target by ion bombardment resulting from the atomic collision processes in the target. At first sight, the process is abstractly very simple; a schematic representation of sputtering is shown in Figure 4.35.

A collisional cascade occurs in the uppermost region of the solid target when a high-energy (order of tens keV) beam of ions hits the surface and its energy is

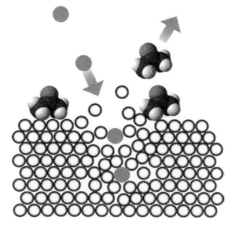

Figure 4.35 Schematic representation of sputtering process.

transferred to the target by means of a billiard-ball-type collision process. Primary ions dissipate their energy by collisions with the target atoms; if these are energetic enough, they collide with other atoms until the energy is completely dissipated. Via these collisions, energy is transferred to the surrounding medium. A fraction of the primary ions' energy may gradually be reflected toward the surface, producing the emission of secondary particles (atoms and molecules). In the context of the sputtering process, two factors influence the information depth related to a secondary particle: penetration depth and attenuation length. The penetration depth is the volume of the collision cascade, while the attenuation length is the attenuation of the flux of upward-moving secondary particles. Both factors have similar orders of magnitude; therefore, the penetration depth of the primary ion must be a relevant physical parameter in the explanation of the information depth, although it depends on the sample's nature and properties. Actually, the point of emission of secondary particles is up to few nanometers away from the point of primary impact, where the residue energy is in the order of tens eV. In this condition, only atoms or molecules placed at the uppermost layer of the sample are emitted. Thus, *SIMS is the mass spectrometry of emerged surface layers.* The information depth is of the order of a monolayer. The majority of sputtered atoms have quite small energies and must originate in a narrow depth range beneath the surface. Therefore, the sputter capability is related to the energy deposited onto the target near the surface. The amount of particles emitted by sputtering is expressed in terms of sputter yield, which determines the rate at which material is removed from the sample during sputtering:

$$\text{Sputter yield } (Y) = \frac{\text{Numbers of removed atoms}}{\text{Number of impinging ions}}$$

4.1.7.4 Ionization (Generating Secondary Ions)

As a consequence of the collision cascade, secondary particles are emitted from the uppermost layer of the sample. In particular, particles will be mainly neutral species, and there are also electrons and positively and negatively charged single atoms or clusters, including molecules. Secondary ions are extracted and analyzed by a mass spectrometer, which provides the detailed chemical analysis of the sample surface. The formation of ions in inorganic targets is influenced by several processes that involve electron transfer between the emitting species and the matrix they are leaving. For example, the ion yield of a metal will be significantly different than that of the same particular element's oxide. Secondary ion yields may vary by several orders of magnitude as a function of the chemical composition of the sample. Moreover, that yield depends on the electronic state of the surface. This phenomenon is called the "matrix effect" and it can be quite acute. For example, the presence of easily ionizable elements, like Na^+, Ag^+, etc. or Cl^-, O^-, etc., will favor, by several orders of magnitude, the intensity of positive or negative secondary ions detected during TOF-SIMS analysis. The ionization probability is one of the elements of the general equation describing the emission of the secondary ions (see Section 4.1.7.2); therefore, *TOF-SIMS analyses are not normally quantitative* because the ionization probability will vary, sample by sample, because of the matrix effect.

In the case of organic targets, SIMS is characterized by a *different ion formation process from most other mass spectrometers*. The ionization mechanism includes the chemistry of the materials, like acid–base reactions, ejection of an electron, molecular rearrangements and transpositions, etc. The study of the processes is very helpful for the chemical analysis of the organic materials as low mass fragments provide useful information about the chemical structure of the target. The precise modeling of the processes involved during ionization is not known, but it probably occurs via collisions due to the energetic recoils inside the cascade. Molecular dynamics simulations have shown that it likely occurs in proximity to or just above the surface. The recent introduction of polyatomic cluster primary beams actually changed this "historical" point of view [2]. In fact, because of the particular characteristics of the interaction between the polyatomic ion beam and the surface (including low penetration depth, high density of deposited energy, and nonlinear effects caused by the time overlap of collision cascades), peculiar effects are observed, particularly that the use of cluster projectiles produces a strong increase of the ionization probability [3] of the sputtered particles, and this positively affects the sensitivity of the technique as well as (due to the increase of the signal-to-noise ratio) its lateral resolution, which, for molecular materials, typically increases by an order of magnitude [4].

4.1.7.5 Construction of TOF-SIMS

The typical setup of a TOF-SIMS instrument is schematized in Figure 4.36. We can select three main sections: the primary particle source, the focusing/rastering optic, and the mass spectrometer.

The primary beam sources that are used in TOF-SIMS are based on several ion production mechanisms (field extraction, plasma ionization, electron bombardment, etc.). After the ion formation, beams are focused, mass-filtered, and pulsed in order to couple the beams with the TOF mass analyzer. State-of-the-art beams can be focused up to the nanometer scale with a pulse duration compatible with high mass resolution (>15000 $\Delta M/M$). The emission brightness, spatial resolution, current production, and user-friendliness will drastically vary for each type of source. The mode of operation of the TOF-SIMS experiment will influence the choice of the source. In the common *static* experimental modality, high-energy primary beams with low current and high brightness, in terms of secondary ion production, are used. Gallium, gold, and bismuth liquid metal sources that raster across the sample region of interest are the

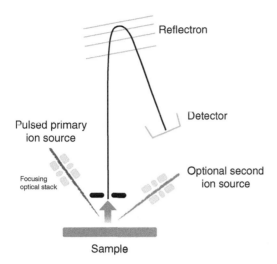

Reflectron

Pulsed primary
ion source

Detector

Focusing
optical stack

Optional second
ion source

Sample

Figure 4.36 Schematic representation of TOF-SIMS instrument.

common analytical primary beams used in *static* TOF-SIMS. *Dynamic* TOF-SIMS analysis typically operates in a dual-beam mode. A low-energy, high-current source is used to produce a crater in the sample, and a second analytical primary ion is rastered on the crater bottom to analyze the emerging surface.

During a TOF-SIMS experiment, samples are located under the primary beam in ultra-high vacuum conditions ($<10^{-8}$ mbar) to allow free passage of the bombarded and emitted ions. A very short pulsed keV primary ion beam strikes the sample and causes the almost instantaneous emission of secondary ions. The pulsing of the primary ion beam provides the start signal for the time measurement. Secondary ions are then accelerated by an electrostatic extractor, and then all the ions of a given polarity enter the TOF section with the same kinetic energy. Since secondary ions are not all emitted with the same energy but with a more or less broad energy distribution, the initial energy spread of some eV for organic molecules and tens of eV for atomic secondary ions limits the mass resolution in a linear TOF analyzer. Therefore, a reflectron-type analyzer, consisting of a combination of energy drift regions, is used in TOF-SIMS instruments to provide secondary ions with energy focusing. In a single-stage reflectron ion mirror system, when starting from the sample, the higher-energy secondary ions will, at a given time, be further along the drift region than the lower energetic once. Ions of higher kinetic energy will penetrate the mirror deeper than low-energy ions. At the point of reversing direction toward the detector, the high-energy ions are trailing behind the low-energy ions. The ions are then accelerated out of the ion mirror, and they leave the drift region toward the detector. By placing the detector in the focusing point, it is possible to obtain mass resolutions of the order of magnitude of 15 000.

4.1.7.6 Analytical Capabilities of TOF-SIMS
4.1.7.6.1 Static TOF-SIMS

The most broadly used application of TOF-SIMS analysis is surface mass spectrometry, acquiring spectra in order to qualitatively characterize the chemistry of the sample. In the case of inorganic targets, the high mass resolution and high surface sensitivity (up to femtomoles) allow us to perform accurate trace analysis.

Typically, in the analytical mode, the analysis is carried out on the sample region of interest without extensive pretreatment. Analysis can be performed by using a large defocused primary beam or by rastering a focused beam over the sample; a low-energy electron beam is used if the sample is an electrical insulator. Generally, it is relatively easy to interpret the spectra using the universal approach to mass spectrum analysis. Considering organic targets, such as biological samples or polymeric materials, though the sputtering is a hard process, molecular information can be gathered by TOF-SIMS since the pioneering experiments performed by Benninghoven and coworkers in 1969 [5]. They discovered a number of different anion species in the spectrum but also molecular cluster ions and organic molecular fragments, like $CH_3(CH_2)_nCOOH^-$, with high intensities. These results revealed that sputtering is much softer than anybody expected at that time. Since then, it has been possible to detect characteristic molecular secondary ions from even larger or more fragile molecular species by ion bombardment (Figure 4.37) [6].

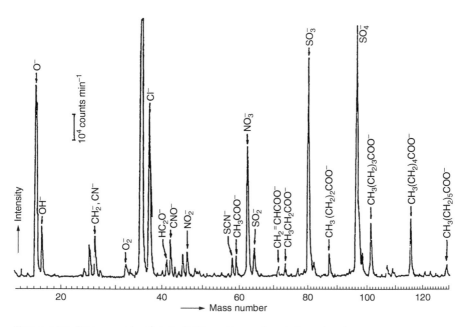

Figure 4.37 First example of static SIMS spectrum of organic molecules.

In their experiments with molecular sub-monolayers onto solids, Benninghoven et al. measured the rate of degradation of several parent-like ions as a function of the primary ion dose, introducing the definition of *damage cross section σ*. For a secondary ion i with intensity I_i and a flux of primary ions φ, the damage cross section σ_i, is expressed by

$$I(\varphi)_i = I_{i0} \exp(-\sigma_i \varphi)$$

A simple definition of σ would be the area per incident particle from where the emission of the specific fragment does not appear. The order of magnitude of damage cross sections measured for small organic molecules under keV ion bombardment indicates an upper dose limit for static SIMS measurements equal to 10^{-14} cm². The equation shows that the value of the intensity is reduced by 1–10% for a dose in the range of 10^{12}–10^{13} ions cm⁻². Accordingly, *this dose range represents the static limit.*

4.1.7.6.2 Visualization of Sample Surface

The high transmission and parallel detection of TOF analyzers, joined to the high focus of ion beams, allows the TOF-SIMS technique to make static imaging analyses of the surfaces [7]. Commercially available TOF-SIMS instrumentation can rapidly collect and store images that contain the full mass spectrum at every image pixel. Without an initial hypothesis about the nature of the sample and the peaks of interest, it is possible to make one mass-resolved image for each peak in the spectrum, obtaining so-called chemical maps. The chemical map is a false-color image showing the lateral distribution of the selected mass (or sum of masses) over the sample. Usually, darker color represents lower-emitting regions and bright color represents higher-emitting regions. For example, Figure 4.38 reports the nano- and microstructure of polymer blend films of polystyrene (PS) and poly(2-vinylpyridine) (P2VP) imaged with TOF-SIMS [8]. By increasing the relative amount of PS,

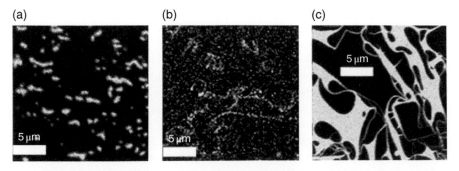

Figure 4.38 TOF-SIMS chemical maps of the sum of polystyrene signals. Sample composition P2VP/PS 1/4 1/5 (a), 1/10 (b), and 1/15 (c).

round-shaped structures, single wires, and entangled wires are detected by TOF-SIMS in imaging mode.

With TOF-SIMS, we can scan very large regions of interest as primary beams can be rastered over millimeters (macro-rastering). Actually, there are virtually no limits in the maximum scan size because modern instruments are able to move the sample under the beam, allowing the investigation of very large areas (on the order of cm^2). The spatial resolution of TOF-SIMS can be defined as its ability to distinguish between two points on the surface. It is limited by the beam size but also depends on the nature of sample. Primary beams can be focused at nanometer scales (<100 nm); however, the actual lateral resolution depends on the emitted particles. The useful lateral resolution, ΔL, is the minimum square area in which N secondary ions of a given mass can be desorbed and detected. This corresponds to the area ΔL^2, which has to be damaged to record N secondary ions. Thus ΔL is related to the secondary ion emission, which is obviously sample dependent, and generally it is not equal to the lateral resolution, which, in contrast, depends on the beam spot size.

For secondary ions with high ionization probability, such as inorganic elements like Na^+ or F^- (detected from glasses, ceramic composites, etc.), the lateral resolution mainly depends on the beam focusing. When the sample under investigation contains low-emitting elements, like organic materials (fibers, polymers, biological samples, etc.), the lateral resolution is strongly affected by the poor signal-to-noise ratio. The acquisition time is also limited by *static* conditions; therefore, pixel binning is often applied to increase the quality of chemical maps at the expense of high lateral resolution. Thus, even if the primary beam is highly focused, the lateral resolution will vary as a function of the ionization probability of the selected mass in the matrix in which it is detected. Moreover, as the ion source becomes increasingly focused, the mass resolution is negatively affected. In order to get a highly focused primary beam, longer time pulses are needed for physical electrostatic reasons; thus, higher beam focus leads to decreased mass resolution. Therefore, it is necessary to balance three experimental conditions – acquisition time, beam focus, and pixel size/binning – to perform high-quality TOF-SIMS chemical mapping.

When TOF-SIMS analysis is performed beyond the *static* limit, 3D depth profiling imaging is obtained. In this operative modality, a high-current ion beam is rastered over a defined squared area on the surface of the sample in order to produce a crater, and then a second highly focused beam is scanned to acquire chemical maps from the crater bottom. Imaging by in-depth sections (by alternate scanning of the two beams) produces a 3D chemical tomography. Chemical tomography is usually represented by means of an isosurface in order to highlight the region of the sample where the intensity of the selected mass was above a chosen threshold.

4.1.7.6.3 Dynamic SIMS: In-Depth Analysis

The experimental method used to perform depth profiling with TOF-SIMS is based on the use of two beams. A low-energy, high-current source is used to produce a crater in the sample, and a second analytical primary ion is rastered on the crater bottom to analyze the emerging surface. TOF-SIMS spectra are collected by the analytical beam, after which the sample undergoes a period of sputtering using the high-current source. Following this, the ion beam is then blanked, and the sample is analyzed again. The process is continued until the desired depth is reached. When insulating samples are being analyzed in depth, a settling time can be interposed between the two beams' etching/analysis time. During this period, low-energy electron flooding is used to compensate the surface potential back to the neutral condition.

Depth profiles are usually represented by plotting the intensity of a selected mass as a function of the sputtering time or number of incident ions. Alternatively, profiles are plotted over the depth when the crater profundity is measured or if the sputter yield of the materials under the energetic ion used for the sputtering is known. Indeed, to be of practical value, the sputter yield is usually converted to $nm\,s^{-1}$:

$$\text{Rate of atoms sputterd} = Y \times \text{rate of arrival of ions} = Y \times \frac{I}{e}$$

where I is the ion beam current (A) and e is the ion/electron charge (C). If ρ ($g\,cm^{-3}$) is the target density having atomic weight w and the raster area is A (cm^2), the etch rate is given by

$$\text{Etch rate}\left(nm \cdot s^{-1}\right) = \frac{IYw}{Ae\rho N}10^{-2}$$

The most important parameter involved in the depth profiling is the depth resolution. It is a measure of the sharpness of an interface brought about by physical or instrumental effects. A generally accepted method for measuring depth resolution is to measure the depth range over which the intensity changes from 16 to 84% of its total values during a step-shaped abrupt intensity change.

During a TOF-SIMS depth profile, the species sputtered away from the material originate from the uppermost layers of surface due to the very short mean free path of atomic species in a solid material; however, it must be stressed that although the penetration depth of the primary ions (that in turn depends on its mass and energy) is usually much deeper, the depth damaged by the ion irradiation (altered layer) is much thicker than the analyzed depth, affecting the depth resolution. In particular, depth resolution, sputtering yields, and secondary ion yields all significantly affect the quality of the depth profiles and are affected by other factors. Experimental conditions (primary ion energy, projectile species, and beam angle of incidence) as well as surface morphology and target chemistry all affect the depth profile quality. In particular, depth resolution is mainly affected by the sample's surface roughness because a rough surface shadows some portions of the sample from the ion beam. Moreover, there are a range of incidence angles between the sputter beam and the sample surface, which will produce a range of sputter yields and therefore broaden the sample interfaces. Thus, the sputtering process can induce surface roughness during the experiment, degrading the resolution as a function of depth because when the collision cascade of the primary beams encounters the material underneath the surface, it is redistributed by ion beam mixing [9].

4.1.7.7 Examples and Spectra Interpretation
4.1.7.7.1 *Speciation Analysis*

TOF-SIMS spectra generally present in the low mass region, starting with peaks related to the elements (positive ions or anions) starting with hydrogen up to, virtually, the highest stable elements constituting the surface (Figure 4.39). TOF high mass resolutions, together with the isotopic abundance distribution, help the users in the correct assignation.

Under static conditions, TOF-SIMS can be used to analyze organic compounds by obtaining molecular information. Typically, a small amount (a few microliters) of a diluted sample solution containing the unknown compounds is deposited onto the surface of an inert substrate and analyzed by means of TOF-SIMS. TOF-SIMS spectra basically present peaks related to elements constituting the compound in the low mass regions, then peaks related to fragmentations, and likely the molecular or quasi-molecular peak (e.g. $M-H^+$, $M-O^-$, $M-Na^+$, etc.). It must be stressed that due to the particular ionization processes in a TOF-SIMS spectrum, *peaks as much as six orders of magnitude less intense than the largest peak in the spectrum can be the most important for the analysis.* When interpreting the spectrum of an unknown compound, the analyst compares it with standard samples analyzed on the user's own instrumentation or compares it with the library of standard spectra that TOF-SIMS instrument manufacturers often supply with their instruments. There are also commercially available databases (e.g. the SurfaceSpectra Static SIMS Library) that the analyst can use; however, many features of the spectra are often not interpretable.

Data produced by electrospray ionization produce a large proportion of even-electron ions compared with odd-electron ions or positive ionized radicals. *The sputtering process is unlikely to produce secondary ions that cannot be compared with other mass spectrometers in terms of pattern fragmentations.* The odd-electron species resulting from the sputtering process can then

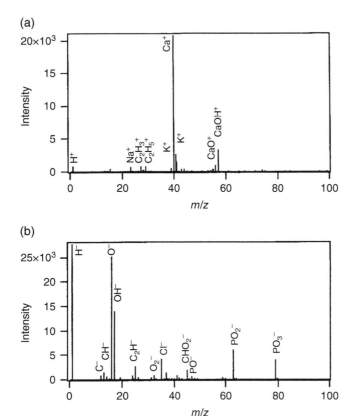

Figure 4.39 Positive (a) and negative (b) TOF-SIMS spectra for a calcium phosphate sample.

undergo subsequent collisions (usually involving hydrogen extraction) to produce a spectrum dominated by even-electron ions. Because the ions undergoing fragmentation and rearrangement are highly vibrationally excited, these species can undertake unusual and peculiar rearrangements and fragmentation in the SIMS experiment with respect to electron impact mass spectrometers.

During the TOF-SIMS experiment, the M^+ ion, after collisions with other organic species or even atomic hydrogen in the matrix, will extract a hydrogen to become MH^+. This is the most common cationization process detected in *static* SIMS. The peak at 267 AMU, which is not present in the electron impact mass spectrometry spectrum, is another typical fragmentation that occurs during the sputtering process in the SIMS experiment. This peak corresponds to the loss of H_2O from the MH^+ peak, according to the mechanism

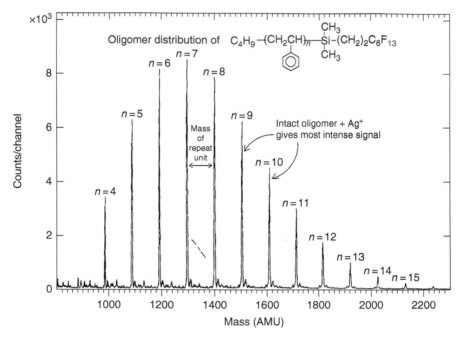

Figure 4.40 TOF-SIMS spectrum for the PTFE-functionalized PS, showing successful reaction termination through the detection of intact silver-cationized oligomers.

proposed by Spool in an extensive study on the interpretation of static secondary ion spectra [10]. In the case of negative ion spectra, fragments can be explained via an anionization process involving electron capture.

TOF-SIMS is also very useful in the investigation of polymeric surfaces, including the study of the effect of polymer degradation. Figure 4.40 shows the acquired molecular weight distribution of a polymer deposited as a sub-monolayer onto silver [11], although, unlike other mass spectrometers, like matrix-assisted laser desorption/ionization, TOF-SIMS is unable to detect the molecular weight of very heavy macromolecules (typically <10 000 AMU).

The TOF-SIMS spectra of polymeric surfaces are characterized by the presence of peaks in molecular fragments formed by the cleavage of the macromolecules present on the sample surface. Leggett and Vickerman proposed an empirical model suggesting that the kinetic energy is transferred from the primary ions [12]; however, from an experimental point of view, basically the analyst identifies the polymer by studying the pattern fragmentation according to its molecular structure (Figure 4.41) [11].

(a)

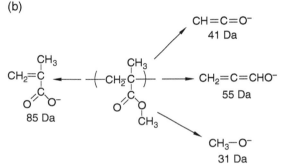

(b)

Figure 4.41 (a) Characteristic positive fragments for poly(methyl methacrylate) (PMMA). (b) Characteristic negative fragments for PMMA.

Questions

- What is SIMS and how is it performed?
- What species are emitted from the bombarded surface in SIMS?
- What is the difference between "static SIMS" and "dynamic SIMS?"
- What information can be drawn from a SIMS analysis?
- What are the main factors influencing the information depth related to a secondary particle?
- What is the "matrix effect" in SIMS?
- Which ion beams are normally used in SIMS?
- Describe what a chemical map is and how it is obtained by SIMS.
- Define the terms "lateral resolution" and "depth resolution" and explain how they can be affected by the experimental conditions of the SIMS analysis.
- Discuss some applications of TOF-SIMS analyses.

References

1 Tuccitto, N., Giamblanco, N., Licciardello, A., and Marletta, G. (2007). Patterning of lactoferrin using functional SAMs of iron complexes. *Chemical Communications* (25): 2621–2623.

2 Havelund, R., Licciardello, A., Bailey, J. et al. (2013). Improving secondary ion mass spectrometry C60 n+ sputter depth profiling of challenging polymers with nitric oxide gas dosing. *Analytical Chemistry* 85 (10): 5064–5070.

3 Tuccitto, N., Torrisi, V., Delfanti, I., and Licciardello, A. (2008). Monoatomic and cluster beam effect on ToF-SIMS spectra of self-assembled monolayers on gold. *Applied Surface Science* 255 (4): 874–876.

4 Kollmer, F. (2004). Cluster primary ion bombardment of organic materials. *Applied Surface Science* 231–232: 153–158.

5. Benninghoven, A. (1969). The mechanism of ion formation and ion emission during sputtering [Zum Mechanismus der Ionenbildung und Ionenemission bei der FestkörperzerstÄubung]. *Zeitschrift für Physik* 220 (2): 159–180.

6 Benninghoven, A. (1994). Surface analysis by secondary ion mass spectrometry (SIMS). *Surface Science* 299–300: 246–260.

7 Torrisi, V., Tuccitto, N., Delfanti, I. et al. (2008). Nano- and microstructured polymer LB layers: a combined AFM/SIMS study. *Applied Surface Science* 255 (4): 1006–1010.

8 Torrisi, V., Licciardello, A., and Marletta, G. (2010). Chemical imaging of self-assembling structures in Langmuir-Blodgett films of polymer blends. *Materials Science and Engineering B: Solid-State Materials for Advanced Technology* 169 (1-3): 49–54.

9 Tuccitto, N., Zappalà, G., Vitale, S. et al. (2016). A transport and reaction model for simulating cluster secondary ion mass spectrometry depth profiles of organic solids. *Journal of Physical Chemistry C* 120 (17): 9263–9269.

10 Spool, A.M. (2004). Interpretation of static secondary ion spectra. *Surface and Interface Analysis* 36 (3): 264–274.

11 Belu, A.M., Graham, D.J., and Castner, D.G. (2003). Time-of-flight secondary ion mass spectrometry: techniques and applications for the characterization of biomaterial surfaces. *Biomaterials* 24 (21): 3635–3653.

12 Leggett, G.J. and Vickerman, J.C. (1992). An empirical model for ion formation from polymer surfaces during analysis by secondary ion mass spectrometry. *International Journal of Mass Spectrometry and Ion Processes* 122 (C): 281–319.

4.2   Analyzers

4.2.1   Time of Flight (TOF)

Anna Bodzon-Kulakowska and Anna Antolak

Department of Biochemistry and Neurobiology, Faculty of Materials Science and Ceramics, AGH University of Science and Technology, Kraków, Poland

4.2.1.1   Introduction

The time-of-flight (TOF) analyzer was designed by W. E. Stephens in 1946, who coined its name as "A pulsed mass spectrometer with time dispersion." The working rule of the TOF analyzer is simple. At the beginning, the ions are either generated (MALDI, SIMS) or introduced (*orthogonal acceleration TOF* [oaTOF]) between two oppositely charged plates and are accelerated toward one of them, in which there is a hole or a slot (Figure 4.42). Then, a focused beam of ions is leaving the exit slot and migrates to the field-free region of the analyzer. The ions move through this region at a particular time, i.e. *time of flight*, which depends on their *m/z* values. In other words, the TOF analyzer separates ions according to the timespan between the ion formation and desorption and their detection. Schematic presentation of the TOF analyzer is shown in Figure 4.42 [1].

As it can be seen, the precise moment of the "takeoff" for all ions has to be precisely defined for correct operation of the TOF. Therefore, this type of analyzer is dedicated for pulsed ion sources, such as MALDI or SIMS. Other combinations of continuous ion sources (such as ESI or ICP) with additional device that ensure pulsed ion supply to the TOF analyzer are also possible. It is

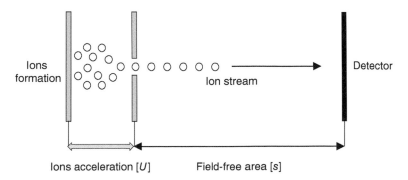

Figure 4.42 Simplified construction of TOF analyzer.

also necessary to generate sufficiently high vacuum inside the device to minimize the number of ion collisions with the gas molecules, which may change the ion flight duration due to the deflection of their flight paths. Also, high vacuum prevents from undesirable fragmentation [2].

4.2.1.2 The Working Rule of TOF Analyzer
The ions generated in the source are under the influence of the accelerating potential V_p and obtain the energy expressed by the formula (z, number of charges; e, elementary charge expressed in Coulombs)

$$E_p = zeU$$

When the ions reach the field-free area, their potential energy becomes converted into the kinetic energy E_k. We assume that all the ions, regardless of their masses, obtain similar kinetic energy E_k:

$$zeU = \frac{mv^2}{2} = E_k$$

A formula for the velocity of a given ion in the field-free region may be obtained by transforming the above equations:

$$v = \sqrt{\frac{2ezU}{m}}$$

If the distance to be traveled by an ion in the field-free region (approximately the length of the analyzer tube) is s, then the time necessary to reach the detector is

$$t = \frac{s}{v} = \frac{s}{\sqrt{\frac{2ezU}{m}}}$$

By transforming the above equation, it can be shown that the TOF for a particular ion is directly proportional to the square root of its m/z:

$$t = \frac{s}{\sqrt{2eU}} \sqrt{\frac{m}{z}}$$

To simplify those considerations, ions of larger masses will move in the field-free area slower than lighter ions, and as a result, they will reach the detector later. What is more, as it could be seen from the above equation, for large m/z values, the time interval between subsequent ion detection decreases (see Example 1). For this reason, proper operation of the analyzer requires measurements of the very short time intervals with an appropriate accuracy, which determines the resolution of the analyses.

Example 1   To determine the time interval between registrations of particular pairs of ions (20 and 21 *m/z*, 200 and 201 *m/z*, 2000 and 2001 *m/z*), the following values can be calculated:

$$\frac{t_{21}}{t_{20}} = \frac{\sqrt{21}}{\sqrt{20}} = \frac{4.58}{4.47} = 1.0246$$

$$\frac{t_{201}}{t_{200}} = \frac{\sqrt{201}}{\sqrt{200}} = \frac{14.18}{14.14} = 1.00282$$

$$\frac{t_{2001}}{t_{2000}} = \frac{\sqrt{2001}}{\sqrt{2000}} = \frac{44.73}{44.72} = 1.000223$$

As can be seen above, the time that elapses from reaching the detector by the neighboring ions becomes shorter for higher masses. Moreover, the above equations indicate that ions of the same molecules, but multiply charged, migrate in a field-free area faster than their singly charged counterparts. For example, in the case of the double-charged ions, it is 1.414 times faster (square root of 2), and in the case of triple-charged ions, it is 1.732 times (square root of 3) [2].

4.2.1.3   Linear Mode of Operation of TOF

In the linear mode of TOF operation, the site for generation/introduction of ions, the analyzer, and the detector are placed coaxially. The transmittance of this analyzer is up to 90%, since potential losses of the ions can only result from their collisions with the molecules present in the analyzer (e.g. due to imperfect vacuum). Hence, this analyzer is characterized by very high sensitivity.

Under perfect conditions, the ions are generated and introduced into the analyzer at the same time and at the same place. Under the influence of the same accelerating potential, ions should gain identical kinetic energy to move through the field-free area at the time proportional to the square root of *m/z*. Accordingly, as shown in Figure 4.43, ions of large masses move slower compared with those of lower masses.

Additionally, TOF analyzer possesses another interesting feature. Metastable ions, i.e. those having an excess of energy, can access the field-free area and easily undergo spontaneous decay. If these ions are singly charged, they may decompose, during their way to the detector, into a neutral fragment and an ion. Both molecules generated in this way have the same speed, so they are recorded by the detector at the same time as non-decayed ions (detector registers the "physical" impact of an ion). Therefore, the linear TOF analyzer can be successfully used for the analysis of metastable ions and high mass molecules. Moreover, this is the only analyzer where neutral particles can be sensed by the

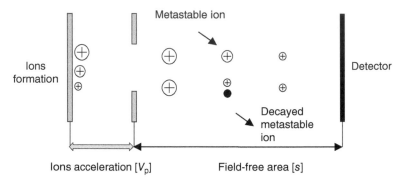

Figure 4.43 The linear mode of TOF. Black dot indicates a neutral molecule.

detector. However, to generate the signal, the breakdown of the metastable ion has to occur in a field-free area.

4.2.1.4 The Spread of the Kinetic Energy Regarding the Ions of the Same Mass

As mentioned above, at least in theory, the kinetic energy of all ions is the same at the beginning of their path through the field-free area. However, in practice, ions of equal masses and charges may slightly differ in kinetic energy. In particular, it is typical for combining a MALDI ion source with a TOF analyzer. These differences in kinetic energy may have several causes.

First of all, in the source, the ions are not generated exactly at the same time. The excitation time is determined by the transfer of excitation energy between the matrix and the molecules of an analyte. For UV lasers it could be between 10 and 50 ns. It means that the differences in time of ion formation may have significant influence on a decrease in resolution (see Example 1) considering the time for ion detection.

Secondly, regarding the place of ion formation, ions can be found at different locations in a gas cloud formed during desorption/ionization process. It means that even if the ions were generated at the same time and had the same kinetic energy, they undergo different effects of acceleration potential. The acceleration potential acts stronger on ions located deeper in the ion source (ion A; Figure 4.44), and those molecules obtain higher kinetic energy than their counterparts localized closer to the field-free region (ion B; Figure 4.44). Ion A enters the analyzer later; however, due to its higher speed, it may reach the detector before ion B.

Thirdly, the initial direction of the ion movements, after their desorption, is random and not necessarily directed along the axis of the analyzer (Figure 4.45). Finally, due to the collisions between the analyte and the matrix, the energy exchange can occur in the ion cloud; thus the initial kinetic energy of the ions can vary. What is more, there is a possibility of undesired fragmentation in the ion cloud.

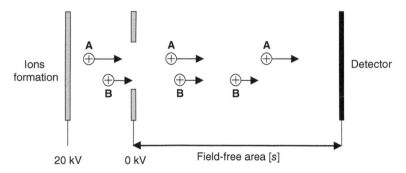

Figure 4.44 Effect of acceleration potential on ions with equal *m/z* values and initial kinetic energies, but situated at different locations in regard to the accelerating electrodes.

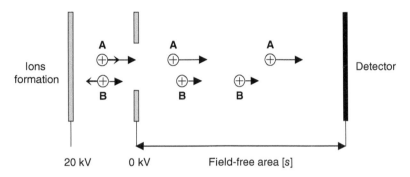

Figure 4.45 Molecules A and B of identical *m/z* values and initial kinetic energy move with the same initial velocity but in opposite directions (additional arrows). The velocity vectors, which the ions obtain as a result of accelerating potential, will have different values. The A particle will move faster than the particle B.

All the factors mentioned above significantly reduce the resolution of the measurements and hinder the accurate determination of *m/z* value for the analyzed substance. However, there are two ways to improve the resolution of the spectrum obtained with the aid of TOF analyzer. It is either to use the reflection mode (reflectron) or a delayed extraction process.

4.2.1.5 Delayed Ion Extraction

Previously, we discussed continuous ion extraction where the ions are accelerated immediately after the ionization process. As discussed above, differences in initial velocities of the ions with equal *m/z* values, as well as their different localization in the gas cloud during the desorption/ionization process, may be the reason for the differences in their kinetic energies. All this contributes to the decrease in the method's resolution. A solution called delayed ion extraction has been introduced to alleviate these effects. Delayed ion extraction results in the separation of two processes: desorption/ionization of the molecules and the process of their acceleration. For this purpose, an additional electrode is used between the ion source and the analyzer (see Figure 4.46) [3].

During the desorption process, the values of electric potentials of both the additional electrode and the plate of the first electrode (where the sample is deposited; Figure 4.46) are the same. As a result, the ions formed in the source are not affected by the electric field. As there is no acceleration, the ions move with the velocities obtained during desorption/ionization process due to the laser impact.

After a delay time, which usually lasts c. 200 ns, the potential of the first electrode increases, and the electric field begins to influence the ions. These ions, which had a lower initial kinetic energy and were not able to move away from the first electrode upon laser shot, would experience the strongest

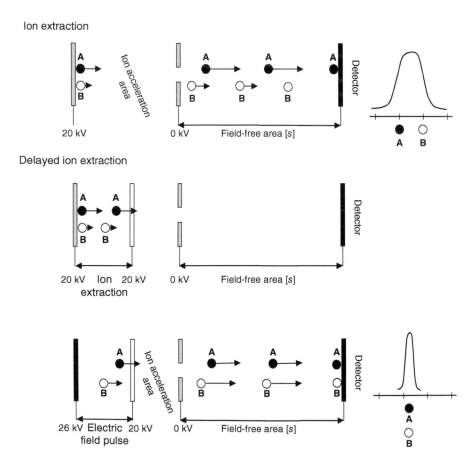

Figure 4.46 Comparison of continuous and delayed ion extraction – see description in the text (ions A and B have the same *m/z* value, but different kinetic energies).

acceleration. In turn, the molecules, which have become distant from the first electrode due to the higher initial kinetic energy, will undergo a lower potential influence and thus will move slower in the field-free area. As a result, the appropriate settings of the delay time and voltages applied at the electrodes allow for the alignment of the kinetic energy of all ions having the same *m/z* values. Unfortunately, ion focusing obtained by this method is mass dependent, so the relevant settings depend on the investigated *m/z* range. Also, the focusing process is less effective for large masses.

In contrast to the reflection mode discussed below, delayed ion extraction does not reduce the sensitivity of the measurement. Additionally, since the ion cloud formed after laser desorption is being dispersed during the delay time, no undesirable fragmentation occurs, which increases the sensitivity of the analysis. Thanks to the delayed ion extraction, the resolution can be improved 3–4 times for the linear analyzers and 2–3 times for the reflectron analyzers.

4.2.1.6 The Reflection Mode

The reflection mode was invented by Mamrin in 1994. The system consists of the set of circular-shaped electrodes called "reflectron," placed outside the field-free area at the end of the analyzer tube. The increasing potential is applied to individual electrodes that stops and reflects the ions at a relevant angle, directing them toward the additional detector (Figure 4.47) [1].

After reaching the reflectron, the ions become decelerated, reflected, and then accelerated again by the voltage applied to the reflectron. This is obtained by the use of a homogeneous electric field applied between the input and the end plate of the reflector. Ions of the same *m/z*, but moving at higher velocities, penetrate deeper into the field than their counterparts (ions with the same *m/z* values) with lower energies. Hence, "fast" ions spend more time in the field area (they travel longer distance) than the ions that move slower.

Due to this difference, molecules with equal *m/z* values, but different kinetic energies (and velocities – for the reasons described earlier), reach the detector at the same time. In other words, the reflectron reduces the

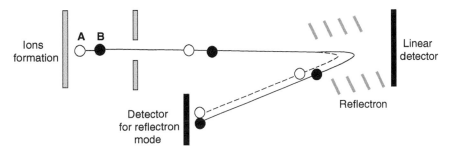

Figure 4.47 The reflection mode of TOF. Ion A (white) and ion B (black) have the same *m/z* value, but different kinetic energies (ion B moves faster).

differences in the TOF of ions having equal m/z values. Therefore, they all reach the detector simultaneously. In an ideal case, by changing the voltage between the reflector plates, it is possible to obtain a spectrum with a resolution depending only on the differences in the time of ion formation. Also, to some extent, the ion beam is being focused, which enables equalization of the angular dispersion resulting from the passage of the ion beam through the slot into the field-free area. In addition, the ion path is being prolonged, which allows for better separation of the ions of similar m/z values. In the case of a linear analyzer, too excessive extension of the linear ion path could result in a substantial loss of the signal due to the angular dispersion of the ion beam.

The use of reflection mode has some limitations. Heavy or light molecules of low energy often do not reach the detector (which is located much far away from the ion source than in the linear mode), or their signal is below the required threshold. Also, in contrast to the linear mode, metastable ions that decompose into a neutral and ionized molecule, before reaching the reflector, would not evoke the signal. What is more, elongation of the ion path prolongs the time for eventual fragmentation in the area between the reflector and the detector, which additionally complicates obtained spectrum. Consequently, the sensitivity is approximately tenfold lower when compared with the linear mode.

Reflection mode allows for the analysis of molecules of masses below 10 kDa, and the resulting resolution reaches about 20 000–30 000 for masses of 3000 Da. Figure 4.48 presents exemplary mass spectra obtained for a peptide in either linear or reflection mode.

4.2.1.7 Orthogonal Acceleration TOF Analyzer

A system in which the TOF analyzer is situated perpendicular to the ion beam enables the use of continuous ion sources (see Figure 4.49). After leaving the source, the ions are focused in order to minimize their dispersion in the direction perpendicular to the beam path. Then, the electric pulse ejects the ion "package" in the direction perpendicular to the TOF analyzer. Such impulse corresponds to the laser ionization process in a MALDI-TOF system. As soon as the detector registers the heaviest ion from the "package," another impulse ejects the next "load" into the TOF analyzer [2].

The advantages of such analyzer are:

- High sensitivity due to the high transmittance of the analyzer.
- High scanning speed.
- High resolution.
- High accuracy in mass determination.

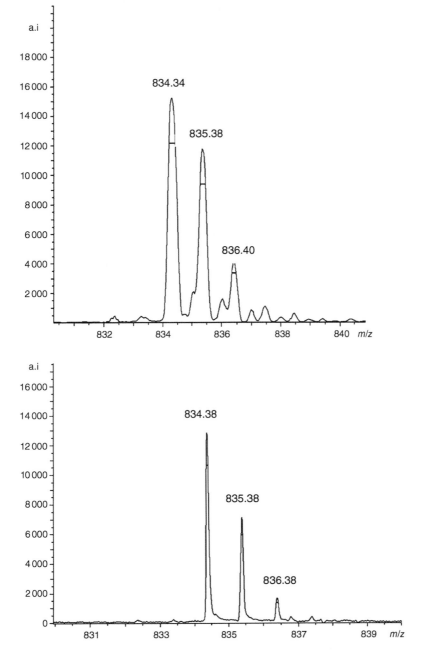

Figure 4.48 The spectrum of a peptide (MW = 833.40): (a) in a linear mode (resolution R_{lin} = 3654) and (b) in the reflection mode (R_{ref} resolution = 10852).

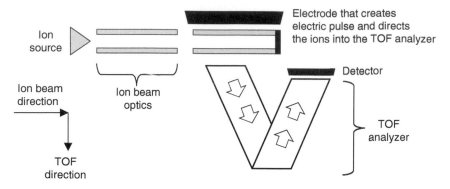

Figure 4.49 An orthogonal TOF analyzer.

4.2.1.8 Summary
In summary, the benefits of a TOF analyzer are as follows:

- Theoretically, the range of the analyzed masses is unlimited.
- As a result of the ionization process (single laser pulse), the whole spectrum can be obtained in tens of microseconds.
- High transmittance of the analyzer results in high sensitivity.
- Linear mode has lower resolution – few thousands – while it has higher sensitivity and allows for the detection of higher molecular weights.
- Reflectron mode is characterized by high resolution but reduced sensitivity when compared with the linear mode. The resolution achieved is tens of thousands (35 000 for insulin [5 734 Da]).
- The TOF construction is not complicated.
- The newest instruments allow for analysis with very high accuracy and the possibility for fragmentation.

The advantages of this TOF analyzer notably diminish its disadvantages, which are as follows:

- Large dimensions of the device.
- High price in comparison with simpler analyzers (e.g. quadrupole).
- The necessity to assure a very high vacuum.
- Limitations in the analysis of low molecular weight substances (use of matrices in the MALDI ion source, which is commonly combined with a TOF analyzer).

Questions

- Describe how the TOF analyzer works.
- Is it possible to detect metastable ions by TOF analyzers?
- Which factors are responsible for lowering the resolution of the TOF analyzers? How is it possible to circumvent this problem?
- What are the limitations of the use of the reflection mode in TOF analyzers?
- What are the advantages of orthogonal TOF analyzers?

References

1 Silberring, J. and Ekman, R. (eds.) (2002). *Mass Spectrometry and Hyphenated Techniques in Neuropeptide Research*. New York: Wiley.
2 Gross, J.H. (2011). *Mass Spectrometry*. Berlin: Springer.
3 Ekman, R., Silberring, J., Westman-Brinkmalm, A.W. et al. (2009). *Mass Spectrometry: Instrumentation, Interpretation, and Applications*. Hoboken, NJ: Wiley.

4.2.2   Ion Mobility Analyzer (IM)

Anna Antolak and Anna Bodzon-Kulakowska

Department of Biochemistry and Neurobiology, Faculty of Materials Science and Ceramics, AGH University of Science and Technology, Kraków, Poland

4.2.2.1   Principle of IM Operation

Ion mobility is defined as a proportionality factor between an ion's drift velocity v_d in a gas and an electric field of a defined strength (E). In other words, this factor is related to the velocity of ions pulled by an electric field along a drift tube, which has been filled with an inert gas. Ion mobility is characteristic for particular substances. Its value depends on the charge, mass, size, and shape of a molecule, as well as on the mass and density of neutral gas used in the analyzer. The temperature and the pressure inside the ion mobility drift tube influence this parameter as well.

For the proper operation, all typical analyzers mentioned so far (ion trap, quadrupole, FT-ICR, Orbitrap) required the presence of a high vacuum. IM is an exception from this rule as it operates at atmospheric pressure. In the IM technique, charged molecules move under the influence of an electric field. The space inside the analyzer is filled with a neutral gas, and the ions are separated as a result of their collisions with gas particles. It means that isobaric ions (having equal masses) with the same ionization fold can be distinguished based on their velocities, which are related to their shapes and dimensions (collision cross section). The smaller cross section of a molecule, the faster it moves through an analyzer under the influence of an electric field, and vice versa. The greater cross section determines the lower mobility under identical conditions. It is an important advantage compared with conventional vacuum analyzers that cannot differentiate between isomers, due to the fact that they possess the same masses. The principle of IM operation is presented in Figure 4.50 [1, 2].

Up to date, three types of analyzers are being used – classic *drift time IMS* and two commercially available analyzers of somewhat more complex constructions: *high field asymmetric waveform ion mobility spectrometer (FAIMS)* and *traveling wave ion guides (TWIG)*.

4.2.2.2   Drift Time IMS

In the simplest approach, IM is composed of the tube filled with neutral gas (nitrogen, helium, or argon), either at atmospheric or slightly reduced pressure. Homogeneous, low-value electric field induces ion movement through the pipe. A time of flight (TOF) that depends on the shape and size of the molecules is registered by a detector. The ions have to be introduced to the analyzer in small "packages." It means that the ion trapping systems (ion traps or trapping funnels) responsible for injection of the ions into the tube should be located in

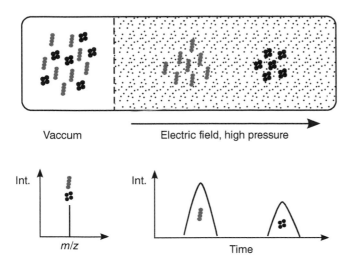

Figure 4.50 Mobility-dependent separation of the ions having the same mass but various shapes.

front of the IMS analyzer. Their role is to ensure high ion capacity for the best sensitivity and to focus ion packets for the best drift resolution. The ions are focused before entering the drift tube to avoid ion losses, and the appropriate ion gate pulses are applied to ensure proper numbers of ions in the ion package. Numerous studies on IMS are conducted to enhance ion transmission and increase the resolution [2].

4.2.2.3 High Field Asymmetric Waveform Ion Mobility Spectrometer (FAIMS)

FAIMS is usually filled with the mixture of nitrogen and helium under the atmospheric pressure (760 Torr). In this analyzer, the ions migrate in the space between two electrodes, to which high-voltage (kV range) asymmetric waveform at radio frequency (RF) is applied perpendicularly to the ions' direction of motion. The movement of ions toward the detector is assured by a gas flow. Asymmetric voltage means that the amplitude of positive voltage is significantly different from the amplitude of a negative voltage. Duration of an impulse is calculated in such a way that an area under the voltage peak for the positive value is equal to the area of a negative value (see Figure 4.51).

Asymmetric high field induces movement of the ions toward one of the electrodes where they become discharged (see Figure 4.51). As there is an atmospheric pressure inside an analyzer, the velocity of ion movement will depend on its shape and will be proportional to the applied voltage. For example, positively charged chloride ion will move fast toward the negatively charged electrode due to its compact shape.

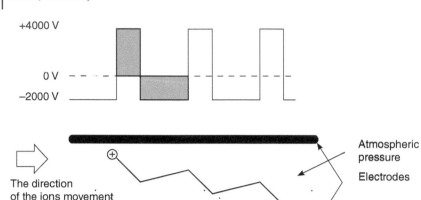

Figure 4.51 An example of an asymmetric alternating field used in FAIMS and the scheme of an ion trajectory generated in the analyzer. Gas flow carries the ions through the analyzer to the detector. Electrodes force the ion movement, depending on their shape, perpendicularly to their route in the analyzer. Additional DC stabilizes trajectory of an ion of particular shape and drives it to the detector.

Small direct current (DC) applied to one of the electrodes will stop the drift of selected ions toward the electrode, and only those ions of a defined shape (not mass!!) achieve stable trajectory, forced by the gas flow. The value of the compensation voltage depends on the ion's shape and is empirically established. In the case of chloride ion, a relatively high-level correction voltage is needed to compensate for its high velocity. In contrast, ions of complex shapes move slower; thus they require an adequately lower voltage. Based on this observation, DC voltage enables separating ions based on their mobility.

Electrodes utilized in FAIMS analyzers have flat or cylindrical design. In the latter case, the ions are being more efficiently focused, as losses resulting from ion repulsion and discharging are minimized. Therefore, it is possible to extend the length of the device; thus separation time can be prolonged, which, in turn, improves resolution. When the flat surface electrodes are used, diffusion of the ions is quite fast. Hence, such analyzer has smaller dimensions. It means that the separation time is reduced, and hence, resolution is worsened in comparison to the cylindrical electrodes.

Besides the shape of electrodes, numerous factors influence the operation of an analyzer and separation effectiveness of the ions characterized by different ion mobility. Selected examples are mentioned below:

- Electric field strength: The higher voltage, the better resolution. However, the use of high voltage involves the risk of an electric arc discharge.
- The distance between electrodes: Bringing the electrodes closer allows for an increase in electric field strength without discharging (Paschen's law) to

certain extent; thus higher resolution can be achieved. A drawback of this solution is that transmittance of the ions is reduced (decreased sensitivity).

• The shape of RF field and its frequency: RF changes should be characterized by a square wave (see Figure 4.51). Deviation from this shape worsens the operational parameters of the analyzer. Better resolution can be achieved, to certain extent, when the more asymmetric field of higher strength is applied.

• Gas composition: A mixture of nitrogen and helium is commonly utilized. The composition of nitrogen and hydrogen may increase the resolution of analyses. However, hydrogen is more challenging in practical use. Better resolution may also be achieved by the addition of solvent vapor (either water or methanol).

• Temperature of the electrodes: In the flat surface analyzers, both electrodes have the same temperature. In contrast, the inner electrode of the cylindrical analyzer should be c. 20 °C cooler than the outer one. Exact values should be empirically established for each analysis.

• Duration of the analysis: Time of the analysis affects the quality of the ion separation in a complex way. Considering the ion movement in the electric field, the longer the separation time, the better the separation of the ions. However, diffusion and ion–ion repulsion would disperse the ion with time – which of course influences the resolution and sensitivity of the analysis. Additionally, long separation time means loss of the ions by collisions with the electrodes, and that is why the duration of the analysis defines the ions that will be detected.

FAIMS analyzer is usually used when it is necessary to increase the signal-to-noise ratio for a given m/z value. In the case of complex sample, several substances with the same m/z value may be present concomitantly. FAIMS analyzer will allow for separation of those substances depending on the molecule's shape (*cross section*). After application of a given compensation voltage, the chosen compound will move through the analyzer, while all contaminants will be discharged on the electrodes.

4.2.2.4 Traveling Wave Ion Guides (TWIG)

The cell of a TWIG analyzer (see Figure 4.52) is filled with nitrogen at 10^{-3} Torr, which flows in the opposite direction to the moving ions (in TOF analyzer the pressure is 10^{-7} Torr; in ICR or Orbitrap it is 10^{-10} Torr). TWIG analyzer is equipped with a set of ring electrodes (*stacked-ring ion guides* [STRIGs]) connected to the RF drive voltage in the alternation of the phase. In addition, traveling wave of DC pulse is applied to the electrodes. This DC wave successively "moves forward" through the ring electrodes. Such combination of the voltages generates specific movement of ions in the analyzer, which resembles "pushing by a sea wave" and is therefore called "voltage surfing" [2].

In brief, a molecule that has a big cross section is being repelled by gas particles, while RF field wave moves it forward. In the case of molecules with a

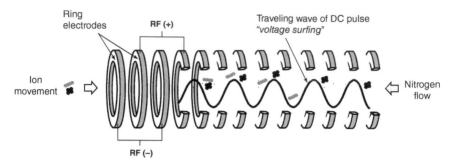

Figure 4.52 Scheme of the TWIG analyzer.

smaller cross section, collisions with the gas particles occur less frequently. Thus, those molecules move more effectively (faster) to the detector. In theory, the velocity of ions of equal masses is a function of their shapes. In fact, the relation between duration of a flight along the analyzer and the molecule's shape is much more complex. For this reason, it is necessary to use standards for calibration. TWIG, as the youngest type of analyzer, is being constantly developed, and it is expected that some technical problems will be solved shortly [2].

4.2.2.5 IM Spectrum

Using an IM, a specific spectrum is generated that resembles a chromatogram, rather than a mass spectrum (see Figure 4.50). In more complex constructions, where IM is combined with classical analyzers, the spectrum contains information regarding the TOF through the IM and m/z value for a particular ion. An example of such design could be Q-TWIG-Q (Xevo^{TM}, Quattro Premier^{TM}) or Q-IMS-oaTOF (Waters Synapt HDMR).

4.2.2.6 Applications

The mass spectrometer is an excellent tool for detection and identification of substances. If the mass of a given substance is known, the technique enables confirmation of its presence in the sample. Miniaturization and simplicity of the system are essential to facilitate application of spectrometers for fast, preliminary analysis of environmental pollution and detection of explosives, warfare gases, or narcotics at the airports, borders, and other public places. IM fit well for miniaturization, as they do not require vacuum for their operation, which makes them suitable for fast analysis. The common applications include military, emergency services, trace detection of explosives, and critical infrastructure protection, to name a few [3].

Ion mobility devices work also in combination with other types of analyzers. When the ions of a given mobility leave IMS tube, other analyzer (e.g. TOF) can accomplish further measurements. This provides novel opportunity to obtain numerous spectra from one separation based on the ion mobility. This enables *m/z* analysis of the molecules that differ in shape (Figure 4.50). For instance, IMS can be connected to the MALDI ion source due to the impulse character of this source. Combining IMS analyzer with orthogonal acceleration time of flight (oaTOF) enables identification of *m/z* of the ions separated based on their mobility. Also, in such an approach, signals from the matrix can be distinguished from substances of a similar mass present in the samples.

Questions

- What is the principle of operation of IM?
- What types of analyzers are currently being used in IM?
- Which factors affect the separation effectiveness of the ions in IM?
- How does a TWIG analyzer work?
- What are the main applications of IM?

References

1 Gross, J.H. (2011). *Mass Spectrometry*. Berlin: Springer.
2 Greaves, J. and Roboz, J. (eds.) (2013). *Mass Spectrometry for the Novice*. Boca Raton: CRC Press.
3 Kiss, A. and Heeren, R.M.A. (2011). Size, weight and position: ion mobility spectrometry and imaging MS combined. *Analytical and Bioanalytical Chemistry* 399 (8): 2623–2634.

4.2.3 Quadrupole Mass Analyzer

Anna Antolak and Anna Bodzon-Kulakowska

Department of Biochemistry and Neurobiology, Faculty of Materials Science and Ceramics AGH, University of Science and Technology, Kraków, Poland

4.2.3.1 Construction and Principles of Operation of a Quadrupole

Quadrupole mass analyzer separates ions according to their m/z by utilizing their stable oscillatory trajectory in the electric field. Wolfgang Paul and Helmut S. Steinwedel described this phenomenon in 1953 for what they were awarded the Nobel Prize in Physics in 1989.

Quadrupole mass analyzer is built of four rods made of metal (molybdenum) or ceramic core covered with a layer of the relevant metal. They are set parallel to each other, as well as to the trajectory of the ion beam, in the form of a square (Figure 4.53). The electric field generated between rods by the applied voltage should have hyperbolic shape. To obtain such characteristic of the field, electrodes with hyperbolic cross sections should be used (Figure 4.53). However, such construction is hard to manufacture, and even minor aberrations in the rods' shape may cause a decrease in resolution and ion transmittance. Therefore, cylindrical rods are being used instead, as they are easier to fabricate in the proper shape. The hyperbolic field is generated in the analyzer when rods are correctly placed, and their radius r is equal to 1.1468 of r_0, which is the distance from the center of an analyzer to the inner edge of the rod [1].

Opposite rods are electrically connected and have the same potential resulting from the combination of direct current (DC) with radio-frequency (RF) voltage (RF of 1–4 MHz), and the sum direct and alternating voltage equals 10^2–10^3 V (Figures 4.54 and 4.55). At any moment, each pair has the potential of the same value but of the opposite polarity. It is expressed as

(a) (b)

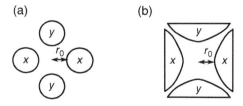

Figure 4.53 Scheme of the quadrupole mass analyzer: (a) cylindrical and (b) hyperbolic. Rods marked as *x* have positive potential applied. *y*, negative potential; r_0, distance from the center of an analyzer to the edge of the rod.

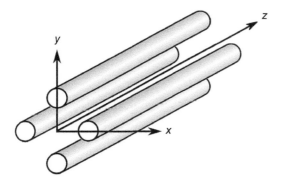

Figure 4.54 Scheme of the cylindrical quadrupole mass analyzer.

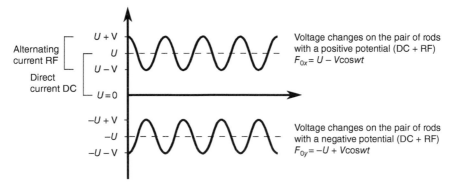

Figure 4.55 Scheme of the changes in voltage applied to rods having positive and negative potentials.

Pair of x rods:

$$\phi_{0x} = U - V\cos\omega t$$

Pair of y rods:

$$\phi_{0y} = -U + V\cos\omega t$$

where

U is DC voltage
$V\cos\omega t$ is RF voltage – voltage of potential V and frequency ω

An ion that is traveling between the rods is being alternately pulled and repelled by them while oscillating on $2r_0$ distance, due to the action of the

(a)

(b)

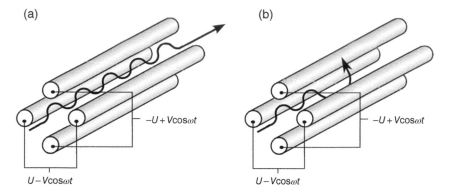

$-U + V\mathrm{cos}\omega t$

$U - V\mathrm{cos}\omega t$

$-U + V\mathrm{cos}\omega t$

$U - V\mathrm{cos}\omega t$

Figure 4.56 Stable trajectory (detectable ion) (a). Unstable trajectory (the ion does not reach analyzer) (b).

electric field, resulting from the superposition of both RF and DC voltages. For the defined values of U, V, and ω, an ion of the given m/z can achieve stable trajectory (Figure 4.56a). Under these conditions, other ions have unstable trajectories and are discharged during collisions with the rods (Figure 4.56b). The mass spectrum is obtained by a continuous change of the V and U values while maintaining their constant ratio, which consecutively permits passage of the ions of particular m/z [2].

4.2.3.2 Behavior of an Ion Inside the Quadrupole

In the beginning, ions from the ion source are directed to the analyzer by a set of focusing lenses. A potential is applied to accelerate the ions so that they could travel through the analyzer to the detector. Sometimes, sufficient momentum is achieved only by repelling the ions bearing the same charge [3].

An ion having the positive charge enters the space between the rods of a quadrupole and moves in the direction parallel to z-axis. As it was mentioned earlier, charged molecule of a given m/z may achieve stable trajectory toward the detector by applying adequate values of DC (U) and alternating voltage (V) of defined frequency (ω). The pairs of rods act as cutoff filters for the molecules having masses higher or lower than the selected m/z, so they do not pass the analyzer.

To visualize the behavior of a molecule under superposition of DC and RF voltages, we should assume that the ions of higher masses respond to DC voltage only. We may say that in such case, the RF voltage changes very fast and does not influence the heavy ion movement. We may say that these ions are "too heavy" to be able to respond to such fluctuations. In turn, ions of low molecular weight will respond to RF voltage only. We may say that these molecules are "agile" enough to respond to voltage oscillations. To simplify further considerations, we assume that the described ions have a positive charge [1].

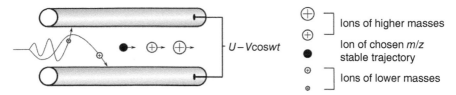

Figure 4.57 A filter for low masses – trajectory of the ions between rods with a positive DC potential applied.

A pair of rods (x) with a positive potential applied (Figures 4.53 and 4.54) acts as a filter for high masses in the quadrupole. It allows molecules of the higher masses to pass to the detector and stops those with masses lower than the mass of a selected ion. This is because the positive DC voltage applied to the electrode repels positively charged ions of higher masses toward the center of the space between the x rods. Thus, those particles may easily migrate through the analyzer.

Ions of the lower masses will strongly respond to the RF voltage, which is also applied to the pair of x rods. Once per each cycle of RF change, the sum of the direct and alternating potentials will force low molecular weight ions to move in the direction of one of the rods. If the mass of such ion is low enough and the potential of the electrode is adequately strong, then the ion will accelerate and collide with the electrode before the total potential changes once again. Collision with the electrode will discharge the ion and eliminate it from the ion beam (Figure 4.57). In summary, the pair of positively charged rods acts as a filter for the ions of higher masses in the quadrupole, because it "pushes" them between the rods so they may pass through the analyzer. Lower masses are not transmitted through the analyzer because superposition of direct and alternating potentials causes their discharge on the rods.

A pair of rods with a negative potential (y) (Figure 4.58) acts as a filter for low masses, in contrary to the pair with the positive charge. It eliminates ions of higher masses than those selected and passes those with adequately low mass. As before, the ions of higher masses will respond to direct negative potential (DC) applied to the pair of electrodes, and heavier ions will be discharged on the rods (positively charged heavy ions will collide with the negatively charged electrode).

Ions of lower masses will also be attracted by the electric field generated by the negative potential. However, in contrast to heavier ions, they will also respond to the field generated by RF potential. As the sum of both potentials will repel those ions once per each cycle, it will allow maintaining these ions between the y rods (Figure 4.58).

Superposition of potential applied to both pairs of rods enables ions of a given m/z to stably oscillate between rods and finally fly to the detector. In turn, ions having lower or higher masses will oscillate with an increasing amplitude until discharged on the rods.

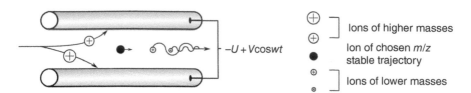

Figure 4.58 A filter for low masses – trajectory of ions between rods with a negative DC potential applied.

4.2.3.3 How Mass Spectrum Is Generated? Changes of *U* and *V*

The movement of ions of a given *m/z* is stable in the *x* and *y* directions for a particular range of *U* and *V* values. For other values of these parameters, the movement of ions will be unstable, and these ions will not pass the analyzer.

Spectrum is being created by passing the ions of successive values of *m/z* through the analyzer due to the changes of *U* and *V* at constant ω so that the *U/V* ratio is always maintained constant. With increasing values of *U* and *V*, the ions of higher *m/z* are able to get to the detector and generate a spectrum. This process is called *scanning of the ion beam* or simply *scanning*. Scans can be taken very quickly as we only manipulate the electric field, and this is one of the advantages of the quadrupole, important during very fast chromatographic separations. The resolution of spectra depends on the *U/V* ratio. The bigger the ratio, the higher resolution is achieved. High resolution may only be achieved when for the given *U* and *V* values only a narrow range of *m/z* is stable [4].

4.2.3.4 Spectrum Quality

The main problem of a quadrupole analyzer is its relatively low resolution across the entire mass range. These analyzers are characterized by a unit resolution. For example, an ion of *m/z* 300 can be distinguished from the signal at *m/z* 301, and an ion of *m/z* 1200 can be distinguished from the peak at *m/z* 1201. The resolution depends on the manufacturing precision of quadrupole's rods and on their distribution in space (to achieve satisfactory results, micrometer accuracy is required). The resolution also depends on the *U/V* ratio. The higher the value, the higher the resolution. On the other hand, transmittance (the number of ions that pass through the analyzer and reach the detector) decreases with increasing *U/V* ratio, which affects the sensitivity of analyzer.

The number of oscillations of the ions passing the quadrupole is also important for the spectra quality. Simulations of the ions' trajectory revealed that for a molecule with a kinetic energy of 10 eV, 100 oscillations are sufficient to achieve satisfying results. Therefore, the resolution can be enhanced by installing an ion reflector at both ends of the quadrupole, which elongates the flight path and thus increases the number of oscillations. A similar effect may be obtained by increasing the frequency of the alternating voltage on the rods or by slowing down the velocity of the ions in the analyzer because the faster they move through the quadrupole, the poorer resolution of the spectrum is achieved.

4.2.3.5 Applications of the Quadrupole Analyzer

The quadrupole analyzers are called mass filters because only the ions of the defined m/z values can pass through the device. In contrast to TOF or magnetic analyzers, the kinetic energy of the ion is irrelevant. It should only be high enough to allow an ion to fly through the quadrupole.

As it was mentioned earlier, the spectrum is generated by scanning the ion beam and allowing only the ions of certain m/z to pass to the detector. It means that the rest of ions are eliminated by discharging them on the rods, which decreases the sensitivity of the measurements. Therefore, the quadrupole mass analyzer is suitable for the analysis of ions within a narrow range of m/z or to monitor only selected ions.

The advantages of quadrupole are:

- Small size (diameter of the rods, 0.5–1.0 cm; length, 5–25 cm).
- High transmittance.
- Fast scanning.
- Relatively low price.

The ability to monitor one or several selected m/z values within the whole range of the analyzed spectrum, along with the possibility of quick switching between the selected m/z values, makes this device an excellent candidate for linking to GC-MS and LC-MS, in both qualitative and quantitative modes. In addition, these analyzers operate at low vacuum (10^{-5} Torr, $6 \cdot 10^{-3}$ Pa), which simplifies their use in tandem mass spectrometry.

The narrow measuring range (50–4000 m/z) predisposes this analyzer for the investigation of organic and inorganic substances of relatively low masses. This range can be extended by a combination of a quadrupole with an ESI ion source (see Section 4.1.3.2). Multiple ionization of the analyte enables analysis of larger molecules (up to 100 000 m/z). Another feature includes successful analysis of the very low mass compounds, even below 50 m/z, which cannot be achieved in ion traps.

4.2.3.6 Quadrupoles, Hexapoles, and Octapoles as Focusing Elements: Ion Guides

Reduction of the U voltage (DC) in relation to the V voltage (RF) in the quadrupole, within the stability range of the ions' trajectory, simultaneously increases m/z range of the passing ions. In an extreme case, the reduction of U to zero (*RF-only quadrupole*) allows for the passage of the entire ion beam through the analyzer. This phenomenon is used to transfer all the ions from one element of the spectrometer to another. In the literature, such mode is denoted as "q." For this type of application, hexapole (six rods arranged, denoted as "h") and octapole (eight rods denoted as "o") are often used instead of a quadrupole. In such configuration, the ions are more efficiently transferred in a wider range of m/z. However, only quadrupole enables analysis of the ions regarding their m/z. Hexapoles and octapoles only allow for transferring and focusing the ions, but not for their analysis [4].

Properly arranged rods connected to the alternating voltage V are "transparent" to neutral molecules, but, at the same time, they are able to focus ions in the center of the analyzer. This setup can be used for connecting the ion source operating at atmospheric pressure with a suitable analyzer. The introduction of such an intermediate step allows for relevant reduction of the pressure between the ion source and the analyzer because neutral gas molecules will "escape" between the rods and could be pumped off. Interestingly, application of a low vacuum (10^{-5} Torr) during operation of the device triggers additional focus of the ion beam in the center between the rods due to the reduction of their kinetic energy (*collisional cooling*). This allows for focusing the ions within the rods (*collisional focusing*) and increases the efficiency of ion transmission.

Apart from focusing and transferring the ions between elements of the spectrometer, quadrupoles, hexapoles, and octapoles are utilized as collision chambers for tandem mass spectrometry (see Chapter 7). Such chambers are often bent-shaped in order to elongate the ion path and to increase fragmentation efficiency. Also, this kind of geometry reduces the level of noise in the mass spectrum as it prevents neutral particles and photons created during fragmentation from reaching the detector.

Questions

- What is the principle of operation of a quadrupole mass analyzer?
- What are the main factors affecting the resolution of the spectra recorded by a quadrupole mass analyzer?
- How is it possible to enhance resolution?
- What is the main factor affecting the sensitivity of the spectra recorded by a quadrupole mass analyzer?
- What are the main advantages of the quadrupole mass analyzer?
- Besides detection of ions, what other applications are possible for the quadrupole?

References

1 Ekman, R. and Silberring, J. (eds.) (2002). *Mass Spectrometry and Hyphenated Techniques in Neuropeptide Research*. New York: Wiley.
2 Ekman, R., Silberring, J., and Westman-Brinkmalm, A.M. (eds.) (2009). *Mass Spectrometry: Instrumentation, Interpretation, and Applications*. New York: Wiley.Kraj, A., Desiderio, D.M., and Nibbering, N.M.
3 Gross, J.H. (2011). *Mass Spectrometry*. Berlin: Springer.
4 Greaves, J. and Roboz, J. (eds.) (2013). *Mass Spectrometry for the Novice*. Boca Raton: CRC Press.

4.2.4   Ion Trap (IT)

Anna Bodzon-Kulakowska and Anna Antolak

Department of Biochemistry and Neurobiology, Faculty of Materials Science and Ceramics, AGH
University of Science and Technology, Kraków, Poland

4.2.4.1   Introduction

In 1960 the inventors of the quadrupole analyzer, Wolfgang Paul and Helmut Steinwedel, modified its construction, designing an ion trap (quadrupole ion trap). An ion trap (Figure 4.59) is composed of three hyperbolic electrodes: two spherical and one torus shaped (a torus-shaped electrode comes from one quadrupole rod coiled in a circle). The diameter of the ion trap is typically about 5 cm, so it has quite small dimension. In this kind of analyzer, ions of different masses may be trapped between the three electrodes due to the formation of a potential well. This is achieved by applying a relevant combination of RF and DC potentials to the electrodes. The ions are moving inside the trap on stable trajectories in a three-dimensional space (z and r directions) (Figure 4.59). In order to detect the ions of specified m/z, their trajectory is destabilized, and, as a result, they are ejected from the trap to the detector. The spectrum is generated by successive destabilization of the trajectory of ions with subsequent m/z. The greatest advantage of an ion trap is its ability to store the ions for a time long enough for additional procedures, such as fragmentation [1, 2].

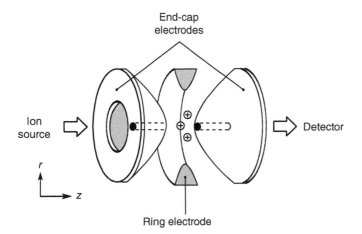

Figure 4.59 Ion trap (ring electrode presented in a cross section).

4.2.4.2 Behavior of an Ion Inside the Ion Trap

At the beginning, the ions generated in the ion source are accelerated by a 5–20 V potential and channeled to the analyzer by the set of lenses and octapoles. Such introduction is important because the quality of the spectrum depends on the number of ions inside the analyzer. Too many charged molecules injected at the same time may decrease resolution due to the repulsion forces between ions (*space charge effect*). This may also lead to the reactions between the ions, thus generating false results. The maximum number of ions inside the trap is about 10^6–10^7. On the other hand, insufficient number of ions would decrease the sensitivity.

At the beginning of the analysis, a package of ions enters the trap due to the negative potential operating for a short time (in a positive ion mode). Next, the ions are ejected from the trap toward the detector to assess their number. This allows for choosing an adequate timespan for the next "portion" of ions to accumulate a sufficient number of them inside the trap. This procedure prevents the analyzer from overloading and increases the sensitivity of measurements.

The analyzer is filled with a neutral gas (usually helium of the highest purity). As soon as the ions enter the trap, they begin to oscillate under the influence of electric field generated by applying the RF voltage to the ring electrode. As a result of collisions with helium particles, the kinetic energy of ions is being dispersed (ion cooling). This enables holding ions inside the trap. Otherwise, they would directly hit the spherical end electrode and discharge.

> Helium purity has significant impact on spectra quality. The high purity helium is expensive, but expenditure of the gas is very low (one bottle is sufficient for many years).

In addition, the use of the gas pressure of 10^{-1} Pa [10^{-4} Torr] compensates for the ions' repulsion and prevents from their dispersion by concentrating them inside the center of the ion trap. In this area, the electric field has the highest symmetry and induces the most stable trajectories of the analyzed ions. Moreover, the ions are concentrated in the center of the analyzer. Upon excitation, they begin their move from this central position, which ensures minimum beam dispersion on the way to the detector. It means that higher sensitivity and resolution can be achieved. The presence of helium plays an important role also in tandem mass spectrometry, as it contributes to the fragmentation of ions due to the collisions of helium atoms with trapped ions (Section 7.3).

Ions trapped inside an analyzer oscillate in a specific manner in the direction of *r*-axis (the trajectory resembles the shape of figure "8") and along the *z*-axis (Figure 4.60). The eight-shaped trajectory is not flat, but spatially curved, in

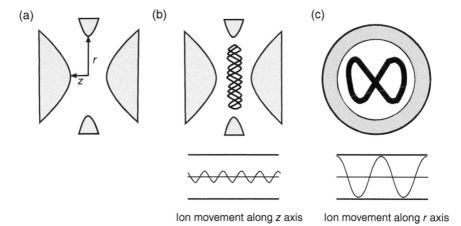

Ion movement along z axis   Ion movement along r axis

Figure 4.60 The movement of ions in the trap: (a) direction of the move, (b) side view of a 100 Da ion move, and (c) the move of ions (view in the plane of the ring electrode).

the same manner as the plane of a lower-value potential generated in the trap by electrodes. As can be seen, the ions move inside a trap along complex trajectories that may intersect at many points. The high density of the ions at intersection points causes deformation of the electric field, which results in deterioration of the spectrum quality due to the widening of the peaks. This effect is associated with the repulsion of ions (space charge effects). Hence, as mentioned above, it is important to accumulate the relevant number of ions inside the trap. Of course, this number varies, depending on the purpose of the analysis – either isolation of ions of a defined m/z value or generation of the full spectrum [3].

4.2.4.3 Analysis of the Ions
Separation of ions regarding their m/z value is obtained by releasing them from the trap due to destabilization of their trajectory. A group of ions ejected from the analyzer is then transferred to a detector. Application of the relevant potentials to the electrodes in a trap results in the formation of the potential well, in which the ions are trapped. One can imagine that the ions are arranged in the form of the layers from the heaviest to the lightest in a bowl (well potential). Tilting the bowl or lowering its edge (changing the RF potential during the analysis) causes "the pouring out" – ejecting the ions of subsequent masses.

Ion trap analysis can be performed in a mode of mass selective instability, resonance ejection, axial modulation, or nonlinear resonance. Those variants are described below [4].

4.2.4.4 Mass Selective Instability Mode

Mass selective instability mode is one of the methods used for obtaining a mass spectrum (Figure 4.61). In this mode, the RF potential is applied to the ring electrode. Its frequency is constant, while the value of the potential can be changed. Spherical electrodes are grounded.

The changes in RF potential during the analysis in a mass selective instability mode are shown in Figure 4.62. At the beginning, the RF potential applied to the ring electrode is equal to zero to remove all ions from the trap before the next analysis. Then, during the introduction of ions into the analyzer, RF potential is kept at a constant level. This allows for setting the value of the potential (*storage voltage*) so that only the ions of m/z higher than the defined value are collected in the trap. This is important especially during the analysis of complex mixtures. Additionally, such handling eliminates signals derived from the air, solvent, or other low molecular weight impurities from the spectrum. Ions of these substances will not be collected in the trap, as they are not able to achieve a stable trajectory under such conditions.

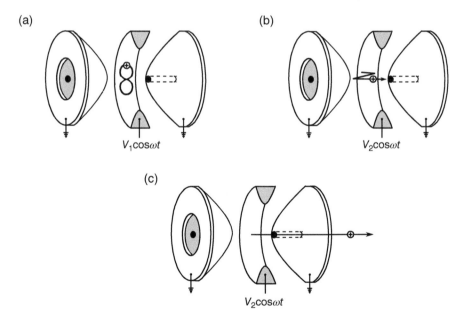

Figure 4.61 Excitation of ions in the mass selective instability mode: (a) trapping – stable trajectory of the ions of given m/z; (b) excitation – unstable trajectory ($V_1 < V_2$); and (c) ejection of the ions from the trap to the detector (Note that the presented trajectories are very simplified. In fact, ion movement is much more complex (see Figure 4.60)). V_1, the potential applied to the ring electrode during ion trapping; ω, the frequency of the potential applied to the ring electrode; V_2, the potential applied to the ring electrode during ion excitation.

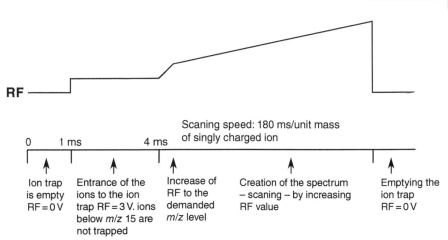

Figure 4.62 Changes of the *V* potential in the mass selective instability mode.

When the trap is filled with a specified group of ions, the value of the *V* potential begins to grow. It triggers more rapid movement of ions of successively increasing values of *m/z*. This leads to destabilization of their trajectory in *z* direction and their ejection into the detector through the hole in one of the spherical electrodes. Detection of the subsequent groups of ejected ions of the same *m/z* enables obtaining the mass spectrum.

The highest mass detectable by using this mode depends on the potential limit that can be applied to the ring electrode (e.g. for the potential of 7000–8000 V, $r_0 = 1$, and a frequency of 1.1 MHz, the maximum measurable value is 650 Da) [1, 4].

4.2.4.5 Resonant Ejection Mode

Oscillation of the ions, in both *z* and *r* directions, takes place at a defined frequency depending on *m/z* value for a given ion. Therefore, it is possible to destabilize ions by applying a defined resonant frequency to the ring electrodes in the trap.

In the resonant ejection mode (Figure 4.63), the ions are trapped as described before (Section 4.2.4.4), i.e. the corresponding RF potential is applied to the ring electrode to collect ions without the low molecular weight impurities. The potential applied to the ring electrode remains constant during further steps. In the next phase, an additional RF potential (hundreds of millivolts), shifted in phase (180°), is applied to each of the spherical electrodes. This may cause the energy transfer to the ion of a given *m/z* due to the resonance. The only condition is that the frequency of RF potential changes corresponds to the oscillation frequency of a given ion in the *z* direction. As a result, an excited ion is ejected from the trap through a hole in a spherical electrode. Time-dependent changes

(a)   (b)

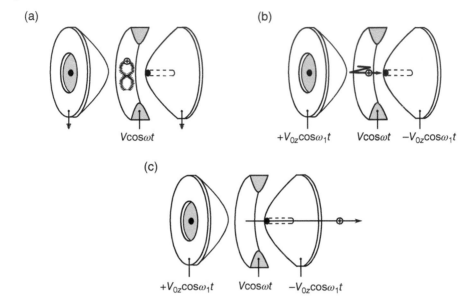

Vcosωt $+V_{0z}cos\omega_1 t$ Vcosωt $-V_{0z}cos\omega_1 t$

(c)

$+V_{0z}cos\omega_1 t$ Vcosωt $-V_{0z}cos\omega_1 t$

Figure 4.63 Excitation of ions in the resonant ejection mode: (a) trapping – stable trajectory of an ion of given *m/z*; oscillation in *z* direction with a frequency dependent on *m/z*; (b) excitation of an ion by applying a potential to the electrodes having frequency equal to the oscillation frequency of the ion in the *z* direction (ω_1); and (c) ejection of ions from the trap toward the detector (Note that the presented trajectories are simplified. In fact, ion movement is much more complex (see Figure 4.60)). V_{0z}, potential applied to spherical electrodes; ω_1, frequency of the potential applied to spherical electrodes; *V*, potential applied to the ring electrode; ω, frequency of the potential applied to the ring electrode.

of the RF potential applied to the spherical electrodes enable ejecting ions of subsequent *m/z* values to the detector and, as a result, to obtain the mass spectrum.

The mode described above allows for collecting ions within defined *m/z* range. This could be achieved by the use of relevant frequencies of RF potential, so the trap can be first emptied from the ions of lower *m/z* values and then from the ions of higher *m/z* value than defined *m/z* range. In the previous mode (Section 4.2.4.4), it was only possible to obtain the spectrum from the lowest to the highest *m/z* value. As resonant ejection mode enables trapping the ions of a defined *m/z* range, this allows for selection of ions for the tandem mass spectrometry.

This method is characterized by higher sensitivity and resolution than the previous one and provides the analysis of a wider range of masses. Its main advantage is the possibility of trapping (and analyzing) ions of a defined *m/z* range [2, 4].

4.2.4.6 Axial Modulation

This mode of operation is a kind of conjunction of the two previous modes. It operates by applying RF potential of about $6V_{(p-p)}$[1] to the spherical electrodes and a frequency equal to a half of the frequency of the potential on the ring electrode. In such a system, for a moment before ejecting a defined ion from the trap, by increasing the potential of the ring electrode, it resonates with the potential applied to the spherical electrodes. This results in better focusing of an ion beam leaving the trap and eliminates the space charge effect (widening of the peaks associated with the repulsion of ions). In the previously discussed mass selective instability mode, signals generated from low masses are widened due to the presence of ions of higher m/z value in the trap. In the axial modulation mode, this problem is eliminated. Such a method allows for increasing the resolution (obtained peaks are not as wide) and broadening the range of the analyzed masses. In addition, this mode provides better sensitivity, as the number of ions inside the trap is increased by an order of magnitude.

4.2.4.7 Nonlinear Resonance

As a result of applying an excitation potential to the spherical electrodes, a weak field (usually dipole or hexapole) appears besides the existing quadrupole field. This increases the resolution of the obtained spectrum but, at the same time, contributes to the loss of ions from the trap due to the nonlinear resonance.

This phenomenon is used in the nonlinear resonance mode. It turns out that the ejection of an ion, as a result of excitation with frequency relevant to induce nonlinear resonance, is very fast. This allows obtaining spectra of good resolution with high scanning speed. The exact explanation of this mode is beyond the scope of this textbook.

4.2.4.8 Linear Ion Trap (LIT)

Construction of a linear ion trap (LIT) is based on the structure of a quadrupole. RF potential is applied to its rods, which allows for focusing ions between the rods and for controlling their move in x and y directions. At the exit of the trap, additional electrodes with appropriate potential are located. They control the move of ions along z-axis (Figure 4.64), which enables entering and trapping ions in the center of the trap between the rods [3,5].

The linear trap can operate in two modes. In the *axial ejection mode*, the ions of certain values of m/z leave the trap in the z direction. This is due to the changes of an alternating voltage (AC) applied additionally to the rods and to the changes of RF voltage focusing ions in the x- and y-axes.

---

1 $V(p-p)$, *peak-to-peak voltage*: for RF current, it is the absolute sum of the amplitudes of the negative and positive voltages applied to the electrode. For example, when the $V_{(p-p)}$ equals 6 V, the maximum positive voltage applied to the electrode is 3 V, and a maximum negative voltage applied to the electrode is -3 V.

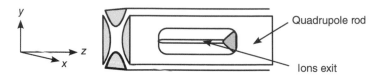

Figure 4.64 Scheme of a linear ion trap and ejection of ions of a defined *m/z* in the *x* and *y* directions.

The Difference Between AC and RF

Alternating current (AC) is a more general term of the alternating voltage. Its value is usually in the range of kilohertz. An alternating voltage marked with radio frequency (RF) corresponds to the alternating voltage of a frequency of c. 3 kHz to 300 GHz.

In the second mode, radial mass selective ion ejection, the change of voltage on the rods allows for ejection of the ions having specified *m/z* in *x* and *y* (radial) directions. This type of trap is equipped with the rods of hyperbolic shape. Additionally, there are slots cut in one pair of the opposing rods through which the excited ions can exit toward the detectors. The combination of two detectors provides an increase in sensitivity. This mode of operation enables selection and keeping the chosen ions, their subsequent fragmentation, and further analysis.

LIT handles higher number of ions when compared with the classic Pauli ion trap (QIT; see Section 4.2.4.2) due to the reduction of repulsion between ions having the same charges (space charge effect). Ions in the LIT oscillate during their move along the quadrupole, and the trajectories of particular ions do not intersect, as it happens in case of QIT analyzer. In addition, the ion cloud is spread along the entire length of the device. Higher amount of the trapped ions provides an improved limit of detection when compared with QIT. LIT is also characterized by faster scanning and simpler construction; however, precision and positioning of the rods remain crucial for its efficient operation.

The linear trap can also be used for the storage of ions, e.g. to transfer them to the analyzers that work in a pulsed mode, such as FT-ICR, Orbitrap, or orthogonal TOF. The ions are introduced into the trap, where collisions with the neutral gas particles reduce their kinetic energy to almost zero (*ion cooling*). Ions can be released from the trap and then delivered to the analyzer at any time (they are usually kept inside the trap for milliseconds to seconds). Also, in the LIT analyzer, fragmentation of an ion can be performed, and the resulting fragments may be forwarded to the next analyzer. Additionally, LIT can be used to protect ultrahigh vacuum in the FT-ICR analyzer.

4.2.4.9 Applications

Application of an ion trap is similar to the quadrupole analyzer. The mass range from 50 to 4000 m/z (sometimes 6000 m/z) makes it useful for the analysis of organic and inorganic substances of relatively low masses. The constructional limitation of an ion trap debilitates analysis of the molecules having masses lower than 50 m/z. As in the case of a quadrupole analyzer, the trap works well in combination with ESI ion source, as multiple ionization of the analyte typical for this source enables analysis of larger molecules, despite the limited analysis range.

The ion trap is a simple and relatively low-priced analyzer of small size, suitable for miniaturization. In contrast to other analyzers, it does not require high vacuum for proper working (pump with $40\,l\,s^{-1}$ flow instead of $250\,l\,s^{-1}$).

Working principles lead to increased sensitivity compared with the quadrupole analyzer. In the ion trap, the ions are collected inside the analyzer, which improves signal-to-noise ratio. Additionally, it is characterized by a higher resolution and scanning speed in comparison with the Q analyzer.

The great advantage of the traps is that they allow for the collection of ions from the defined m/z range by destabilizing all others. Not only this is used in the tandem mass spectrometry but also in the studies of the reaction kinetics between ions and gas molecules. Collection of ions increases sensitivity of the analysis for substances at low concentrations in complex mixtures. This increase in the sensitivity is achieved by setting trap parameters in such a manner that only ions from defined m/z range obtain stable trajectory. The abundant substance is not trapped, which eliminates the repulsion effects between positively charged ions (space charge effect) and improves quality of the resulting spectrum. Due to the operating characteristics, the analyzer can be linked to the sources operating in continuous way (e.g. EI, ESI), as well as in the pulsed mode (such as MALDI or DIOS) [4].

Questions

- What is an ion trap and how it works?
- What is the major drawback of introducing too many ions inside the ion trap (more than 10^7)?
- List and describe the various modes of operation for an ion trap.
- Which modes of operation allow to trap ions of a defined m/z range?
- Why linear ion trap can handle higher number of ions if compared with the classic Pauli ion trap?
- What other usages linear ion traps can have besides detection?
- What are the applications of ion traps?
- What are the advantages of ion traps over quadrupole analyzers?

References

1 Ekman, R. and Silberring, J. (eds.) (2002). *Mass Spectrometry and Hyphenated Techniques in Neuropeptide Research*. New York: Wiley.

2 Ekman, R., Silberring, J., and Westman-Brinkmalm, A.M. (eds.) (2009). *Mass Spectrometry: Instrumentation, Interpretation, and Applications*. New York: Wiley.Kraj, A., Desiderio, D.M., Nibbering, N.M.

3 Gross, J.H. (2011). *Mass Spectrometry*. Berlin: Springer.

4 March, R.E. and Todd, J.F.J. (eds.) (2005). *Practical Aspect of Ion Trap Mass Spectrometry*. Boca Raton: CRC Press.

5 Greaves, J. and Roboz, J. (eds.) (2013). *Mass Spectrometry for the Novice*. Boca Raton: CRC Press.

4.2.5   High-Resolution Mass Spectrometry

Piotr Stefanowicz and Zbigniew Szewczuk

Faculty of Chemistry, University of Wrocław, Wrocław, Poland

4.2.5.1   Introduction

One of the most important tasks of mass spectrometry is to determine the chemical formula of an investigated compound. Elemental composition of a given molecule can be determined based on a very accurate determination of its molecular mass. The number of possible atomic compositions to be assigned to the analyzed compound significantly decreases with increasing mass accuracy. Therefore, mass spectrometers characterized by high resolution and high precision (better than 1 ppm) are greatly desirable for accurate analysis of higher mass molecules, such as proteins, metabolites, and petroleum fractions in their complex mixtures. Very high resolving power is also needed to separate isobaric mass peaks (Figure 4.65).

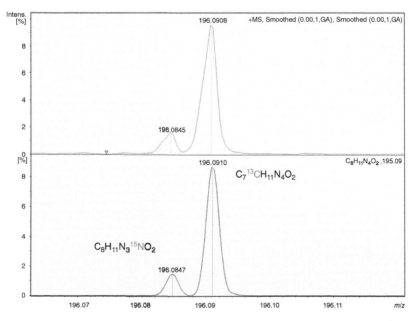

Figure 4.65 Distribution of isobaric peaks in the isotopic cluster of peaks [M + H]$^+$ of caffeine recorded using ICR mass spectrometer (upper panel) and calculated (lower panel). The cluster must contain only one heavy atom. The most abundant signals are separated peaks corresponding to $C_8H_{11}N_3{}^{15}NO_2$ (*m/z* 196.0847) and $C_7{}^{13}CH_{11}N_4O_2$ (*m/z* 196.0910). *Source*: From the authors' collection.

4.2.6   Ion Cyclotron Resonance (ICR)

Piotr Stefanowicz and Zbigniew Szewczuk

Faculty of Chemistry, University of Wrocław, Wrocław, Poland

4.2.6.1   Introduction

Currently, Fourier transform ion cyclotron resonance mass spectrometry (FT-ICR-MS) offers the highest resolution (more than 500 000 FWHM), resolving power, and accuracy than any other mass analyzer. Although the application of a cyclotron analyzer for mass detection was already developed in 1949 by Hippie et al. [1], the application of FT-ICR mass spectrometers was first demonstrated in 1973 by Comisarow and Marshall [2]. They used Fourier transform to determine mass-to-charge (*m/z*) ratio of ions based on their cyclotron frequencies in a fixed magnetic field.

A characteristic element of all ion cyclotron resonance (ICR) instruments is the liquid helium-cooled superconducting magnet producing homogeneous magnetic field in the range of 3–18 T (Tesla). The magnitude of the applied magnetic field strength has a significant impact not only on the resolution of the analyzer but also on its cost. There is an ICR accumulative chamber inside the magnet, acting, at the same time, as an ion analyzer and detector. The charged particles rotate under the influence of a magnetic field, at a frequency related to their *m/z* ratios. Measurement of the *m/z* is based on the analysis of the cyclotron frequencies of the ions flying near the detection plates.

4.2.6.2   Cyclotron Frequency

The trajectories of ions are curved in a magnetic field, and the radius of the trajectory depends on the intensity of the field and ion velocity. If the magnetic field is strong (7 T or more) and the ion velocity is low, the radius of the trajectory becomes small (<1 mm). Therefore, the ions can be trapped on a circular trajectory in the ion cyclotron.

If an ion with mass *m* is rotating in a magnetic field *B* with a velocity *v*, there are two forces acting on the ion:

Lorentz force (centripetal):

$$F = zvB \tag{1}$$

Centrifugal force:

$$F = \frac{mv^2}{r} \tag{2}$$

where *r* is the radius of circular path.

If these two forces are balanced, the trajectory of the ion is stabile:

$$zvB = \frac{mv^2}{r} \tag{3}$$

Dividing both sides of the above equation by v,

$$zB = \frac{mv}{r}, \quad \text{i.e.}: \frac{v}{r} = \frac{zB}{m} \tag{4}$$

Frequency formula is given by

$$f = \frac{v}{2\pi r}, \quad \text{i.e.}: f = \frac{1}{2\pi} \cdot \frac{v}{r} \tag{5}$$

Next, incorporating the above equation to Formula (4),

$$f = \frac{z}{m} \cdot \frac{B}{2\pi} \tag{6}$$

According to this equation, in a constant and uniform magnetic field B, all ions can move on circular orbits with characteristic cyclotron frequencies.

When using SI units, i.e. f in Hz (Hertz) and B in T (Tesla), we obtain the formula

$$f = \frac{1{,}535611 \times 10^7 \, B}{m/z} \tag{7}$$

Therefore, the cyclotron frequency of the rotating ion is inversely proportional to its m/z ratio and depends on the intensity of the magnetic field B, which is a constant value for a given ICR spectrometer.

For example, for the 7 T magnetic field, the cyclotron frequency is 3.84 MHz at 28 Th and 26.8 kHz at 4000 Th. The frequency range is thus very large.

It is important to note that the frequency (f) depends on the m/z ratio and is independent on the velocity and on the radius of circular path. An increase of the kinetic energy of the ion spinning in the cyclotron does not affect its cyclotron frequency, although it increases the radius of the trajectory of motion.

4.2.6.3 ICR: Principles of Operation

The frequencies of the trapped ions must be determined in order to obtain a mass spectrum. The relatively broad dynamic range of modern FT-MS spectrometers (around 10^5) allows simultaneous analysis of up to one million charged particles. In the case of a larger number of ions present in the chamber, there may be adverse Coulomb repulsions between ions having the same charge (space charge effect), which leads to ion suppression. FT-MS analyzers are very sensitive; theoretically it is possible to register MS spectra for

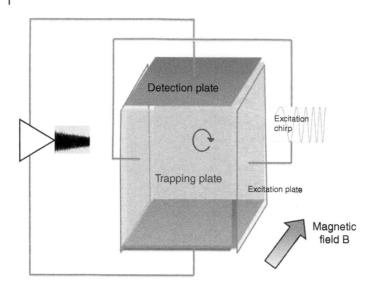

Figure 4.66 Schematic depiction of the ICR chamber. The mass analysis involves three steps: (a) ion formation and storage using trapping plates, (b) excitation of the trapped ions by an external broad frequency range pulse, and (c) detection of the ions by measuring their image current.

10 ions circulating in the ICR chamber. An important advantage of the FT-ICR-MS method is the possibility of relatively long (multi-minute) ion analysis. This allows for studying some reactions taking place in the chamber, including ion reactions with neutral molecules or fragmentation reactions with electron capture (electron capture dissociation [ECD]; see Section 7.4.5).

The ICR chamber has the shape of a cube or cylinder and contains three pairs of electrodes oriented perpendicularly to each other, which are called trapping plates, excitation plates, and detection plates, respectively (Figure 4.66).

4.2.6.4   Injection of Ions into the ICR Cell

ICR mass spectrometers may contain any type of ion source, although the instruments utilizing ESI, MALDI, or APCI ionization are most commonly produced. Most high-resolution mass spectrometers are equipped with a quadrupole mass filter and a fragmentation chamber that allows to perform MS/MS experiments. The generated ions are directed through the focusing lenses into the ICR chamber. There is a very low pressure inside the chamber, around 10^{-12} Ba, allowing the ions to move around in the chamber without collisions, even for a few minutes. This requires the use of very efficient turbomolecular pumps. The excited ions with m/z of 100 Th can cover a distance of about 30 km in 1 second. The ions are introduced into the ICR cell through a hole in one of the trapping plates, oriented parallel to the magnetic field lines.

4.2.6.5 Trapping Electrodes

The trapping electrodes are responsible for the maintenance of ions inside the ICR chamber by using a small electrical potential. This induces the oscillating ion movement between trapping motions. The oscillating and cyclotron motions of ions lead to the generation of the third type of motion, which is a magnetron motion. The magnetron motion makes precise ion analysis difficult, which is why various methods are used to minimize this problem. The trapping plates also allow to remove unnecessary ions from the chamber by changing the voltage, after the measurement is completed.

4.2.6.6 Excitation Electrodes

Ions concentrated in the center of the ICR chamber rotate at a cyclotron frequency according to their *m/z* values. Unfortunately, direct measurement of these frequencies is impossible because their trajectory radii are too small to be sensed by the detection plates and the ions are scattered at various places on the orbits.

In order to detect passing ions, it is necessary to excite them onto an orbit with a larger radius and to organize them. The excited "ion packs" circulating at higher orbits approach the detection plates, which makes them measurable.

In order to excite the rotating ions, a variable radio-frequency (RF) voltage is applied to the excitation plates (chirp). If the frequency of the alternating electric field at some point corresponds to the cyclotron frequency of the analyzed ion, the latter is excited. Ions that differ in the *m/z* value are therefore excited at different frequencies.

At the moment of excitation, all ions with the same *m/z* value absorb energy from the alternating electric field between the excitation plates, which will increase their kinetic energy. The excited ions move in a spiral motion onto an orbit with a larger radius, keeping their cyclotron frequency. The radius of this orbit depends on the duration and amplitude of the changing potential. However, if the ions are excited too much, they will obtain a radius that exceeds the dimensions of the ICR chamber and will be expelled. Excitation of ions causes their ordering, i.e. ions having identical *m/z* values move close to each other, creating ion clouds. After ion excitation, the electric field on the excitation plates is switched off, and then the detection of ions by means of detection electrodes is possible. After some time, the ions will return to the original orbit, and another excitation will be necessary for further measurement.

4.2.6.7 Detection Electrodes and Fourier Transform

Ions of a specific cyclotron frequency induce an image current that is detected by a pair of electrodes in the ICR chamber. The detection electrodes are

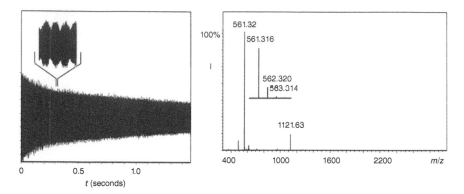

561.32
100%
561.316

I

562.320
563.314

1121.63

400   1000   1600   2200   *m/z*

0   0.5   1.0
t (seconds)

Figure 4.67 Graph of changes in current intensity on detection plates generated by ion movement in a cyclotron (FID) and MS spectrum calculated by means of Fourier transform. *Source*: From the author's collection.

connected to ground through a resistor. When the positive ion cloud approaches one of the detection plates, the electrons are attracted to this electrode plate from ground. Then as the ions circulate toward the second detection plate, the electrons travel back to the second electrode plate. This motion of electrons moving back and forth between the two plates produces a small but detectable AC. This migration of electrons is amplified and detected as a sinusoidal image current, which will be analyzed in terms of its frequency and amplitude. Because all ions are excited by the RF chirp, the image current represents a super position of many components of the cyclotron frequencies with different amplitudes of all the ions in the cell (Figure 4.67). The overall signal is very complicated and fades over time as the ions relax and return to their stable circular orbits near the center of the cell. This image current can be converted into a mass spectrum using the Fourier transform.

Simultaneous analysis of all cyclotron frequencies allows for registering high-resolution mass spectrum in a relatively short time. Although the analysis time (<1 second per spectrum) is not as short as in other MS analyzers, it is possible to use FT-MS as a liquid chromatography detector (LC-FT-MS technique). An important advantage of FT-MS is the ability to analyze substances within a wide range of *m/z* values (in most spectrometers from 100 to 4000 Th). However, it should be emphasized that at higher masses (above 2000 Th), the resolution is rapidly decreasing, especially in the case of the analyzers equipped with magnets generating magnetic fields of relatively low intensity.

The cyclotron frequency ranges from a few kHz to several MHz, which enables it to be precisely measured using standard electronic methods. The ion of a given *m/z* value has the same cyclotron frequency regardless of its kinetic energy, which eliminates the need to adjust ion kinetic energy (in other types of analyzers such systems are often used to increase resolution).

In FT-ICR-MS, the ions are not detected by hitting an electron multiplier but only by passing near detection plates during rotation in a magnetic field. Therefore, the m/z values are resolved only by the ion cyclotron resonance frequency but not in time or space, which provides much better stability. Consequently, different ions are detected simultaneously during the detection interval. Because the ions are measured nondestructively, repeated experiments may be summed up to improve the signal-to-noise ratio. This provides an increase in the sensitivity and resolution. The resolution can also be improved by increasing the magnetic field (strength of the magnet) and by elongating detection time.

4.2.6.8 FT-ICR Properties as *m/z* Analyzer

In ICR cells, the sustained off-resonance irradiation (SORI) can be performed. In this method, the ions trapped in the ICR chamber are excited by means of the variable electric field. The excitation uses the frequency shifted in relation to the resonance frequency. This allows to increase the kinetic energy of ions in a controlled manner, thus preventing their release from the ICR chamber. At the same time, collision gas (usually argon) is introduced into the cyclotron chamber via an electronically controlled pulse valve. Collisions of ions with argon atoms lead (similarly to the classical CID) to the gradual increase of the oscillatory energy, followed by fragmentation of trapped ions. The mechanism of fragmentation and diagrams of ion decay are very similar to those observed in CID conditions (Section 7.4.2).

An important advantage of ICR mass spectrometers is the possibility of carrying out ECD, which allows fragmentation of polypeptides. The mechanism of this process is similar to the ETD method (Section 7.4.6). The electrons, however, are not arising from a free radical medium (e.g. fluoranthene) but are emitted by a heated cathode instead. The energy of electrons is low, i.e. below 1 eV. Electron uptake by the multiply protonated peptide cation causes fragmentation of $N-C_\alpha$ bonds, yielding N-terminal c and C-terminal z fragments. In contrast to CID (leading to the formation of mainly b and y fragmentation ions), ECD is a nonergodic process [3], i.e. the cleavage happens prior to intramolecular energy redistribution. ECD generally results in the cleavage of a wider range of peptide backbone bonds than CID, with less dependence on peptide composition. As a result, many labile side-chain groups are preserved. This allows for sequencing of peptides containing labile posttranslational modifications, including glycations [4] (Figure 4.68) and N-phosphorylations (undergoing complete dissociation under CID conditions) [5]. In addition, ECD fragmentation method has proved useful in localizing amino acid residues, in which hydrogen atoms in peptide bonds have been exchanged for deuterium atoms [6].

The rate of H/D isotopic exchange of the amide protons of individual amino acid residues depends on the spatial structure of the protein. Therefore, studies

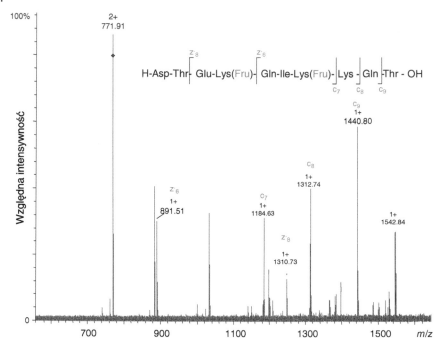

Figure 4.68 ECD-MS spectrum of a double glycated peptide. *Source*: From the authors collection; Fru, fructose.

on the kinetics of this process, and in particular the determination of exchangeable amino acid residues, can provide important information on the protein conformation. Unlike CID [7], ECD proceeds with a relatively low degree of hydrogen scrambling, which makes it a method of choice for localization of deuterated amides within peptide chain.

It should be noted that electron capture neutralizes one positive charge within the ion, which means that only ions with multiple positive charges can be analyzed by this method. This makes the sequencing of short peptides impractical, as the small number of basic amino acid residues does not form repeatedly charged ions. Another limitation of this method is its uselessness in the spectrometers with ion sources in which mostly singly charged ions are generated, including MALDI.

In recent years, Bruker Daltonics has launched an ultra-high-resolution solariX mass spectrometer containing an improved ICR chamber with dynamic harmonization of the electric field (ParaCell chamber), developed by Nikolaev et al. [8] (Figure 4.69).This resulted in a very high resolution of the analyzer (over 10 000 000 FWHM) and exceptional accuracy of m/z measurement (error <1 ppm), despite the use of a superconducting magnet generating a relatively small magnetic field (7 T). The use of high-resolution solariX not only enabled the analysis of isotopic peak distributions in high molecular proteins [9] but

Figure 4.69 The ICR cell (ParaCell®) applied in Bruker's solariX XR Fourier transform mass spectrometer [11].

also the study of subtle cluster distributions of individual "isotope peaks" containing isobaric peaks derived from various heavy isotope atom combinations of ^{13}C, ^{15}N, ^{17}O, ^{18}O, ^{2}H, ^{33}S, and ^{34}S. Analyzing the ions of $[M+2H]^{2+}$ substance P on the solariX spectrometer, Nikolayev et al. successfully separated all isobaric peaks in the isotopic cluster A + 3: ^{15}N^{34}S (m/z 675.867 834), ^{13}C^{15}N$_2$ (m/z 675.870 143), ^{13}C^{34}S (m/z 675.870 951), ^{13}C$_2$^{15}N (m/z 675.873 268), ^{13}C^{18}O (m/z 675.875 185), ^{13}C$_3$ (m/z 675.876 382), and ^{2}H^{13}C$_2$ (m/z 675.877 895) [10].

The most important advantage of the FT-ICR analyzer is the highest recorded mass resolution of all mass spectrometers (above 500 000). The accuracy of measurements using an internal standard (1 ppm) is also high. Unlike other analyzers (where increasing sensitivity usually leads to a decrease in resolution and vice versa), the FT-ICR analyzers enable high-resolution measurements of small amounts of complex mixtures with very good resolution. Another advantage of this analyzer is the ability to conduct a nondestructive ion detection, resulting in a relatively long measurement, which enables precise determination of all cyclotron frequencies and testing various reactions, including multiple fragmentation.

A serious drawback is the need of a high-intensity magnetic field produced by a superconducting magnet cooled with liquid helium, as well as the requirement to achieve a high vacuum, which is associated with high price of the apparatus and high costs of its maintenance (consumption of liquid helium about 100 l for 3 months, liquid nitrogen about 100 l per week). A yet another disadvantage of the analyzer is also the decrease in resolution at high m/z values, which is particularly observable in ICR spectrometers equipped with magnets with a low magnetic field, and also the presence of some artifacts, such as harmonics and sidebands present in the mass spectra.

4.2.7 Orbitrap

Piotr Stefanowicz and Zbigniew Szewczuk

Faculty of Chemistry, University of Wrocław, Wrocław, Poland

Orbitrap is the mass analyzer developed during the last 20 years. This instrument measures m/z values based on the frequency of their oscillations utilizing only electrostatic field for ion trapping. Orbitrap provides high mass accuracy and resolving power and a large space charge capacity. Although the resolution and accuracy of this analyzer are not as high as in the case ICR FT-MS spectrometer discussed above, Orbitrap is more compact, less costly, and easier to maintain and does not require expensive cryogenic liquids.

4.2.7.1 History of Development and Principles of Operation

Orbitrap is a result of an evolution of the Kingdon trap – a device used for storage of ions in a vacuum chamber over relatively long period of time (a few seconds). The Kingdon trap was composed of an outer cylindrical electrode, two end-cap electrodes, and thin-wire central electrode. The ions orbiting around central electrode were trapped on the stable trajectory, because of the equilibration of electrostatic and centrifugal forces. The end-cap electrodes produced electrostatic potential used to achieve ion trapping in the axial direction. Although the Kingdon trap was a storage device, it did not allow for obtaining mass spectra. In 1981 Knight modified Kingdon trap by dividing the external electrode and changing its shape, which created quadrupole field in the axis of a trap. In the modified trap the axial harmonic oscillations of ions were observed. The analysis of the resonance of ions inside the Knight trap, in theory, allowed measurements of m/z, but because of distortion of the quadrupolar nature of the axial potential, no mass spectra were reported for the device. In 2000 Makarov significantly improved Knight's design by optimizing the shape of electrodes in order to generate perfect quadrupole field in axial direction. The Orbitrap mass analyzer is composed of three electrodes (Figure 4.70): the axial spindle-shaped internal one (1) surrounded by two outer barrel-like electrodes (2A and 2B) separated by a thin layer of insulator.

Like in Kingdon trap, the stability of ions' trajectories is a result of equilibration of electrostatic and centrifugal forces; however due to the shape of electrodes, the axial electric field has a perfectly quadrupolar character. Electrostatic potential U in this device depends on the two cylindrical coordinates, r and z, which describe the distance of ion from the analyzer axis and axial position of the ion, respectively, and is given by Eq. (1):

$$U(r,z) = \frac{k}{2}\left(z^2 - \frac{r^2}{2}\right) + \frac{k}{2} \cdot (R_m)^2 \cdot \ln\left[\frac{r}{R_m}\right] + C \tag{1}$$

where R_m is characteristic radius, C is constant value, and k is field curvature [12].

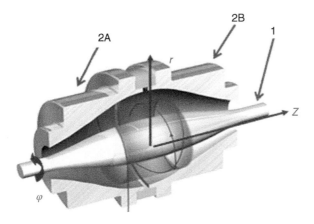

Figure 4.70 Cross section of the Orbitrap mass analyzer.

Equation (1) demonstrates that the ion's motion in the Orbitrap can be formally divided into two independent components: orbital motion and, resulting from harmonic potential, axial motion. The potential describing oscillations of an ion depends on coordinate z only. The frequency ω of axial oscillations is given by Eq. (2):

$$\omega = \sqrt{\left(\frac{z}{m}\right) \cdot k} \tag{2}$$

where z is ion's charge, m ion mass, and k constant value. Determination of ω frequencies allows calculation of m/z values for individual ions in the Orbitrap.

4.2.7.2 Analyzing Ions in the Orbitrap

The ions generated in the ion source (e.g. ESI) undergo collisional cooling to decrease their kinetic energy. Dimension of the ion package should not exceed a few mm, and their transfer to the analyzer must be performed in a very short time interval (<100–200 ns). Those factors are crucial for the stability and coherence of the ions' motion. The ions collected in the ion trap (usually a linear ion trap or C-trap; Section 4.2.4) are simultaneously injected into the mass analyzer. Injection is accompanied with a gradual increase of potential applied to the central electrode (up to 3500 V), resulting in a decrease in diameter of ionic cloud. Ions are injected far from the equatorial plane of Orbitrap, which leads to axial oscillations of the ion cloud (along z-axis), without a need for any additional excitation. The injected ions should have proper kinetic energy to allow their radial trapping. The trapped

(a)

(b)

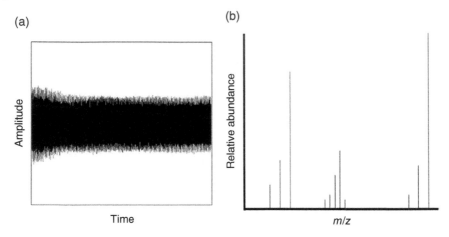

Figure 4.71 Graph of changes in the image current generated by ion movement in the Orbitrap mass analyzer (a) and MS spectrum calculated by means of Fourier transform after converting frequencies to *m/z* according to Eq. (2) (b).

ions in the Orbitrap oscillate synchronically along *z*-axis. The external electrode is divided into two insulated parts (2A and 2B; Figure 4.70). The axial motion of ions induces the difference of potential between parts 2A and 2B. This signal is amplified and subjected to Fourier transform, allowing determination of all frequencies of the oscillating ions (Figure 4.71). These frequencies are transformed into *m/z* values based on Eq. (2). The *k* value is obtained empirically by a calibration of instrument using a series of standards with known *m/z* ratios.

4.2.7.3 Orbitrap Properties as *m/z* Analyzer

Resolution of the Orbitrap is in the range of 150 000–500 000 and is comparable with ICR FT spectrometers only. The example of spectrum is presented in igure 4.72. The analyzer is also characterized by high accuracy (1–5 ppm), and calibration remains stable for a long time (drift of *m/z* in 6 hours is lower than 5 ppm). Internal standard further increases the accuracy of *m/z* measurements, thus reducing the error to 100 ppb. Similarly to ICR FT-MS, Orbitrap is losing resolution when *m/z* ratio increases. However, this effect is less pronounced than in the case of ICR FT-MS, especially for the high field version of Orbitrap. Therefore, for *m/z* > 2000, the resolution of Orbitrap may be higher than that achieved with magnetic FT-MS.

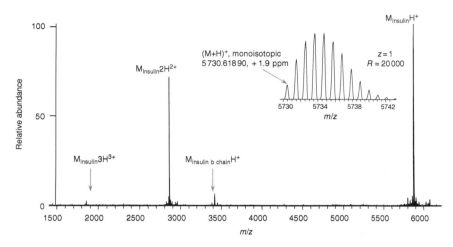

Figure 4.72 Mass spectrum of bovine insulin measured over the extended mass range, up to *m/z* 6000.

m/z range of commercially available Orbitrap is 50–6000 *m/z*. However, it has been suggested that after relatively simple changes in the instrument's hardware and configuration, this value may be expanded up to 10 000 *m/z* [13].

Dynamic range although this parameter is lower than in the case of triple quadrupole instruments and is equal to approximately 4 orders of magnitude (e.g. Q-Exactive 5000), which allows application of Orbitrap in quantitative analysis.

Scanning speed 1–2 seconds per scan at maximal resolving power of the analyzer; however this parameter can be significantly increased if the analysis does not require full resolution of the instrument. For example, 0.4 seconds scan cycle time corresponds to the resolution of c.30 000.

Sensitivity of the Orbitrap is one of the highest among the currently used mass spectrometers, allowing detection of attomolar quantities of analytes.

4.2.7.4 Analytical and Proteomic Applications of Orbitrap

Orbitrap is often used as a component of hybrid mass spectrometers. The combination of this analyzer with a quadrupole ion trap offers an instrument with MS^n capability and high resolution, as well as high mass accuracy characterized by an unprecedented power in structural elucidation. These features make the Orbitrap very useful for identification of unknown compounds in untargeted metabolomics research.

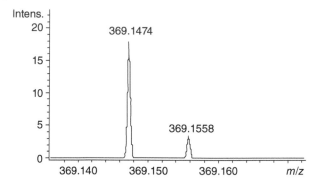

Figure 4.73 ESI-MS spectrum of 5-(2-ethoxy-5-((4-methylpiperidin-1-yl)sulfonyl) phenyl)-3-methyl-1,2,4-oxadiazole.

The most common application of this instrument remains proteomics [14], where besides mass accuracy, the high efficiency of fragmentation by both CID and ECD is key, allowing sequencing of long peptides, as well as an efficient analysis of posttranslational modifications. Recently, it has been demonstrated that Orbitrap, after minor modifications, may be used to investigate native protein assemblies like proteasome, chaperones, or virus capsids [13]. Although the molecular masses of such structures are in the range of megadaltons, modified Orbitrap instruments, including MS/MS experiments, allowed their full characteristics. In these experiments, a relatively low decrease in resolution with simultaneous increase of m/z for the Orbitrap analyzer is beneficial.

High-resolution mass spectrometer allows extraction of the ions with precisely defined m/z. This feature initiated further applications in the analysis of food contaminants [15] and in anti-doping control [16]. The selectivity in this approach is comparable with LC-MS/MS method, but the data on the mass of precursor ion and fragment ions are not necessary for development of analytical method. Therefore, the already archived data may be reinvestigated even long time after acquisition. The main benefit of employing the Orbitrap in such measurements is the combination of high resolving power and mass accuracy higher than in the case of time-of-flight (TOF) analyzers. On the other hand, the resolution of TOF instruments is constant, while in the case of Orbitrap the resolution depends on the acquisition time (the longer the acquisition time, the higher resolution).

Questions

- What is the resolution of a mass analyzer required to resolve two isobaric peaks: $C_{16}{}^{13}CH_{24}N_3O_4{}^{34}S$ and $C_{14}{}^{13}C_3H_{24}N_3O_4S$ (heavy isotopes are underlined) [17].
- Please identify the isobaric peaks in the isotopic cluster A + 3 on the narrowed part of the ESI-MS spectrum of 5-(2-ethoxy-5-((4-methylpiperidin-1-yl)

sulfonyl)phenyl)-3-methyl-1,2,4-oxadiazole $(C_{17}H_{23}N_3O_4S + H)^+$ presented in Figure 4.73.

What other isobaric components are present in the isotopic cluster $A + 3$? Why are they not visible in the spectrum?

- Emulate the ultra-high-resolution spectrum of glutathione (ECG, monopro-tonated ions). Pay attention to the isotopic peak clusters.
- A recently developed set of isobaric tags TMT 10-plex allows parallel multi-plexed quantification of proteins in 10 independent samples [18]. 10-Plex multiplexing is achieved by HRMS based on a difference between ^{13}C and ^{15}N isotopes. The structures of reporter ions are given below. Assuming that the resolution of Orbitrap mass analyzer is equal to 60 000 FWHM, evaluate whether separation of isobaric reporter ions is possible using this mass analyzer.

TMT reagent   Labeled peptide

CID dissociation of TMT-labeled peptide (R = peptide)

Isobaric reporter ions formed by collision-induced dissociation of TMT-labeled peptides.

- Why is it advantageous to have high magnetic fields in cyclotrons?
- Why is it advantageous to have long-time ion analysis?
- How does the cyclotron work?
- Which ion sources are compatible with ICR?
- What is the function of the trapping electrodes in ICR?
- What are the parameters that control the radius of the ions orbiting inside ICR?
- What is the main advantage of FT-MS?
- How is it possible to improve the resolution in FT-ICR-MS?
- What is the sustained off-resonance irradiation (SORI)?
- What is the difference between ECD and CID?
- What are the main advantages and drawbacks of FT-ICR analyzers?

References

1 Hipple, J.A., Sommer, H., and Thomas, H.A. (1949). A precise method of determining the Faraday by magnetic resonance. *Physical Review 76*: 1877–1878.
2 Comisarow, M.B. and Marshall, A.G. (1974). Fourier transform ion cyclotron resonance spectroscopy. *Chemical Physics Letters 25*: 282–283.
3 Zubarev, R.A., Kelleher, N.L., and McLafferty, F.W. (1998). Electron capture dissociation of multiply charged protein cations: a nonergodic process. *Journal of the American Chemical Society 120*: 3265–3266.
4 Stefanowicz, P., Kijewska, M., and Szewczuk, Z. (2009). Sequencing of peptide-derived Amadori products by the electron capture dissociation method. *Journal of Mass Spectrometry 44*: 1047–1052.
5 Kowalewska, K., Stefanowicz, P., Ruman, T. et al. (2010). Electron capture dissociation mass spectrometric analysis of lysine-phosphorylated peptides. *Bioscience Reports 30*: 433–443.
6 Stefanowicz, P., Petry-Podgórska, I., Kowalewska, K. et al. (2010). Electrospray ionization mass spectrometry as a method for studying the high-pressure denaturation of proteins. *Bioscience Reports 29*: 91–99.
7 Ferguson, P.L. and Konermann, L. (2008). Nonuniform isotope patterns produced by collision-induced dissociation of homogeneously labeled ubiquitin: implications for spatially resolved hydrogen/deuterium exchange ESI-MS studies. *Analytical Chemistry 80*: 4078–4086.
8 Nikolaev, E.N., Boldin, I.A., Jertz, R., and Baykut, G. (2011). Initial experimental characterization of a new ultra-high resolution FT ICR cell with dynamic harmonization. *Journal of the American Society for Mass Spectrometry 22*: 1125–1133.

9 Li, H., Wolff, J.J., van Orden, S.L., and Loo, J.A. (2014). Native top-down electrospray ionization-mass spectrometry of 158 kDa protein complex by high-resolution Fourier transform ion cyclotron resonance mass spectrometry. *Analytical Chemistry 86*: 317–320.

10 Nikolaev, E.N., Jertz, R., Grigoryev, A., and Baykut, G. (2012). Fine structure in isotopic peak distributions measured using a dynamically harmonized Fourier transform ion cyclotron resonance cell at 7 T. *Analytical Chemistry 84*: 2275–2283.

11 BRUKER. Innovation with Integrity. http://www.bruker.com/fileadmin/ user_upload/8-PDF-Docs/Separations_MassSpectrometry/Literature/ Brochures/1816259_solarix_XR_brochure_07-2013_eBook.pdf (accessed 28 January 2019).

12 Hu, Q., Noll, R.J., Li, H. et al. (2005). The Orbitrap: a new mass spectrometer. *Journal of Mass Spectrometry 40*: 430–443.

13 Rose, R.J., Damoc, E., Denisov, E. et al. (2012). High-sensitivity Orbitrap mass analysis of intact macromolecular assemblies. *Nature Methods 9* (11): 1084–1086.

14 Scigelova, M. and Makarov, A. (2006). Orbitrap mass analyzer: overview and applications in proteomics. *Proteomics 6*: 1–2.

15 Kellmann, M., Muenster, H., Zomer, P., and Mol, H. (2009). Full Scan MS in comprehensive qualitative and quantitative residue analysis in food and feed matrices: how much resolving power is required? *Journal of the American Society for Mass Spectrometry 20*: 1464–1476.

16 Scigelova, M. and Makarov, A. (2009). Advances in bioanalytical LC–MS using the Orbitrap™ mass analyzer. *Bioanalysis 1*: 741–754.

17 SIS. Exact masses of the elements and isotopic abundances. http://www. sisweb.com/referenc/source/exactmas.htm (accessed 28 January 2019).

18 Pichler, P., Kocher, T., Holzmann, J. et al. (2010). Peptide Labeling with Isobaric Tags Yields Higher Identification Rates Using iTRAQ 4-Plex Compared to TMT 6-Plex and iTRAQ 8-Plex on LTQ Orbitrap. *Analytical Chemistry 82*: 6549–6558.

4.2.8 Hybrid Mass Spectrometers

Giuseppe Di Natale

CNR Institute of Crystallography (IC), Secondary Site, Catania, Italy

4.2.8.1 A Brief Comparison of Mass Analyzers

The basic principles reported in the previous chapters and describing the separation of ions according to their mass-to-charge (m/z) ratio pointed out some advantages and limitations of common mass spectrometers. For example, although the high mass detection range represents the main strength of TOF instruments in comparison with other analyzers [1], TOF analyzers also offer other important properties. Indeed, in this case, all ions generated in source are directed toward the detector and analyzed, implicating a high duty cycle value, while the short time spent by each ion during its transport through the mass spectrometer reduces its collisions both with the residual gas molecules and with other parts of the device, thus increasing ion transmission. The high transmission efficiency and the high duty cycle are therefore responsible for the high sensitivity of TOF mass spectrometers. In addition, implementation of the delayed pulsed extraction and reflectron in the modern MALDI-TOF mass spectrometers provides high-resolution analysis (FWHM > 10 000) and very good mass accuracy [1].

Compared with TOF instruments, the ions analyzed by quadrupoles or ion traps (ITs) require more time to be transported from the source to the detector, due to the higher gas pressure and lower kinetic energy of the ions, whose transmission is therefore reduced. Moreover, quadrupoles and IT analyzers are devices that use the stability of trajectories in oscillating electric fields to separate ions according to their m/z ratios. Therefore, voltage parameters applied to the analyzer have to be changed each time when ions of various m/z values are detected (full scan mode). Under these conditions, the duty cycle of the spectrometer decreases proportionally to the total time spent to detect each ion, producing a lower sensitivity than in the case of TOF instruments. However, it is important to remember that sensitivity of a quadrupole is influenced by the data acquisition mode. Indeed, when detection of selected ions (single ion monitoring [SIM]) is applied, the analyzer separates ions with a specific m/z ratio only. This method is clearly more sensitive than the full scan mode as the time of signal integration is longer [1, 2].

Looking at the features of IT analyzers, separation of ions based on the stability of the trajectories in oscillating electric fields has an important functional outcome. An ion generated in the source can be selected and fragmented in the same analyzer, and the produced fragments can be isolated and fragmented again. This process can be repeated, allowing MS^n experiments to be performed. This represents a great advantage of ITs in comparison with other analyzers. On the other hand, sensitivity of each consecutive fragmentation step decreases, and the total time of the analysis increases accordingly. For this reason, such sequential procedures are applied when large amounts of sample are available. Usually, for LC-MS experiments, MS^3 is acceptable without losing too much information. Ion trap analyzers can yield information regarding molecular structure and are highly efficient in the structural analysis [3].

In the last decade, analyzers that employ trapping of ions in electrostatic field were developed. These devices, called Orbitrap mass analyzers, use the Fourier transform algorithm to obtain mass spectra with very high mass resolution (FWHM > 150 000 at m/z 600) and mass accuracy (<2 ppm) [4], thus achieving performance of double-focusing instruments.

The main drawback of Orbitrap analyzer is the scan speed. In the first commercially available Orbitrap instrument, the normal acquisition cycle was one second. Nevertheless, the advances in electronics enabled the development of novel Orbitrap analyzers with improved acquisition rates [4]. Moreover, it is important to note that stand-alone Orbitrap analyzer cannot perform ion fragmentation; therefore additional fragmentation devices must be implemented in Orbitrap spectrometers.

From the considerations reported above, it turned out that there is no single mass analyzer that provides fast acquisition rate, high sensitivity, high selectivity, MS^n capabilities, and accurate mass measurement. The effort to exploit the strengths of each analyzer while avoiding their weaknesses prompted the development of hybrid mass spectrometers that combine different analyzers in sequence. The development of many early hybrid instruments was also motivated by the desire for MS/MS spectral acquisition. The coupling of two physically distinct devices can affect the type and quality of data obtained, depending on the type of analyzer used in the first and second stages of analysis. In this chapter, several hybrid mass spectrometers will be described. The following discussion will be structured according to the strengths and weaknesses of each analyzer described above.

4.2.8.2 Triple Quadrupoles

As already mentioned, quadrupoles (see Section 4.2.3) can select ions having specific m/z ratio by adjusting the values of the direct potential (U) and the RF voltage (V). It is also possible to stabilize the trajectory of all ions across an adequate mass range when the quadrupole operates in RF-only mode (U is equal to zero) [1, 2]. In view of their properties, it is possible to combine more quadrupoles using some of them as analyzers (Q), while the RF-only device can be used for all ions as a guide as well as a collision cell (q). In this context, triple quadrupole (QqQ) instruments are of significant interest (Figure 4.74a). Four main MS/MS scan modes are particularly used in QqQ instruments (triple quad): (1) product ion scan, (2) precursor ion scan, (3) neutral loss scan, and (4) selected reaction monitoring (SRM) [1].

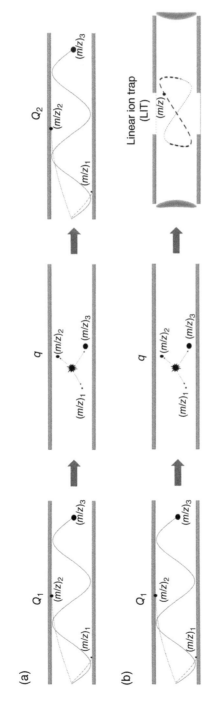

Figure 4.74 (a) Schematic view of a triple quadrupole instrument. Depending on the scan mode used, the ions with different m/z values ((m/z)₁, (m/z)₂, (m/z)₃, etc.) can be scanned or selected in Q_1 analyzer and fragmented (CID) in the RF-only device/cell collision (q). Product ions can be scanned or selected in the Q_2 analyzer. (b) Schematic view of a quadrupole–linear ion trap hybrid instrument. The Q_1q part of the instrument is generally used for ion fragmentation, while LIT operates in the scan mode, providing greater sensitivity and speed of analysis than the quadrupole operating under the same conditions.

The most common scanning mode is the product ion scan (1). It is based on selecting an ion (parent ion) with a chosen m/z ratio in the first analyzer (Q_1). This ion collides inside the central quadrupole (q), which serves as a collision cell and undergoes fragmentation. The reaction products (fragment ions or daughter ions) are analyzed by the second analyzer (Q_2). During the process, the second analyzer (Q_2) focuses on fragment ions generated during collisions. It is important to note that those fragments belong to the selected parent ion and serves as a specific "fingerprint" of a given molecule, thus allowing its identification.

During precursor ion scan (2), the second analyzer (Q_2) focuses on a particular product ion of interest *AFTER* collision, while the first mass analyzer (Q_1) scans the m/z ratios. This mode is used to reveal all parent ions (i.e. all molecules in the analyzed mixture) that generate the same daughter ion. This mode can be used for screening for all compounds having similar chemical structure.

In the neutral loss scan (3), both first (Q_1) and second mass analyzers (Q_2) operate with a constant mass offset. When precursor ion is transmitted through the first mass analyzer, this ion is recorded by Q_2 if it yields a product corresponding to the loss of a specific fragment (specific group) from the precursor ion, e.g. water molecule, phosphate, carbon dioxide, etc., after leaving collision cell (q). This loss provides identification of molecules having specific structural features.

In the SRM (SRM scan mode), the first (Q_1) and second mass analyzers (Q_2) are both focused on the selected ions with preset m/z, and the instrument is tuned to transmit only the ions derived from fragmentation. It is important to note that fragmentation pathway is much more indicative than the parent m/z value to identify a specific molecule; therefore, SRM mode is more selective than SIM mode used in single quadrupoles. The higher sensitivity obtained by using this method compared with SIM is due to the improved signal-to-noise ratio, which is characteristic for fragment mass spectra, and the high duty cycle due to the detection of only one specific ion. In the case that m/z values for multiple ions are preset to monitor multiple reactions, the term "multiple reaction monitoring (MRM)" is used. An additional asset of using SRM is significantly lower occupancy of the hard disk space for data acquisition. This mode is also used to increase sensitivity of the analysis for low-abundant species.

It is important to note that the same operational mode may lead to reduced sensitivity when other mass analyzers are used. For example, the SRM scan mode can be performed in an IT, but the ion selection mechanism is different. Namely, ions of different masses produced in the source are collected together inside the trap and are consecutively expelled when a specific m/z ion is selected. Therefore, the duty cycle of the spectrometer decreases depending on the fraction (m/z range) of the ions selected among all the ions inside the trap.

For this reason, QqQ are the main analyzers employed for quantitative analysis of individual molecular species [5, 6]. A yet another obstacle for quantitative application of the IT is a possibility to saturate the device, in which case the relation between number of ions and detected output is no longer linear.

On the other hand, the employment of QqQ for qualitative analysis of complex mixtures is very limited. Separation methods coupled with the mass spectrometer, such as HPLC, are necessary before analyzing complex samples. QqQ, as well as IT analyzers, can be interfaced with HPLC using the ESI source as LC-MS method. Nevertheless, an important requirement for the LC-MS is the high scanning speed of mass analyzer in order to acquire the maximum number of mass spectra and identify analytes consecutively eluted during the chromatographic run. This is particularly important during nano-LC-MS/MS runs, where a particular ion may last no longer than 20 seconds. In the case of analytical UPLC linked to MS, individual peaks may only be c. 2–3 seconds broad. Identification of several compounds by a QqQ system implicates that the first and third quadrupoles (Q_1 and Q_2) operate in the scan mode, while fragmentation occurs in the central quadrupole (q). As already discussed above, when mass spectrometer is used in the full scan mode, the duty cycle of the spectrometer decreases proportionally to the total analysis time devoted to each ion. Increasing this time also increases the sensitivity, enabling identification of samples having low abundances. On the other hand, an increased acquisition time reduces the number of chromatographic points acquired, and thus analytical information may be lost. Therefore, QqQ may not be the best choice when very complex mixtures are investigated.

4.2.8.3 Q-IT

The performance of a QqQ can be improved by changing the features of the last quadrupole (Q_2). As a matter of fact, an interesting evolution of the QqQ instruments has led to the replacement of the last quadrupole with an IT analyzer (Figure 4.74b) [7]. The Q_1q segment of the instrument is operated as a conventional QqQ for CID experiments. Performing CID in the quadrupole (q) has the advantage of efficiently fragmenting ions (without sacrificing the loss of sensitivity and the loss of lower m/z fragment ions, typical for ITs [8]), along with having higher-energy collisional activation. Using the IT for the second stage of mass analysis provides greater sensitivity and higher scanning speed of analysis, as compared with a quadrupole mass analyzer used in the scan mode. Moreover, the IT replacing the last quadrupole can also trap low molecular mass ions, which might have been lost otherwise during fragmentation, due to the low mass cutoff.

4.2.8.4 Q-Orbitrap

Subsequent instrument developments are focused on the improvement of high resolution, mass accuracy, and sensitivity of the IT analyzer. Q-Exactive instruments (Thermo Scientific) [4] involve the combination of the Orbitrap mass

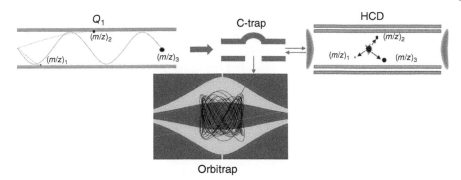

Figure 4.75 Schematic view of the quadrupole/Orbitrap hybrid mass spectrometer. The Q_1 quadrupole is used for isolation of precursor ions. C-trap device allows for accumulation of the ions before injection into the Orbitrap analyzer. Orbitrap analyzer is employed for detection of full scans and MS/MS spectra after HCD fragmentation. The high-energy collision-induced dissociation (HCD) provides collisional dissociation similar to that offered by a quadrupole (q) but with more detailed structural information than that obtained by the low-energy CID fragmentation.

analyzer with a quadrupole analyzer (Figure 4.75). The mass filtering quadrupole is used for isolation of precursors, while Orbitrap is employed for detection of full scans and MS/MS spectra using a multipole for higher-energy collision-induced dissociation (HCD). The combination of fast and efficient quadrupole isolation with HCD fragmentation [9] and Orbitrap detection provides improved data quality. Nowadays, Q-Orbitrap mass spectrometers are suitable for quantitative mass measurements, with performance comparable with that of the commonly used MS/MS QqQ [10].

4.2.8.5 Q-TOF
As mentioned above, an important feature of the TOF analyzer lies in its scanning speed. Orthogonal acceleration TOF (oaTOF) [11] was recently developed for coupling TOF analyzer with continuous ionization sources (ESI source). This, in turn, enabled the development of HPLC/oaTOF instruments [12].

The oaTOF technique gained a significant role in the investigations of complex mixtures obtained from tissues, body fluids, or cell cultures studied in the genomics, transcriptomics, proteomics, lipidomics, and metabolomics fields [13]. Direct introduction of ions, generated in the source, into the orthogonal accelerator has some limitations, including low MS/MS efficiency. As a matter of fact, it is important to consider that TOF instruments typically operate at "high" ion kinetic energies, whereas "low" ion kinetic energies are typical for quadrupole mass filters and ITs [1]. Therefore, performing collision experiments (CID) at the kinetic energy levels of TOF instruments reduces MS/MS efficiency. Combining a quadrupole analyzer with a TOF instrument in a Q-TOF configuration is advantageous because of the higher MS/MS efficiency demonstrated for QqQ instruments and the speed and sensitivity of analysis typical for a TOF analyzer [14, 15]. The most common feature of these instruments includes a quadrupole analyzer Q_1 and a quadrupolar collision cell q, followed by an oaTOF (Figure 4.76a). The resulting hybrid instrument provides good performance of the TOF analyzer regarding resolution and mass accuracy, but a limited mass range due to the low transmission of the quadrupoles.

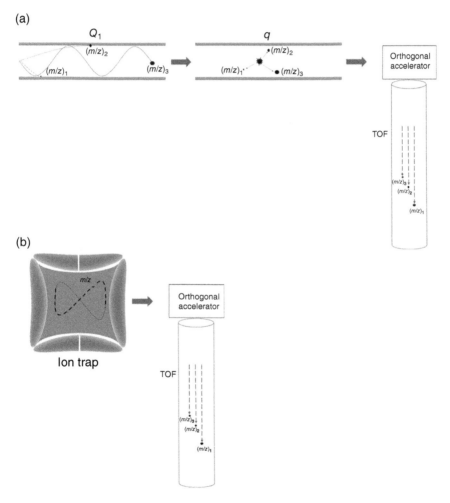

Figure 4.76 Schematic view of hybrid TOF instrument. (a) Q-TOF hybrid. The "low" ion kinetic energies typical of quadrupole enable efficient collision experiments CID. TOF analyzer provides fast analysis, high resolving power, and sensitivity. (b) Ion trap analyzer combined with a TOF instrument (Trap-TOF hybrid; scheme in box with broken line). MS^n capabilities of ion traps provide structural information of unknown molecules. Product ion detection by TOF analyzer provides higher mass accuracy and much higher resolving power than ion trap analyzer.

4.2.8.6 IT-TOF

Several hybrid instruments have been proposed more recently, combining TOF and IT analyzers in the IT-TOF configuration (Figure 4.76b). In some cases, the quadrupole IT is a conventional three-dimensional ion trap (Q-IT) [16], while in others it is a linear quadrupole ion trap (LIT) [17]. As already discussed, the key feature of the IT analyzer is MS/MS (and MS^n) capabilities. However, as noted previously, ITs possess modest resolving power and mass measurement accuracy. These limitations can be overcome by coupling an IT with a TOF analyzer, thereby providing better mass accuracy and much higher resolving power for the analysis of product ions. Therefore, the IT is used to accumulate ions and to perform ion selection and activation in MS/MS experiments, while the TOF analyzer is used for mass analysis instead of the classical ion ejection methods used in ITs. It is important to highlight that the fast scan speed of TOF analyzer makes hybrid IT-TOF setup compatible with the very fast LC investigations. The MS^n and accurate mass measurements are routinely employed to rapidly confirm the identification of expected metabolites or to elucidate the structures of uncommon or unknown metabolites. These features make the LC IT-TOF spectrometer a very powerful analytical tool for unambiguous identification of molecules [18].

4.2.8.7 IT-Orbitrap

The most recent type of hybrid instruments is the trap–trap combination where two different types of trapping mass spectrometers are combined (Figure 4.77). These instruments do not provide the speed of analysis demonstrated by the TOF-based instruments, sacrificing this capability for superior mass measurement accuracy and mass resolving power. The general goal of this configuration is to achieve even better mass accuracy and resolving power

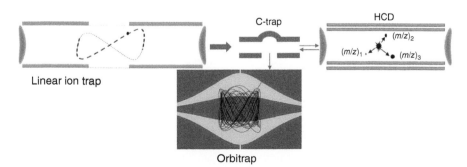

Figure 4.77 Schematic view of linear ion trap (LIT)/Orbitrap hybrid mass spectrometer. LIT as the first stage of analysis provides multiple low resolving power MS/MS spectra. C-trap device allows accumulation of ions before injection into the Orbitrap analyzer. Orbitrap analyzer is employed for detection of full scans and MS/MS spectra. Implementation of trap/Orbitrap hybrid instrument with high-energy collision-induced dissociation (HCD) provides more structural information of unknown compounds.

than that obtained by employing TOF analyzer for the second stage of mass analysis in the MS/MS experiment. Thus, the second trap is not an IT but an Orbitrap mass analyzer [4]. The high-resolution and accurate mass (HR/AM) detection offered by the Orbitrap is combined with the sensitive ion detection, precursor isolation, and fragmentation capabilities associated with linear ion trap (LIT). A unique attribute of having the LIT as the first stage of analysis is the ability to obtain multiple and low-resolution MS/MS spectra using just the LIT, while the Orbitrap is acquiring a separate spectrum at high resolving power and mass accuracy. In the last decade, trap–trap hybrid instruments had quickly become important analytical tools in the development of the proteomics field, as they provide the speed and sensitivity required for liquid chromatography (LC)-compatible full scan and MS/MS detection of peptides. Moreover, one of the main advantages of HR/AM systems is that they allow the estimation of accurate charge states and the mass determination of the multiply charged species, thus simplifying database searching [19] and increasing accuracy of identification.

Questions

- Which are the main outcomes of the high and low ion kinetic energies required in TOF and quadrupole/ion trap analyzers, respectively?
- Which mass analyzer provides fast acquisition rate, high sensitivity, high selectivity, MS^n capabilities, and accurate mass measurement?
- Which instrumental requirements brought to the development of hybrid mass spectrometers?
- Which hybrid instruments are mainly employed for quantitative analysis of individual molecular species?
- When complex mixtures are investigated, which type of hybrid instrument could be used? Which features these instruments should provide?
- Which mass analyzers provide the higher MS/MS efficiency?
- Which hybrid instruments sacrifice the speed of analysis for superior mass measurement accuracy and mass resolving power?

References

1 de Hoffmann, E. and Stroobant, V. (2007). *Mass Spectrometry: Principles and Applications*. Chichester: Wiley.
2 Dawson, P.H. (1976). *Quadrupole Mass Spectrometry and Its Applications*. Elsevier.

3 Wong, S.H. and Cooks, R.G. (1997). Ion trap mass spectrometry. *Curr. Sep. 16*: 85–92.

4 Eliuk, S. and Makarov, A. (2015). Evolution of orbitrap mass spectrometry instrumentation. *Annual Review of Analytical Chemistry 8* (1): 61–80.

5 Zhao, Y. and Brasier, A.R. (2013). Applications of selected reaction monitoring (SRM)-mass spectrometry (MS) for quantitative measurement of signaling pathways. *Methods 61* (3): 313–322.

6 Lange, V., Picotti, P., Domon, B., and Aebersold, R. (2008). Selected reaction monitoring for quantitative proteomics: a tutorial. *Molecular Systems Biology 4*: 222.

7 Hager, J.W. (1999). Performance optimization and fringing field modifications of a 24-mm long RF-only quadrupole mass spectrometer. *Rapid Communications in Mass Spectrometry 13* (8): 740–748.

8 Major, F.G. and Dehmelt, H.G. (1968). Exchange-collision technique for the RF spectroscopy of stored ions. *Physical Review 170* (1): 91–107.

9 Olsen, J.V., Macek, B., Lange, O. et al. (2007). Higher-energy C-trap dissociation for peptide modification analysis. *Nature Methods 4* (9): 709–712.

10 Fedorova, G., Randak, T., Lindberg, R.H., and Grabic, R. (2013). Comparison of the quantitative performance of a Q-Exactive high-resolution mass spectrometer with that of a triple quadrupole tandem mass spectrometer for the analysis of illicit drugs in wastewater. *Rapid Communications in Mass Spectrometry 27* (15): 1751–1762.

11 Guilhaus, M., Selby, D., and Mlynski, V. (2000). Orthogonal acceleration time-of-flight mass spectrometry. *Mass Spectrometry Reviews 19* (2): 65–107.

12 Lacorte, S. and Fernandez-Alba, A.R. (2006). Time of flight mass spectrometry applied to the liquid chromatographic analysis of pesticides in water and food. *Mass Spectrometry Reviews 25* (6): 866–880.

13 Girolamo, F., Lante, I., Muraca, M., and Putignani, L. (2013). The role of mass spectrometry in the "Omics" Era. *Current Organic Chemistry 17* (23): 2891–2905.

14 Glish, G.L. and Goeringer, D.E. (1984). Tandem quadrupole/time-of-flight instrument for mass spectrometry/mass spectrometry. *Analytical Chemistry 56*: 2291–2295.

15 Glish, G.L., McLuckey, S.A., and McKown, H.S. (1987). Improved performance of a tandem quadrupole/time-of-flight mass spectrometer. *Instrumentation Science and Technology 16* (1): 191–206.

16 Martin, R.L. and Brancia, F.L. (2003). Analysis of high mass peptides using a novel matrix-assisted laser desorption/ionisation quadrupole ion trap time-of-flight mass spectrometer. *Rapid Communications in Mass Spectrometry 17* (12): 1358–1365.

17 Campbell, J.M., Collings, B.A., and Douglas, D.J. (1998). A new linear ion trap time-of-flight system with tandem mass spectrometry capabilities. *Rapid Communications in Mass Spectrometry 12* (20): 1463–1474.

18 Liu, Z.Y. (2012). An introduction to hybrid ion trap/time-of-flight mass spectrometry coupled with liquid chromatography applied to drug metabolism studies. *Journal of Mass Spectrometry 47* (12): 1627–1642.

19 Clauser, K.R., Baker, P., and Burlingame, A.L. (1999). Role of accurate mass measurement (+/- 10 ppm) in protein identification strategies employing MS or MS/MS and database searching. *Analytical Chemistry 71* (14): 2871–2882.

4.2.9 Sector Instruments

Anna Antolak and Anna Bodzon-Kulakowska

Department of Biochemistry and Neurobiology, Faculty of Materials Science and Ceramics, AGH University of Science and Technology, Kraków, Poland

4.2.9.1 Introduction

The first commercially available spectrometers with magnetic and electrostatic analyzers (also known as the magnetic sector and the electrostatic sector) appeared in the 1950s [1, 2]. At first, they were cumbersome and difficult to handle, but over time they had been developed and became more approachable. A few years ago, they were the basic analyzers in mass spectrometry laboratories, especially in high-resolution measurements. Nowadays, these devices are being replaced by smaller, cheaper, and easier-to-handle types of analyzers. However, this type of sector instruments is still used in quantitative measurements of high accuracies – such as isotope ratios, anti-doping control, analysis of toxic substances and its derivatives, etc. The magnetic and electromagnetic–magnetic analyzers use the phenomenon of deflection of the ion trajectory in the magnetic and electrostatic field for the separation of ions with different m/z values.

4.2.9.2 Rule of Operation of Magnetic Analyzer (B)

In the magnetic analyzer (see Figure 4.78), the charged particles leave the ion source and are accelerated by the potential of 2000–8000 V. They therefore move with the kinetic energy of

$$E_k = \frac{mv^2}{2} \tag{1}$$

where

m is ion mass
v is ion velocity, in which kinetic energy corresponds to the potential energy of the charged particle in the electric field, expressed as $E_{el} = zeU$
z is ion charge
e is elementary charge expressed in Coulombs
U is accelerating potential

Hence

$$zeU = \frac{mv^2}{2} \tag{2}$$

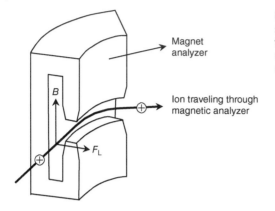

Figure 4.78 Magnetic field: simplified diagram of the direction of forces acting on ion moving in the magnetic field. *Source*: Figure modified from Ref. [1].

Magnet analyzer

Ion traveling through magnetic analyzer

The velocity vector v of the ions reaching the analyzer is perpendicular to the direction of the lines of the magnetic field with induction B. Under these conditions, each of charged particles is affected by the Lorentz force F_L, perpendicular to the magnetic field lines and to the direction of the ion move (right-hand rule). This force depends on the induction of magnetic field (B), ion charge, and velocity:

$$F_L = qv \cdot B, \quad F_L = zev \cdot B \tag{3}$$

The ion path became curved under the influence of Lorentz's force. This force acts here as a centripetal force described by the equation

$$F_c = \frac{mv^2}{r} \tag{4}$$

where

r is the radius of the ion trajectory in the magnetic field. Thus, we may combine above equations to obtain

$$F_L = F_c, \quad zevB = \frac{mv^2}{r} \tag{5}$$

By transforming the above formula, we get an expression that presents the principle of the magnetic analyzer operation:

$$r = \frac{mv}{zeB} \tag{6}$$

This formula enables calculation of the trajectory radius of an ion with given mass, velocity (i.e. momentum), and charge, moving in the magnetic field of magnetic induction B.

At a given magnetic induction, ions of uniform charge and momentum will move along an arc of the same radius r. It means that the magnetic analyzer separates ions due to their momentum and charge. The momentum of an ion of a particular charge is related to its mass, which ultimately determines the m/z value of the analyzed molecule.

The shape of the analyzer predetermines the available ion trajectory. To collect ions of subsequent m/z values, the strength of magnetic field should be gradually changed. This is illustrated by the following equation obtained by substituting v determined in Formula (6) to the Eq. (2):

$$\frac{m}{z} = \frac{B^2 r^2 e}{2U} \tag{7}$$

For a given B, only the ions with the specific m/z ratio will reach the detector moving along the path that corresponds to the curvature r of the analyzer (Figure 4.79). In order to obtain the mass spectrum (the entire range of m/z), the value of the magnetic field strength should be changed (magnetic scanning).

Another method of analysis in the magnetic sector is electrical scanning, where B remains constant while the accelerating potential U (see Eq. (7)) is alternated. Thus, ions with corresponding m/z values are sequentially focused at the same point. The disadvantage of this system is that sensitivity declines with decreasing voltage. This method is used in high-resolution analyses where high measurement stability is required and the measuring range does not exceed tens of Daltons. Both modes are well described by Eq. (7). If U is constant, magnetic scanning is performed, and when B is constant, electrical scanning occurs. Since the path of all the ions is in a curvature of the same

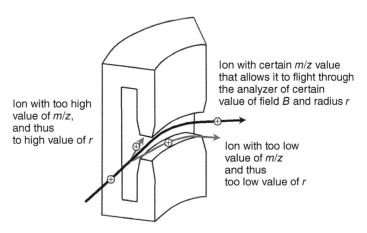

Ion with too high value of m/z, and thus to high value of r

Ion with certain m/z value that allows it to flight through the analyzer of certain value of field B and radius r

Ion with too low value of m/z and thus too low value of r

Figure 4.79 Illustration of an ion move with different m/z inside the magnetic sector with a given radius r and magnetic field value B.

radius r and the analysis of ions of subsequent m/z is distributed over time, only one detector, e.g. an electron duplicator, is sufficient to detect the molecules.

Magnetic analyzers work well with "continuous" ion sources such as EI, CI, or ESI. The resolution of the magnetic sector R varies between 2000 and 7000 depending on the radius value. The limit is related to the fact that the ions entering the analyzer do not have uniform energy; hence the ions with the same m/z may have slightly different momentum, which lowers the resolution of the measurements [2].

To Summarize
1) Magnetic analyzer separates ions depending on their momentum and charge.
2) For a given magnetic field strength value B, an ion of charge z and momentum (mv) will move along the path of radius r.
3) The ions are separated according to whether their trajectory corresponds to the radius of the analyzer arc. Charged particles moving on trajectories of the different radius will be neutralized on the analyzer walls and will not reach the detector.

4.2.9.3 Electrostatic Sector (*E*)
The electrostatic sector consists of two metal plates profiled in the shape of a fragment of a circle. Between those plates, a constant potential difference is maintained. The velocity vector v of the ions entering the analyzer is perpendicular to the direction of the electric field.

As a result of an electrostatic attraction expressed by the equation (E is the intensity of the electric field)

$$F_e = zeE \tag{8}$$

which plays a role of centripetal force

$$F_c = \frac{mv^2}{r_e} \tag{9}$$

molecules move on a circle trajectory of radius r_e:

$$zeE = \frac{mv^2}{r_e} \tag{10}$$

The kinetic energy of the ion entering the analyzer arises from the acceleration potential and is described by the equation

$$E_k = \frac{mv^2}{2} = zeU \tag{11}$$

By inserting the formula for v calculated from Eq. (11) to Formula (10), we obtain a simple relation:

$$r_e = \frac{2U}{E} \qquad (12)$$

It follows that ions accelerated by the potential U and passing through the homogeneous electric field E have the same radius of curvature r_e, irrespectively of their m/z. As previously shown, the kinetic energy is related to the acceleration potential U ($E_k = zeU$). After appropriate transformation of the above formula and inserting it into the Eq. (12), the following relation is obtained:

$$r_e = \frac{2E_k}{zeE} \qquad (13)$$

From the above formula it can be seen that in the electrostatic sector with constant electric field intensity, the radius of the ion path depends only on its kinetic energy, not on its m/z. The ions with defined E_k for which the curvature of the trajectory corresponds to the curvature of the analyzer will pass through it (see Figure 4.80). Two ions of the same m/z will be focused even when they slightly differ in kinetic energies. This allows for increasing the resolution of the measurements. If these energies are significantly different, then the ions will hit the analyzer walls, and they will not reach the detector [2].

To Summarize
1) The electrostatic sector separates the ions according to their kinetic energy E_k, irrespectively of the m/z value.
2) For a given value of the electric field E, the ion of charge q and the kinetic energy E_k will move along with a strictly defined radius r.
3) The electrostatic sector reduces the kinetic energy distribution of ions with the same m/z, thus increasing the resolution of the measurements.

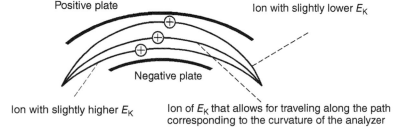

Figure 4.80 Electrostatic sector: a simplified diagram showing the distribution of ions depending on their kinetic energy.

4.2.9.4 Mass Spectrometers with Magnetic and Electrostatic Sector

Continuing the works of the J.J. Thomson and F.W. Aston, many other scholars, e.g. A.J. Dempster, J. Mattauch, R. Herzog, K.T. Bainbridge, and A.O. Nier, contributed to the development of mass spectrometry and improvement of the resolution of measurement due to the construction of double-focusing instruments. As a result, before 1930, mass spectrometry was well recognized as a technique for ion separation as a function of their mass and charge.

If a magnetic analyzer is not preceded by an electrostatic sector, even ions of the same mass and charge are being dispersed after leaving the ion source, due to the differences in their kinetic energy. Consequently, ions with the same mass but slightly different kinetic energies are not focused at one point, and the obtained spectra are characterized by low resolution. The use of electrostatic and magnetic sectors in one spectrometer significantly improves the resolution of measured spectra as a result of focusing a beam of ions of different kinetic energy and angular deviation. Thus, in double-focusing instruments, ions of equal kinetic energy access the magnetic sector; therefore obtained resolution is 10 times higher than in a single magnetic analyzer and can reach even 100 000. The separation of ions in the double-focusing analyzers is often performed at constant acceleration voltage and constant electric field, with alternating magnetic field. Thus, depending on B value, the ions with the corresponding m/z values are focused. For effective focusing, magnetic induction should increase with the increase the mass of analyzed substances. It is also possible to manipulate the electric field at constant parameters of the magnetic field. Such a way of operation is used for accurate measurements of the m/z value of the analyte at high resolution.

Different types of electrical and magnetic setups are achievable, but in each case, the primary purpose of the analysis is to separate the ions according to their m/z value with the highest possible resolution. In simple geometry, the electrostatic sector precedes the magnetic analyzer and ions separated based on the kinetic energies, and m/z values are focused at one detector. In such setup, the angular dispersion of ions in the electrostatic is compensated in the magnetic analyzer. Also, differences in translation energy of ions may be neutralized by a combination of both sectors.

In reverse geometry, the magnetic analyzer precedes the electrostatic sector (Matsuda geometry). The magnetic analyzer lowers the level of ion current generated in the source, passing for further analysis only a beam of molecules with the same m/z values. The advantage of inverted geometry spectrometers is the sensitivity improvement and noise reduction.

Currently, the limit of m/z that can be analyzed using the magnetic and electrical sectors is about 16 000. In the separation sectors, high vacuum (about 10^{-5} Pa) is applied so that the ion trajectory is not disturbed by the collisions with gas molecules. Mass spectrometers with this type of analyzer are characterized by very high repeatability and wide dynamic range and are the best for quantitative analysis. The disadvantage is its high price and complex service, requiring high qualifications of the operator [1].

Questions

- For which purposes magnetic and electrostatic sector analyzers are still used nowadays?
- How do magnetic analyzers separate ions?
- What is the disadvantage of using the electrical scanning in a magnetic sector?
- What is the resolution of the magnetic sector?
- How the electrostatic sector separates the ions?
- What is a "double-focusing" instrument? What resolution can be achieved in such kind of instruments?

References

1 Gross, J.H. (2011). *Mass Spectrometry*. Berlin: Springer.
2 Silberring, J. and Ekman, R. (eds.) (2002). *Mass Spectrometry and Hyphenated Techniques in Neuropeptide Research*. New York: Wiley.

4.3   Ion Detectors

Piotr Suder

Department of Biochemistry and Neurobiology, Faculty of Materials Science and Ceramics,
AGH University of Science and Technology, Kraków, Poland

4.3.1   Introduction

For years of mass spectrometer development, the instruments were constantly modified to reach better analytical capabilities, and their construction was supplied by new ion sources, analyzers, and vacuum technology. One of the most important, but usually undervalued part of mass spectrometry (MS), is ion detector. This element is critical for sensitivity of the entire instrument. At the beginning of MS, photosensitive plates or Faraday cups were routinely used. Currently, in majority of mass spectrometers, electron multipliers are used as basic detectors due to their cost efficiency, sensitivity, and reliability. Electron multipliers can convert kinetic energy of the ion into electron beam, generating measurable current. Other commonly used detectors (microchannel, photomultiplier, channel multiplier) are similar in their principles of operation to a typical electron multiplier, which will be discussed later.

Lastly, especially in the MS imaging, there are attempts to replace typical detectors with other solutions. The major disadvantage of current technology is its inability of simultaneous mass spectra acquisition from the 2D structures (like tissue sections). Instead, point after point must be taken consecutively to complete entire analysis. There are some efforts commenced to design a detector capable of working like a CCD matrix in photo cameras that can acquire mass spectra from hundreds of points at the same time. The working name of such solution is Medipix or Timepix pixelated detector, invented by the AMOLF Institute (Amsterdam, The Netherlands).

Other interesting types of ion detection are solutions used in the ion cyclotron resonance (ICR) chambers and Orbitraps. In these instruments, the process occurs by the measurement of cyclotron frequencies of the ions traveling in the ICR chamber or by the measurement of oscillation frequencies of ions spinning around central electrode in the Orbitrap. Detailed principles of ions' measurement in ICR analyzer were given in Section 4.2.6. For details concerning Orbitrap, see Section 4.2.7.

4.3.2   Electron Multiplier

Up to date, the most popular type of detector is electron multiplier. It consists of a few electrodes (usually from 10 to 20) called dynodes. An ion leaving analyzer hits the first dynode (so-called conversion dynode or conversion electrode). Kinetic energy of the ion is sufficient to remove few electrons from the electrode surface. As a consequence, a simply measurable signal coming from moving electrons (electric signal) rather than moving ions is generated. Emerging electrons consecutively hit the following dynodes. Multiplier construction and currents on the dynodes allow for generation of the electron cascade traveling through dynodes. Every dynode enhances the electric signal, generating so-called secondary electrons. Using this way, the initial signal coming from a single ion (and a single electron from conversion dynode) can be amplified 1×10^4 to 1×10^9 times, depending on electric potentials on the following dynodes. Amplification can be adjustable, depending on the operator's requests.

A variant of the detector described above is the channel electron multiplier. Basis of its operation is the same as of a typical electron multiplier, but its construction is slightly different. It looks like a small funnel made of glass, with inside walls covered with lead oxide and tin. There are no distinct dynodes in this construction; their role is provided by the whole internal surface of the funnel. The ion hits conversion electrode or directly metal surface inside the funnel, thus initiating the electron cascade. Amplification of the cascade occurs, while the electron traveling along the detector expels additional electrons from the metal-coated surface in the funnel. In the lower part of the funnel, the device measuring electron current is located. Comparison of the construction schemes of both detectors is presented in Figure 4.81.

4.3.3 Microchannel Detector

This is another variant of an electron multiplier having the same theoretical basis of operation, but its construction is completely different. Microchannel detector consists of thousands of channel multipliers in a form of tiny tubes collected in a plate resembling a honeycomb. The effective area being a target for ions is about 80% of the detector surface, where the hitting ion is capable to

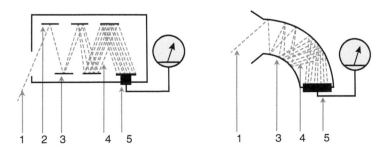

Figure 4.81 Comparison between dynode-based electron multiplier (left) and channel electron multiplier (right): (1) ion beam; (2) conversion dynode; (3) dynode; (4) electron cascade; (5) electron current measuring device.

initiate electron cascade. Every miniaturized channel has its own device measuring electron current. Total signal from all devices is converted into a mass spectrum. Such construction has a number of advantages. It is possible to simultaneously measure ions at various m/z values. Thus, the time of data processing is shorter in comparison with typical detectors, where data handling is sequential. It is also possible to measure m/z of the ion without precise focusing of the ion beam. This is particularly useful in the TOF-based constructions. Unluckily, such approach is less sensitive than conventional ion focusing. Additionally, microchannel detector is able to register ion series in the very short time lapses (even in nanosecond intervals), so it is routinely used in MALDI-TOF systems, where ion series can be generated in microsecond steps (depending on the applied laser frequency). Of course, microchannel detector has also a set of drawbacks. The most important are the relatively short time between maintenances and high price.

4.3.4 Medipix/Timepix Detector

This type of detector is currently tested for commercial purposes but is marketably unavailable yet; however information shows some promising effects of collaboration between the Amsterdam Scientific Instruments (http://www.amscins.com/applications/mass-spectrometry) and Omics2image (www.omics2image.com) to implement this innovation in commercial mass spectrometers. This type of detector may be a solution to the problems with imaging MS. To make an ion image of the surface, it is necessary to acquire MS spectrum pixel after pixel received from the sample. This stepwise procedure substantially increases time necessary to complete the analysis. For instance, an ion image of the tissue section from $1\,cm^2$ at poor resolution $300\,dpi$ requires spectra acquisition from about $14\,000$ points (118×118 points). Even using fast-scanning MALDI-TOF mass spectrometer, equipped with semiconductor laser of $1000\,Hz$ frequency, the time consumed per 1 pixel is about 1 second. Hence, the analysis will be completed after $14\,000$ seconds (almost 4 hours). As can be seen from the example, the major drawback of imaging is the time necessary for finalizing the analysis using a typical approach. An idea staying behind the Medipix detector is simultaneous acquisition of mass spectra from hundred points instead of analyzing point after point [1]. If it would be possible to simultaneously acquire spectra from 1000 points, the analysis would last 14 seconds instead of 4 hours. Currently, efforts are undertaken to design a detector able to acquire hundreds or thousands of spectra instantaneously. Such detectors have few layers. The basic coating is luminescent, catching ions traveling from the sample. In the recent construction, the role of luminescent layer is replaced by the modified microchannel detector. This device transforms ions into the electron cascade measured by CMOS matrix located just under the detector. Electronics measures ions' intensities and

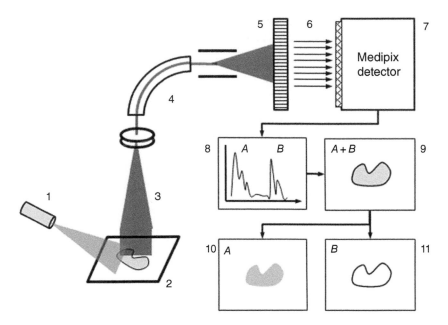

Figure 4.82 Concept of the Medipix detector operation. (1) ion source; (2) sample; (3) secondary ion beam; (4) ion focusing devices; (5) microchannel detector converting ions into electrons; (6) electron beam; (7) Medipix detector containing CMOS matrix; (8) mass spectra from scanned area; (9–11) ion images reconstructed from data acquired by the detector.

transforms them into a set of MS spectra in a form of 2D density image for every detected ion. In such construction, in a principle similar to digital camera, the problem of slow point-by-point spectra acquisition is eliminated. Up to date, it is possible to simultaneously measure $1\,mm^2$ of the surface with resolution 512×512 pixels. Basis of the concept for this detector is shown in Figure 4.82.

4.3.5 Ion Detection in ICR and Orbitrap-Based Mass Spectrometers

Ion detection in the ICR and in Orbitrap analyzers is distinct from those described previously for conventional mass spectrometers. In both analyzers, the ions travel on stable trajectories. For ICR, such trajectory can be easily disturbed by electric field generated by excitation electrodes. Ions, absorbing energy, change their trajectory, thus increasing oscillation radius. Then, they can dissipate energy and simultaneously interact with acquisition electrodes. Their oscillation frequency is a measure of m/z ratio. Acquired data are recalculated by the Fourier transformation into mass spectrum suitable for interpretation. Such kind of data acquisition allows for a much more accurate measurement than in other types of mass spectrometers, where detection is based on multipliers. Resolution of measurements in FT-ICR mass spectrometers reaches a level of 1×10^7 due to direct measurement of electric current, which is much more stable than conversion electrodes. This is also the reason why high-resolution analyses in sector instruments are performed using electric scan instead of magnetic scan. Further details of FT-ICR system can be found in Section 4.2.6.8.

The Orbitrap-based MS ion detection is similar, in its principles, to FT-ICR. Ions introduced into the Orbitrap chamber travel on circular orbits that oscillate along the central electrode. Oscillation frequency is a measure of m/z and is detected by outer electrodes. Similarly to ICR, oscillation frequency is recalculated to m/z and presented on a conventional mass spectrum. Resolution of the Orbitrap analyzer is reaching 1×10^6. Details about Orbitrap analyzers are given in Section 4.2.7.

Questions

- What are the most common detectors currently used and how do they work?
- What kind of solutions is currently investigated to generate data from 2D structures?
- What are the main advantages and disadvantages of the microchannel detectors?

Reference

1 Kiss, A., Smith, D.F., and Jungmann, J.H. (2013). Heeren RM Cluster secondary ion mass spectrometry microscope mode mass spectrometry imaging. *Rapid Communications in Mass Spectrometry* 27: 2745–2750.

5

Hyphenated Techniques

5.1   Gas Chromatography Combined with Mass Spectrometry (GC-MS)

Anna Drabik

Department of Biochemistry and Neurobiology, Faculty of Materials Science and Ceramics, AGH University of Science and Technology, Kraków, Poland

5.1.1   Introduction

Gas chromatography (GC) is a universal separation technique used for identification of complex mixtures. Separated substances should possess one basic property: high volatility. They also ought to be thermally stable, and their boiling point should not exceed the limit of c. 350 °C (662 °F); otherwise they will not be able to evaporate in the inlet (liner) of the injector (Figure 5.1). There are however chemical modifications called derivatizations converting nonvolatile substances into the more volatile derivatives that can be separated and quantified by GC-MS. Usually, particles having molecular weight greater than 1000 Da are thermally labile and cannot be separated using GC technique.

Gas chromatography, as its name suggests, usually utilizes He, H_2, N_2, or Ar as mobile phases. Carrier gas flows through the injector and column to the detector, simultaneously transporting separated molecules of the analyzed sample (Figure 5.2) to the column and detector. The sample is introduced into the chromatograph with the aid of the carrier gas stream and is typically injected in the liquid form using the microsyringe in the volume range of 1–10 µl. The injector is protected with a septum and is also heated in a controlled manner to evaporate constituents of the sample. All nonvolatile impurities and solid particles remain in the liner, which protects the system from eventual clogging. In GC, temperature is the separating medium (corresponding to solvent polarity in liquid chromatography [LC]), and the carrier gas plays a role as a transporting

Mass Spectrometry: An Applied Approach, Second Edition. Edited by Marek Smoluch, Giuseppe Grasso, Piotr Suder, and Jerzy Silberring.

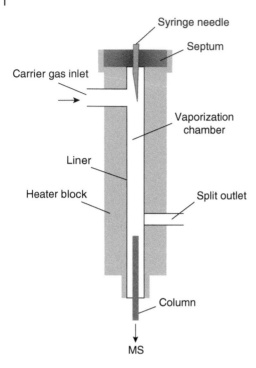

Figure 5.1 Scheme of the sample injector.

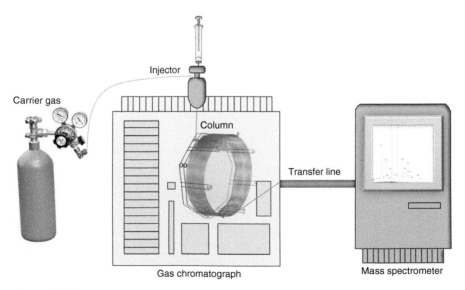

Figure 5.2 Coupling the gas chromatograph with a mass spectrometer.

system, in a similar manner to the mobile phase in LC. Often, when the gas chromatograph is combined with mass spectrometer as a detector, the injector in the split mode is utilized to remove substantial part of the sample to avoid system overloading. For example, at a $101\,ml\,min^{-1}$ of the total carrier gas flow and the split ratio of 100 : 1, only $1\,ml\,min^{-1}$ of gas is actually entering the col-umn. This split also protects high vacuum required in the mass spectrometer.

Modern GC exploits capillary columns, usually characterized as the open tubular. There is, however, a possibility of enabling packed capillary columns of much bigger internal diameter. Currently, they are rarely adapted for coupling with mass spectrometers [1]. Open tubular columns are manufactured of the fused silica coated with polyimide sheath. Typically, their length is 10–50 m (33–165 ft) and internal diameter 250 μm. Capillary columns are considered to have high resolution, and the number of theoretical plates can reach even 5000 per 1 m. Packed columns (typical dimensions 1.5 m × 4 mm) are filled with solid stationary phase, and capillary columns have stationary phase bonded inside the capillary, as described in Table 5.1. Thus, capillary columns are characterized by a continuous stationary phase in the form of a film attached to their inner walls. The stationary phase should ensure separations at high temperatures (c. 350 °C); low bleeding, i.e. no extra signals from the coating (column leaking); and high resolution [1].

Separation of the components is achieved by the differences in displacement of the analyte molecules that are partitioned between the carrier gas stream (mobile phase) and the polymer coating (stationary phase). Sample constitu-ents are eluted with the aid of temperature gradient, which affects forces of their interactions with the stationary phase. This means that the elution gradi-ent is formed by increasing the temperature, in contrast to the gradient applied in high performance liquid chromatography (HPLC), where mobile phase consists of solvents of various polarity/ionic strength/pH. The column is kept under strictly controlled temperature conditions provided by the heated oven. The most volatile compounds are eluted earliest. Subsequently, the less volatile molecules are leaving the column, and at the end of the gradient, the least volatile molecules (usually of largest masses) are observed.

5.1.2 Detectors

There is a broad variety of detectors to be applied in GC. The biggest advantage offers coupling with the mass spectrometer where, during single experiment, the operator receives information about the constituent's structure. The most com-mon GC detector still remains flame ionization detector (FID) because of the lower cost and simplicity of operation. However, there are many other methods to iden-tify the analyte eluted form GC column, also without flow splitting mode (Table 5.2) [2]. The ideal detector should be characterized by high sensitivity ($10^{-12}\,g\,ml^{-1}$) and a linear dynamic range of about five orders of magnitude. Detector sensitivity or

Table 5.1 Examples of the stationary phases.

Stationary phase type	Structure	Application
Polysiloxanes		Universal; medium hydrophobic phase
Polyethylene glycol (PEG)		Separation of acidic compounds
Chiral phase		Separation of isomers

minimum detectable concentration is defined as the minimum concentration of the analyte passing through the sensor that can be explicitly discriminated from noise, where the signal-to-noise (S/N) ratio is greater than 3.

There are three different forms of detector responses to the signal: normal, differential, and integral (Figure 5.3). A device with an integral logarithmic response generates an output linearly related to the solute concentration. A normal/linear sensor provides response that follows the Gaussian profile of the eluted peak. The *area* of the peak is proportional to the total content of the analyte, whereas in the integral response, the *amplitude* of the peak is proportional to the total analyte content. The differential response is often used to

Table 5.2 Examples of various GC detectors.

Types of detectors used in gas chromatography (GC)

Flame ionization detector (FID)	Most popular GC detector; organic compound analysis
Thermal conductivity detector (TCD)	Universal detector; exerts proportional correlation of concentration and thermal conductivity
Atomic emission detector (AED)	Selective for measuring elemental contents in the sample
Electron capture detector (ECD)	Specific for substances containing chlorine
Flame photometric detector (FPD)	Specific for substances containing sulfur and nitrogen
Photoionization detector (PID)	Enables analysis of aromatic and unsaturated hydrocarbons
Nitrogen phosphorus detector (NPD)	Facilitates analysis of organic compounds containing nitrogen and phosphorus
Sulfur/nitrogen chemiluminescence detector (SCD/NCD)	Supports analysis of molecules with sulfur or nitrogen

Figure 5.3 Types of detector response.

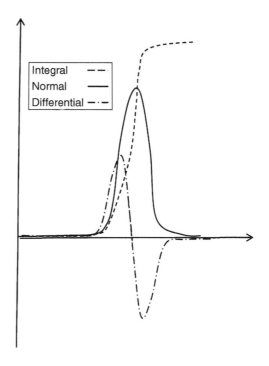

identify the retention time of the analyte, which is the point where the signal crosses from positive through zero to negative values. The signal is amplified in the detector and plotted against time, which gives a rise to chromatogram.

A routinely utilized ion source in GC-MS connections is electron impact (EI) ionization (Section 4.1.1). Moreover, there is a possibility of enabling chemical ionization (CI) (Section 4.1.2). Fragmentation observed during EI ionization process provides a vast number of signals that allow for reliable identification of the molecules by utilizing spectra libraries (e.g. National Institute of Standards and Technology (NIST)). It should be emphasized that mass spectrometry (MS) spectra obtained by EI are unique and can be considered as "fingerprints" of the molecules. Usually, EI ionization causes almost complete fragmentation of the molecules, and often molecular ion is not seen along the spectrum (Section 4.1.1). A typical potential applied to the filament is 70 mV. Reduction of the voltage applied to the filament may help in such cases, at a cost of sensitivity. Application of the CI protects from pronounced fragmentation; however it is less reproducible.

Along with the ion source, also the MS analyzer is of a great matter. The most commonly employed analyzers are quadrupole and ion trap (Sections 4.2.3 and 4.2.4). Nowadays, even a combination of the analyzers like in the case of triple quadrupoles (QqQ), orthogonal accelerating TOF, or Orbitrap can result in a much higher resolution in comparison with standard GC-MS systems.

5.1.3 Chemical Modifications: Derivatization

Applications of the GC-MS method are very comprehensive, nevertheless restricted to the analysis of volatile compounds up to c. 1 kDa. To enable separation of nonvolatile molecules using GC technique, chemical modification of an analyte should be applied. Derivatization is changing the aeriform properties by replacing polar groups like –OH, –SH, or –NH$_2$ with various residues listed in Table 5.3 [1]. This procedure also influences the stability of substances or even their detection limits (chlorine derivatives may be observed in electron capture detection [ECD] detector). The volatility of particular compounds strictly depends on their molecular weight and interactions between their polar groups. Therefore, by their modifications, one can moderate those properties that are critical for their behavior in GC (Table 5.3).

5.1.4 GC-MS Analysis

Figures 5.4 and 5.5 represent examples of a typical GC-MS analysis. Total ion current (TIC) is registered for each individual spectrum, summed up, and presented in form of a total ion chromatogram (TIC) as a function of retention times of individual compounds or spectrum number (Figure 5.4a). If, however, one is studying the analyte of a known molecular weight, e.g. *m/z* 106, the extracted ion chromatogram (EIC) might be generated to show elution profile and mass spectrum for the selected *m/z* value (Figure 5.5a). It can be obtained as a single

Table 5.3 The most popular derivatization techniques.

Derivatization type	Characterization	Derivatization reagent
Silylation	Replacement of an acidic hydrogen in the compound with alkylsilyl group; can also serve to enhance mass spectrometric properties of derivatives by producing more favorable diagnostic fragmentation patterns for structure investigations	Trimethylbromosilane (TMBS) N,O-Bis (trimethylsilyl) trifluoroacetamide (BSTFA)
Alkylation and arylation	Replacement of an active hydrogen in R-COOH, R-OH, R-SH, and R-NH$_2$ with an alkyl or aryl	Methyl triflate; tetramethylammonium hydroxide (TMAH)
Acylation	Conversion of compounds with active hydrogens such as –OH, –SH, and –NH into esters, thioesters, and amines, respectively. Carbonyl groups adjacent to halogenated carbons enhance ECD response	Acetic anhydride, trifluoroacetic anhydride, boron trichloride – methanol
Esterification	Derivatization of acids. Underivatized acids tend to tail because of the adsorption and nonspecific interactions with the column	Boron trifluoride – methanol

measurement (Figure 5.4b) or as an average of few subsequent scans (Figures 5.4c and 5.5b), which is recommended to fulfill reliability and quality requirements.

It can be seen from Figures 5.4 and 5.5 that EIC may be beneficial for easier identification of the compounds. This is also a common method to simplify measurements by programming selected ion chromatograms prior to measurements. For example, when the operator is interested in searching for the presence of a defined molecule, e.g. cocaine, the system may be preset to scan for the ion(s) arising from this particular compound. Also, several ions can be traced by this method (multiple ion chromatograms). This routine provides much faster scanning speed and higher sensitivity and saves space on the hard disk. There is also a certain risk in applying such procedure. If accidentally the background will increase, e.g. due to signal instability or impurities/coeluting compounds, this anomaly, observed as a "peak," will be overlooked in this mode, as only narrow part of the chromatogram/spectrum is monitored.

5.1.5 Two-Dimensional Gas Chromatography Linked to Mass Spectrometry 2D GC-MS

For the analysis of very complex mixtures that are composed of numerous different substances and characterized with the same boiling point, such as food or petrol, two-dimensional (2D) gas chromatographs are utilized. 2D GC-MS consists of two columns serially connected by means of a modulator and finally

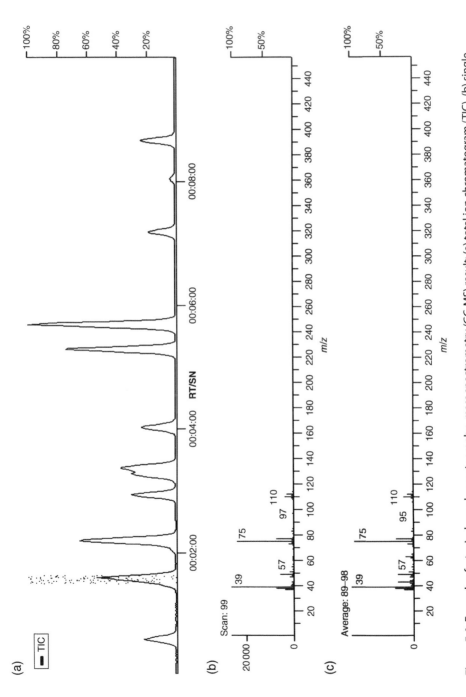

Figure 5.4 Example of a typical gas chromatography–mass spectrometry (GC–MS) result: (a) total ion chromatogram (TIC), (b) single measurement scan 99, and (c) an average of MS scans 89–98.

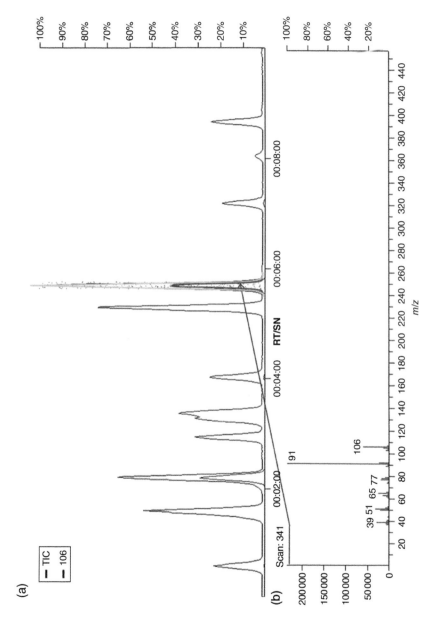

Figure 5.5 (a) TIC and (b) extracted ion chromatogram (EIC). MS spectrum 341 of $m/z = 106$.

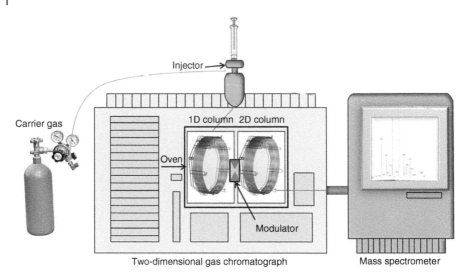

Figure 5.6 Two-dimensional gas chromatography combined with mass spectrometry.

coupled to the MS detector. Two columns are typically of different inner diameters and lengths, where the first column usually is much longer and has larger inner diameter, as compared with the second dimension. Integrated columns allow for the separation of molecules based on their boiling point or sublimation temperature in the first dimension and according to the hydrophobic properties in the second dimension (Figure 5.6). The analyte eluted from the first column is trapped in the sample loop by means of liquid nitrogen and, after switching the valve and reheating, is transferred to the second column [3]. As the analyte arrives in the modulator, it undergoes compression into the sharp peak and is subsequently injected into the second column. Finally, the temperature is raised for separation of the species in the second dimension, after which all sample constituents are consecutively transferred to the mass spectrometer.

The main drawback of the 2D GC-MS coupling is the lack of sufficient scanning speed of the mass spectrometer due to a very fast elution of components from the column. The problem may be circumvented by applying longer columns in the first dimension (from 5 to 15 m), but this may result in peak broadening.

Contradictory to 2D GC-MS, fast or ultrafast separations often used in industrial laboratories for the high-throughput analyses are accomplished by application of short columns, reduced stationary phase thickness, elevated temperature during separation, increased carrier gas flow, or reduced capillary column diameter (Figure 5.7).

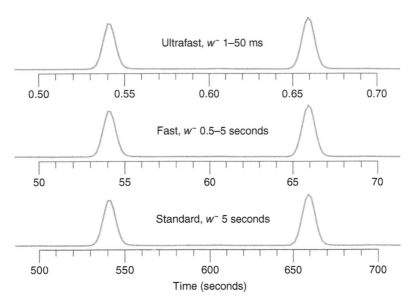

Figure 5.7 Chromatograms obtained with the use of different column lengths. Shorter elution times require fast-operating MS analyzer and detector.

GC-MS coupling enables quantitative analysis of the analyte after applying the calibration curve and/or internal standard. Quantitation, which is superior in case of EI/CI, is also characterized by the highest stability of measurements. The technique, in various setups, has many possible applications, including analysis of carboxylic acids, aldehydes, ketones, amides, esters, ethers, halides, hydrocarbons, glycols, pesticides, phenols, steroids, monosaccharides, steroids, and narcotics, as well as sulfur, nitrogen, and phosphorus derivatives. All those described properties of GC-MS make this technique one of the most widespread analytical methods routinely employed in the quality control laboratories and research centers. GC-MS facilitates identification of complex mixtures in many different disciplines like metabolomics, environmental protection, toxicology, anti-doping control, forensic research, space research, archeology, art, and geochemistry.

Questions

- What properties are required for a substance to be analyzed by GC?
- Which mobile phases are normally used in GC?
- What is the separating medium in GC?
- Why in GC-MS is convenient to operate the injector in the split mode?
- What are the capillary columns? Describe some examples.

- What is the most common GC detector? Why is it so widely used?
- What kind of ion source can be used in GC-MS?
- What are the most popular derivatizations techniques?
- What is 2D GC-MS? What is its major drawback?
- Give some examples of GC-MS applications.

References

1 Niessen, W.M.A. (2001). *Current Practice of Gas Chromatography-Mass Spectrometry*. Boca Raton: CRC Press.
2 Linskens, H.F. and Jackson, J.F. (2012). *Gas Chromatography/Mass Spectrometry*. Berlin: Springer.
3 Stashenko, E. and Martinez, J.R. (2014). Gas chromatography-mass spectrometry. In: *Advances in Gas Chromatography* (ed. G. Xinghua), 1–38. Rijeka: InTech.

5.2 Liquid Chromatography Linked to Mass Spectrometry (LC-MS)

5.2.1 Introduction

Francesco Bellia

CNR Institute of Biostructures and Bioimaging, Catania, Italy

Liquid chromatography (LC) represents the most commonly used separation technique coupled to several types of mass spectrometers (LC-MS). LC-MS is largely applied to obtain qualitative and quantitative data related to the components of complex samples, ranging from chemical to biological samples. Analytical approach differs from the preparative methods, in which the goal is to purify and separately collect one or several components of a mixture. In the latter case, the separation of molecular components represents a critical step to detect and structurally characterize them through mass spectrometry (MS) analysis.

Numerous books and reviews on the theoretical basis of LC are being published every year [1–3]. The aim of this chapter is to summarize the fundamental features of modern LC systems and to address the most important issues related to LC-MS analysis.

5.2.2 Introduction to Liquid Chromatography

Anna Drabik

Department of Biochemistry and Neurobiology, Faculty of Materials Science and Ceramics,
AGH University of Science and Technology, Kraków, Poland

All the chromatographic systems have a common scheme that is depicted in Figure 5.8.

In the classical high performance liquid chromatography (HPLC) system, the sample is injected into a column by the aid of pumps, which produce a controlled flow of eluents. After separation, the analytes reach specific detector and/or can be collected at different time intervals for further processing.

The main core of the chromatographic process is the analytical column. Components of the mixture applied on the column are partitioned between the stationary phase (a highly porous material) and a mobile phase (mixed solvents of various polarities). The different dynamic equilibrium that each component of the mixture establishes during the chromatographic process is responsible for its adsorption/desorption to/from the stationary phase and thus separation. In other words, the separation mostly depends on different affinities of components for the stationary and the mobile phase. The former eluted

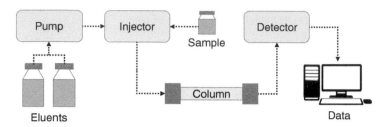

Figure 5.8 The simplified scheme of a chromatographic system.

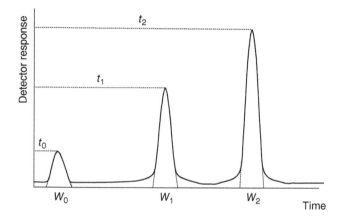

Figure 5.9 Schematic representation of a chromatogram for a sample containing two retained components W_1, W_2, and unretained fraction W_0.

compound is strongly interacting with the eluent, whereas the latter is bound to the column filling. The ideal outcome of a two-component mixture is reported in Figure 5.9; all the molecules that do not interact with the column elute at t_0; t_1 and t_2 are the retention times of the resolved components, whereas W_1 and W_2 represent the peak width. Total volume eluted up to the time t_0 is called *void volume* W_0 and indicates the moment when the component mixture appears on the chromatogram. For each column type and composition of the mobile phase, the retention time depends on the physicochemical features of the particular component, whereas the area under the peak is proportional (within a proper range) to the concentration of the molecule and therefore is used to obtain quantitative information about the analytes.

To optimize the chromatographic separation of studied sample, the balance between three parameters has to be kept – efficiency (a), selectivity (b) (which is a separation factor), and retention (c) (column capacity factor) Eq. (1):

$$R = \overbrace{\frac{1}{4}\sqrt{N}}^{a} \times \overbrace{\frac{\propto - 1}{\propto}}^{b} \times \overbrace{\frac{k}{1+k}}^{c}. \tag{1}$$

While the important concepts such as retention factor (k), selectivity (α), and efficiency/number of theoretical plates (N) can be easily retrieved from chemical tables, here we want to focus on the main kinetic parameter describing the separation between two peaks, that is, the resolution (R), also defined as a distance between the peak centers divided by the peak width measured at the ½ of the peak height (Eq. 2):

$$R = \frac{2(t_2 - t_1)}{W_2 - W_1} \tag{2}$$

Based on this equation, the greater the distance and the narrower the peak width, the better the resolution obtained. Usually, resolution is the parameter mostly considered by the experimentalist as the factor to be optimized and as an indicator of the results' quality. Such a refining process fundamentally concerns the choice of the column, composition of the eluents, and their flow rate. There are commercially available software assisting in gradient planning for maximal resolution. On the other hand, such a gradient adjustment is effective for several components only. Biological mixtures are very complex, and we might always expect coelution of various components in a single fraction, even using the best available column and carefully adjusted mobile phases. Therefore, MS is a very appropriate detection method of choice, as it can simultaneously detect, identify, and quantify molecules in complex samples.

5.2.3 Types of Detectors

Anna Drabik

Department of Biochemistry and Neurobiology, Faculty of Materials Science and Ceramics, AGH University of Science and Technology, Kraków, Poland

Several systems can be used to detect molecules that elute from the chromatographic column. The role of the detector is to provide both qualitative and quantitative information. The ability of the analytes to differently respond in the detector can be used to identify or characterize the sample constituents. If the intensity of the detector signal is proportional to the amount of the molecules present in the sample, quantitative analyses can also be carried out. This, however, requires preparation of the calibration curve, which is a common procedure in quantitative analysis. Internal standard should also be considered, as it compensates the results for eventual fluctuations in sample preparation. Internal standard should preferentially be identical with the compound to be measured, ideally the same molecule containing stable isotope.

The most common detectors linked to an LC system are based on the following:

- Ultraviolet–visible (UV/Vis)
- Diode array
- Light scattering

- Refractive index
- Fluorescence
- Mass spectrometry

UV/Vis and diode array detectors are the most commonly used systems, because they are able to detect a huge variety of molecules. They also have a wide linear dynamic range, and the intensity of UV absorption is proportional to the concentration of the analytes. Standard UV/Vis can only monitor few channels (wavelengths) at the same time, while diode array detector simultaneously scans over the entire wavelength range at every scan event. For this reason, diode arrays are considered to be much more suitable for simultaneous characterizing and quantifying several classes of compounds coexisting in the sample. Moreover, the comparison between the spectrum of the analytes and those of a spectra library allows for identification of unknown compounds in the sample. Diode array detector and parallel measurement in the broad UV/Vis range eliminate fluctuations that may arise during individual scans at preset wavelengths and substantially shorten analysis time. Moreover, ratio of various wavelengths, and in particular its second derivative, may serve as a purity criterion.

Detection by measuring light scattering and refractive index is important for compounds having little or no UV absorption, such as alcohols, sugars, saccharides, fatty acids, and polymers. Monitoring the refractive index requires isocratic separations, whereas light scattering systems (mainly based on the evaporative light scattering detector (ELSD) technique) can be used with gradient separations as presented in Figure 5.10.

Detecting the analytes using fluorescence requires specific structure of the compounds. Often the fluorescence measurements require derivatization of the components before (pre-column) or after (post-column) chromatographic separation. The linear dynamic range is narrower than those of UV/Vis and diode array, but high sensitivity and selectivity are the two features making this detection system valued.

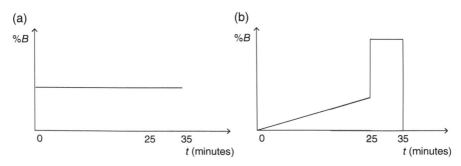

Figure 5.10 Isocratic (a) and gradient (b) chromatographic separation.

The capability of MS to detect compounds is already described in other chapters of this book. The detection based on the mass-to-charge (m/z) values measurement makes MS very selective and unambiguous. Moreover, the tandem MS (MS/MS) approach is fundamental to explicitly identify the components of a complex mixture, such as peptides released from digested proteins, or a mixture of drugs in the seized sample. Therefore, MS detection is unique, as it provides definite identification and quantitation of the compounds at high sensitivity.

The chromatographic resolution is strictly related to the scanning speed of the mass spectrometer. Highly resolved peaks are narrow and need fast scanning speeds. This is why many mass spectrometers characterized with different sensitivities can be used, depending on the number of sample components that need to be analyzed (ultrafast 0.1–50 ms; fast 0.05–1 seconds; standard ~3 seconds).

Detectors can also be used in various combinations (series or parallel) in order to get advantages of the specific information of each acquisition type for the same chromatographic separation. This multi-detection approach is increasingly popular in drug discovery or other screening-type applications. For example, identification and quantitation of peptides formed by the enzymatic hydrolysis, such as human insulin [4] or amylin [5], where ESI-MS detector has been used to identify peptide sequences, whereas UV measurement was helpful to monitor the concentration changes of all hydrolytic fragments.

5.2.4 Chromatographic Columns

Anna Drabik

Department of Biochemistry and Neurobiology, Faculty of Materials Science and Ceramics, AGH University of Science and Technology, Kraków, Poland

The choice of the column is the starting point of any chromatographic analysis. The particles forming the stationary phase have to be quasi-spherically shaped, highly porous, and homogenously packed into the column to reduce the diffusion effects. As the diffusion process contributes to peak broadening, it also negatively affects the resolution and sensitivity. The best way to improve the performance of LC columns is to reduce particles size, as it increases the total surface area of the stationary phase for the same column volume. The drawback of this process is an increased back-pressure, inversely proportional to the square of particle diameter (d), as shown in Eq. (3):

$$P = \frac{v\eta LR}{d^2} \tag{3}$$

The back-pressure (P) is also directly proportional to the flow rate (v) and viscosity (η) of the mobile phase, as well as to the column length (L) and column

Table 5.4 Typical dimensions of common high performance liquid chromatography (HPLC) columns.

Column type	Internal diameter (ID) (mm)	Length (mm)	Flow ($\mu l\,min^{-1}$)	Sample volume (μl)
Preparative	>4.6	50–250	500–30 000	200–50
Analytical	1.0–4.6	15–250	100–1 000	500
Capillary	0.1–1.0	100–200	1–10	5–25
Nano	<0.1	200–350	0.2–0.7	0.5–5

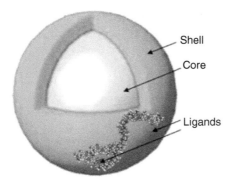

Figure 5.11 Core–shell resin.

resistance (R), a factor that depends on the packing and on the porosity of the stationary phase.

Typical internal diameters and lengths of the most common HPLC columns are reported in Table 5.4.

The HPLC takes advantage of particles 1–10 μm in diameter and the back-pressure can reach 200–300 bars, while the flow rate of the mobile phase ranges between 0.1 and 1 ml min^{-1}. In order to improve separation efficiency, a use of semi-porous particles (core–shell technology) has been applied (Figure 5.11). The permeability is confined to the outer shell of these particles, being greatly reduced with respect to the entirely porous particles having similar dimensions, so that the separation efficacy is clearly improved.

Development of the particles having diameters of less than 2 μm has boosted the modern era of ultra-HPLC (UHPLC). These chromatographic systems retain much higher resolution than classical HPLC instruments and utilize pumps reaching 1000 bar. The flow rate is up to 0.2–10 ml min^{-1} used for columns with dimensions of 50–250 mm, respectively. On the other hand, the entire separation is much faster and completed within a couple of minutes. Therefore, peaks are very narrow, which is an important feature when UPLC is coupled to a mass spectrometer, when a much faster scanning speed is required.

Particle size, flow rate, and column dimension represent common aspects of the LC process. Chemical interactions of the analytes with the stationary phase relay on the chemical features of the adsorbent materials that, in turn, determines the type of separation process. The most commonly used separation modes and the features of the related stationary phases are reported in Table 5.5.

Table 5.5 Main features of the most used chromatographic modes.

Separation mode	Adsorbent material	Functional groups	Eluent
Normal phase (NP)	Silica	$-OH$, $-R$, $-CN$, $-NH_2$	Nonpolar
Reversed phase (RP)	Hydrocarbon-linked silica	C_4, C_8, C_{12}, C_{18}, $-NH_2$, $-CN$, $-phenyl$	Polar
Ion exchange (IE) (SCX or SAX)	Polymer	NH_3^+, NR_2H^+, SO_3^-, carboxymethyl	Salt containing
Size exclusion (SEC)	Polymer	Gel permeation	Buffer or water
Hydrophilic interaction (HILIC)	Silica	$-NH_2$, $-CN$, $-CONH_2$	Aqueous buffer with polar solvent

Majority of the stationary phases packed into chromatographic columns are still silica based. The polar feature of silanol groups at the surface of this material is exploited in adsorption chromatography using nonpolar organic eluents (normal phase [NP]). NP is generally applied for fast purification of mixtures by using low-pressure columns and is particularly used in organic chemistry.

Hydrocarbon chains covalently linked to the silica form a nonpolar surface. The reversed-phase (RP) chromatography implies the use of water and organic solvents (mainly acetonitrile and/or methanol) as eluents. The most used chain length contains 18 carbon atoms (C_{18}), but packings with lower numbers of carbon atoms (C_{12}, C_8, C_4) are also available for separation of more hydrophobic components (large peptides and proteins). Modification of the RP materials with several functional groups ($-NH_2$, $-CN$, $-phenyl$, etc.) improves selectivity of complex mixtures. RP is the most popular separation mode because of the versatility of this technique, mostly due to the compatibility of mobile phases with MS (volatility); therefore majority of the available HPLC and UHPLC columns are packed with RP materials. Care should be taken to protect silica-based column packings from extreme pHs (especially above 9), which can irreversibly destroy particles. Isocratic elution should avoid phosphates, because they may crystalize in the heated capillary (electrospray), leading to its permanent clogging. Trifluoroacetic acid (TFA) as an ion-pairing reagent should be replaced by formic acid (FA), which does not cause ion suppression effect, affecting the sensitivity.

The ion exchange chromatography (IEC) is exploited when the separation process is based on the different charges of the solutes. Positive or negative ions can electrostatically interact with the stationary phase, resulting in separation by means of the anion or cation exchange chromatography, respectively. Strong and weak ion exchange resins can be chosen. The number of charges on a strong ion exchanger remains constant regardless of the buffer pH (strong cation exchange [SCX], strong anion exchange [SAX]). Quaternary ammonium (Q), sulfonate (SO_3^-), and sulfopropyl (SP) are common examples of strong ion exchangers. Initial experiments with unknown samples should not utilize such packings, as the applied material might be irreversibly retained on the column. In contrast, when working with weak ion exchange resins, such as diethylaminoethyl (DEAE) or carboxymethyl (CM) resins, it is important to work within a specific pH range. The major obstacle of such stationary phases is compatibility of solvents with MS ion sources. A typical separation using an ion exchanger utilizes NaCl gradient, which cannot be used in a typical electrospray ionization (ESI) source. Thus, a gradient formed of other volatile solvents at increasing concentration need to be applied, such as NH_4HCO_3 or a pH gradient.

In the size exclusion chromatography (SEC), solutes penetrate the channels formed by a porous packing material, and there are no interactions between the sample components and the stationary phase (in the ideal case). Molecules are separated on the basis of their dimensions relative to the pore size. Larger analytes do not enter the small channels and therefore elute first, whereas smaller molecules penetrate the channels and elute later. In this type of chromatography, the only role of the eluent is to drive sample components through the stationary phase.

5.2.5 Chromatographic Separation and Quantitation Using MS as a Detector

Anna Drabik

Department of Biochemistry and Neurobiology, Faculty of Materials Science and Ceramics, AGH University of Science and Technology, Kraków, Poland

In order to set up chromatographic separation, many parameters need to be taken into account. The assembly with MS detector, especially through an *online* connection, requires specific care on many important details, in particular when capillary LC is planned to be used.

Composition of the mobile phase and its flow rate are the main factors that need to be considered. In the case of the RP chromatography, all the solvents typically used for the separation (water, methanol, acetonitrile) are compatible with the ESI ion source. It is really important to pay attention on the purity of the solvents. Those used for an LC-MS setup should be free from contaminants that may interfere with MS but are acceptable for HPLC alone. If a buffered solution is unavoidable in a water-based eluent, volatile salts like ammonium acetate, ammonium formate, or ammonium bicarbonate are more appropriate than phosphate or acetate buffers containing sodium or potassium ions, as they may irreversibly clog MS system. A yet another problem may arise from the eluent containers. Organic solvents are sold in bottles where the caps are also protected by an additional seal. This seal, under influence of the solvent, is a source of released polymer(s), giving strong signal in the mass spectrometer. It is highly advised to test chemicals (vendor) before analysis, as such polymer signal remains in the LC-MS system for a very long time. Salt solutions (also volatile) should be filtered, in particular for the nano-LC applications.

Important components of the eluent are the ion-pairing agents (commonly TFA, FA, or other fluorinated carboxylic acids). They are also quite important for the chromatographic separation because they suppress the ionization of residual silanols of the stationary phase, thus counteracting nonspecific interactions between the analytes and the stationary phase. The major effect is a general peak sharpening. On the other hand, TFA is responsible for ion suppression in the MS system. A balance between these two opposite effects of coupling agent on the efficiency of HPLC separation has to be properly found. As a general rule, if the ionization of sample components represents the critical step of the LC-MS analysis, FA is better choice than TFA, whereas TFA is preferable when the resolution of the mixture component is the main goal. A mixture of FA and TFA can also be used to compromise between chromatographic and spectrometric performances.

The compatibility of LC and MS systems also concerns the flow rate. The separation of mixtures, especially containing small molecules (<1 kDa), is generally improved by accelerating the flow rate. Such a procedure has a deleterious effect on the sensitivity of MS detector. To overcome this drawback, it is preferable whenever possible to improve the resolution by properly designing the gradient system. The substitution of columns having a diameter of 4.6 mm with those of 1 or 2 mm diameter leads to the decrease of the flow rate (from 1 to 0.2–0.05 ml min^{-1}). On the other hand, the lower the flow rate, the better the sensitivity (improved evaporation and ionization in the ESI source), and the lower the solvent consumption. Moreover, too high flow rate may lead to electrical shortcuts in the instrument and to its destruction. Capillary and nano-LC columns represent the best option to obtain higher sensitivity because of the smaller diameters. For example, identification of peptide components of a reaction mixture is the final step of proteomic analysis and represents one of the main scientific fields that have taken advantage of the high sensitivity of modern UHPLC-HRMS systems. Application of column having diameter 1–2 mm clearly involves the presence of small-sized particle (up to 2 μm) inside the column. The back-pressure of the column reaches values as high as 200–300 bar. However, the signal-to-noise ratio as well as the resolution is greatly improved, as compared with the larger columns, also because of the high scanning speed of the MS analyzers. Roughly, replacement of an analytical column with a nano-column may increase sensitivity by a factor of 5000.

Regarding quantification of the analytes, it is commonly accepted that the absolute intensities produced by MS measurements are not uniformly proportional to the concentration of the detected compounds because of the influence of many physical and technical issues, such as the spray formation, auxiliary gases, efficiency, ion suppression, ionization efficiency, flow rate of the eluent, vacuum, etc.

Using internal standards having chemical properties resembling close similarity to those of the analytes is the best way to obtain quantitative and accurate results. Stable isotope-labeled species of the analytes are perfect candidates in quantitative MS experiments (the procedure is called stable isotope dilution). Difficulties to perform the calibration process when complex biological samples are used further support the importance of the isotope dilution method.

Proper chromatographic and mass spectrometric parameters that allow best detection, identification, and quantification of the molecules strictly depend on their chemical and physical features, the performance of the instruments, and the complexity of the sample. For these reasons, only general guidelines can be drawn for method optimization.

The first important step is optimization of the MS parameters to detect the compound(s). Direct infusion of the analytes, possibly dissolved in the eluents used for the LC separation, is a desirable starting point, providing there are only few components in the sample. The compound may raise some difficulties to form charged ions. If the spectrometric tuning has little or no effect on the ionization process, chemical derivatization of the analytes represents an important strategy: introducing a chargeable group often lowers the detection limit. Extreme care should be taken to apply as little synthetic compound as possible, followed by thorough washing, to avoid contamination of transfer lines. The same applies to the LC system. As a rule, a separate column is used to test system performance.

Chromatographic method should be evaluated using several dilutions of standard compound to calculate the detection limit. Samples mixed with internal standard should be analyzed in order to test the ion suppression of the matrix. This effect can be avoided or significantly decreased by changing elution gradient and/or monitoring fragment ions instead of the intact molecule.

Sensitivity of the ions formed by MS/MS system is generally lower than that obtained by a targeted measurement of the analytes (selected reaction monitoring [SRM], multiple reaction monitoring [MRM]). Also the specificity of the detection is higher, and a signal-to-noise ratio is improved, which significantly increases detection limit.

5.2.6 Construction of an Interface Linking Liquid Chromatograph to the Mass Spectrometer

Anna Drabik

Department of Biochemistry and Neurobiology, Faculty of Materials Science and Ceramics, AGH University of Science and Technology, Kraków, Poland

5.2.6.1 Introduction

As LC analysis is carried out in solution, ionization sources, such as ESI and atmospheric pressure chemical ionization (APCI), can be linked to the LC system by means of an *online* connection. On the contrary, the matrix-assisted

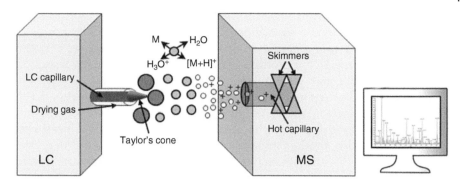

Figure 5.12 Scheme of the electrospray ionization (ESI) interface (positive ion mode).

laser desorption/ionization (MALDI) source needs an *offline* connection to the chromatographic apparatus (Section 5.2.6.5).

The most important aspect to pay attention to is the flow rate of the solvents, which has already been emphasized earlier in this chapter. As a general rule, the lower the flow rate, the higher the sensitivity. This means that flow rates of $0.1–1\,ml\,min^{-1}$ can be used, but the best performance in terms of sensitivity is reached when the flow rate is as low as $100–500\,nl\,min^{-1}$. In these cases, a nanoESI is coupled for direct infusion to the MS.

For the LC-MS analyses, it is fundamental to remove the solvent from the chromatography eluent to produce ions in the gas phase and to force decreasing size of charged droplets. In the second step, charging of the analyte molecule takes place. The solvent undergoes evaporation and dry ions enter the instrument. The most important steps during ESI process that lead to the transfer of the sample ions from the LC column into the gas phase within the mass spectrometer are (i) generation of charged droplets, (ii) desolvation of the droplets, and finally (iii) generation of the ions (Figure 5.12) [6]. It is also important to create suitable pH conditions for efficient ionization of the molecules in the positive or negative ion modes.

There are three major options to achieve this goal under atmospheric pressure, namely, ESI, APCI, or atmospheric pressure photoionization (APPI). These techniques belong to the "soft ionization" methods, which means that they generate ions, which do not spontaneously undergo fragmentation in the ion source. To distinguish such species that are formed by addition or subtraction of protons, from electron impact ionization (Section 4.1.1) generating molecular ions, they are referred to as pseudomolecular ions.

5.2.6.2 ESI Interface
As mentioned above, during ESI (and related techniques) analysis, the sample is transferred to the source with the aid of the mobile phase, and the scheme of the ESI interface is shown in Figure 5.12. The LC capillary tip is

surrounded by the drying gas stream (usually nitrogen at a flow $5–15\,l\,min^{-1}$), where the desolvation process begins. The Taylor cone is formed by the eluent surface deformation in the electric field (potential differences between LC capillary tip and the hot capillary can reach up to $5–8\,kV$). The stream of dry ions reaches the mass spectrometer because of another electric potential difference, this time between two ends of the hot capillary, while neutral particles and other contaminants are captured by skimmers. As a result, analyte ions are transferred to the analyzer where they can be measured. Skimmers play also another important role, as they maintain differential pumping inside the instrument. By adjusting potentials between heated capillary and skimmers, it is possible to force *in-source* fragmentation. Because all ions will be fragmented in this way, it is necessary to inject pure substance to avoid generation of a complex mixture of fragment ions arising from various molecules.

5.2.6.3 APCI Connection to MS

APCI takes place as a result of the charge transfer reactions in the vapor phase to produce ions. This interface has similar construction to ESI, except that a high potential (corona discharge) is applied to the spraying region, i.e. between LC needle and MS inlet. Such setup produces electric discharge that effectively ionizes surrounding gas molecules and adds charges to the mobile-phase species, which are subsequently transferred by proton transfer or charge exchange to the analyte molecules (Figure 5.13). APCI source is applied for the analysis of small molecules that do not undergo thermal decomposition. It also accepts higher flow rates than typical ESI interfaces.

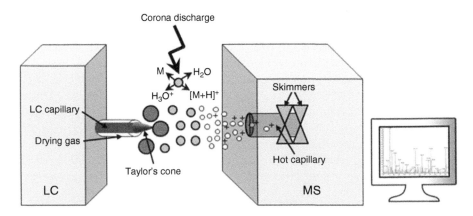

Figure 5.13 Scheme of the atmospheric pressure chemical ionization (APCI) interface.

5.2.6.4 APPI Interface

APPI connection with the chromatographic column for identification of hardly ionizable molecules, like weakly polar or nonpolar analytes, enables irradiation in the spraying zone (Figure 5.14). Ionization of the desolvated droplets is enforced by photoirradiation, typically with the aid of a krypton lamp (10 eV). Analyte molecules absorb photons, and when the applied energy exceeds ionization potential of the species, they turn into ions. This process, likewise ESI, can be performed in both positive and negative modes (see also Chapter 4.1.3.6).

5.2.6.5 LC Connection to MALDI-MS

Coupling LC to MALDI-MS is a typical example of the *offline* connection type (Figure 5.15). It can be easily automated once the spotter is assembled to the LC capillary outlet. Spotter or fraction collector is designed for an automatic

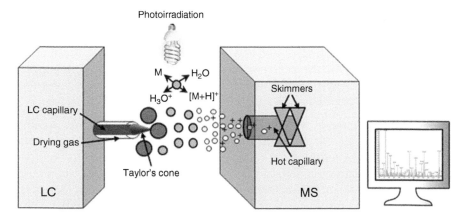

Figure 5.14 Scheme of the atmospheric pressure photoionization (APPI) interface.

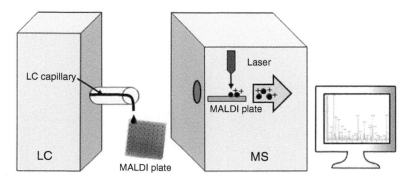

Figure 5.15 Scheme of the liquid chromatography–matrix-assisted laser desorption/ionization (LC-MALDI) interface.

handling of the eluent to deposit fractions on the MALDI target plate. Minute amounts (usually 5–50 µl) of LC eluent can be accurately spotted. Matrix solution is also automatically added. This allows for preparation of target plates without the need of the operator's contribution. This interface enables very high sample capacity, while multiple target plates can be spotted. Samples can be spotted at various time intervals, and therefore the time of MS analysis can be reduced. Piezoelectric spotters are also applied, dispensing much smaller nanoliter volumes.

5.2.6.6 Multidimensional Separations

Analyzing complex mixtures requires more sophisticated separation techniques to be involved. Multidimensional chromatography enables separation of multiple constituents in the sample, even when physicochemical properties of the species are very similar. Sequential connection of two (or sometimes even more) chromatographic columns allows for separation of even thousands of compounds [7]. The highest complexity of the analyzed mixtures is characteristic for biological samples, for example, proteomic analyses of potential biomarkers. Therefore, the most common multidimensional LC connections are SCX/RP (Figure 5.16) and SCX/HILIC [8]. There are however few rules that are essential for an effective analysis [9]:

- Application of multiple columns decreases detection limit, and consequently, information about less abundant compounds might be lost. Hence, sample amount should be enlarged.
- The order of connected columns is of importance, and the system should begin with the column characterized with the highest capacity and diameter, while the second-dimension column should be considered having the highest resolving power.
- The last column should work under conditions (pH, salt concentration) using eluent (mobile phase) compatible with mass spectrometer.

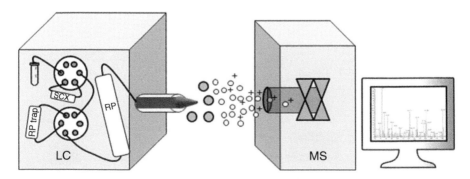

Figure 5.16 Multidimensional liquid chromatography combined with mass spectrometry.

Hyphenated analytical techniques for separation and identification of various molecules provide high potential for analyses of complex samples. The advantage is also the possibility of automatization of each such system. During multidimensional LC-MS/MS studies, one needs to remember that the number of obtained data and analysis time need to be considered for further validation (bioinformatics, statistics, etc.).

Questions

- Which are the main components of a LC-MS system?
- How resolution is defined in chromatography?
- Which are the main differences among the most common detectors linked to an LC system?
- How do the particle size and porosity affect the resolution?
- Which are the main pros and cons of FA and TFA as ion-pairing agents?
- Describe the most important features and benefits of an UHPLC-HRMS system.
- Which are the main differences and the common features of ESI and APCI interfaces?
- Which are the most common multidimensional LC connections?

References

1 Fornstedt, T., Forssén, P., and Westerlund, D. (2015). Basic HPLC theory and definitions: retention, thermodynamics, selectivity, zone spreading, kinetics, and resolution. In: *Analytical Separation Science* (ed. J.L. Anderson, A. Berthod, V. Pino and A. Stalcup). Weinheim: Wiley-VCH.
2 Snyder, L.R., Kirkland, J.J., and Dolan, J.W. (2009). *Introduction to Modern Liquid Chromatography*, 3e. Chichester: Wiley.
3 Sahu, P.K., Ramisetti, N.R., Cecchi, T. et al. (2018). An overview of experimental designs in HPLC method development and validation. *Journal of Pharmaceutical and Biomedical Analysis* 147: 590–611.
4 Bellia, F., Pietropaolo, A., and Grasso, G. (2013). Formation of insulin fragments by insulin-degrading enzyme: the role of zinc(II) and cystine bridges. *Journal of Mass Spectrometry* 48 (2): 135–140.
5 Bellia, F. and Grasso, G. (2014). The role of copper(II) and zinc(II) in the degradation of human and murine IAPP by insulin-degrading enzyme. *Journal of Mass Spectrometry* 49 (4): 274–279.
6 Gelpi, E. (2002). Interfaces for coupled liquid-phase separation/mass spectrometry techniques. An update on recent developments. *Journal of Mass Spectrometry* 37: 241–253.

7 Drabik, A., Bodzon-Kulakowska, A., Suder, P. et al. (2017). Glycosylation changes in serum proteins identify patients with pancreatic cancer. *Journal of Proteome Research* 4: 1436–1444.

8 Drabik, A., Ciołczyk-Wierzbicka, D., Dulińska-Litewka, J. et al. (2014). A comparative study of glycoproteomes in androgen-sensitive and -independent prostate cancer cell lines. *Molecular and Cellular Biochemistry* 386: 189–198.

9 Noga, M., Sucharski, F., Suder, P., and Silberring, J. (2007). A practical guide to nano-LC troubleshooting. *Journal of Separation Science* 30: 2179–2189.

5.3 Capillary Electrophoresis Linked to Mass Spectrometry

Przemysław Mielczarek[1] and Jerzy Silberring[1,2]

[1] Department of Biochemistry and Neurobiology, Faculty of Materials Science and Ceramics, AGH University of Science and Technology, Kraków, Poland
[2] Centre of Polymer and Carbon Materials, Polish Academy of Sciences, Zabrze, Poland

5.3.1 Introduction

Capillary electrophoresis (CE) belongs to the electromigration techniques that have been widely used in the analysis of various chemical compounds. In practice, this technique was applied for separation of various substances, such as proteins, amino acids, fragments of DNA, and polymers, but found its usefulness also for the analysis of bacteria, fungi, viruses, or red blood cells. The advantage of CE is its high resolution. Taking into account high sensitivity of mass spectrometry (MS), connection of these two analytical techniques contributes to the popularity of this method in diagnostics, research, and monitoring in medical sciences. Another feature of this technique is the short analysis time and the minimal consumption of chemical reagents, resulting in a relatively low cost of analysis and the ability to carry out a large number of analyses.

CE, including the most common capillary zone electrophoresis (CZE), utilizes the differences in electrophoretic mobility of separated compounds. The CE system consists of a fused silica capillary filled with a suitable buffer solution. Both ends of the capillary are immersed in buffer reservoirs, between which a high voltage of the order of 5–60 kV is applied. The inner diameter of the applied capillaries varies between 5 and 100 μm, and the outside diameter is between 200 and 400 μm. The schematic of such a device is shown in Figure 5.17.

The flow of ions in solution is driven by an electric field created by the difference in potentials applied at the ends of the capillary. The ability of ions to move in the electric field is called electrophoretic mobility (μe). Positively charged ions (cations) move toward the cathode, while negatively charged ions (anions) move toward the anode. The electric field does not affect electrically neutral molecules. Electrophoretic mobility of ions is dependent on their charge-to-size ratio. Small ions with high charge travel faster toward the electrode than large ions with low charge. Besides electrophoretic mobility, an electroosmotic flow (EOF) can be observed, depending on the buffer composition and the inner surface of the fused silica capillary. EOF is based on the flow of entire solution present inside the capillary. In the case of unmodified fused silica capillaries, a negative charge resulting from the presence of silanol groups appears on the inner walls. Cations that are present in the buffer form a

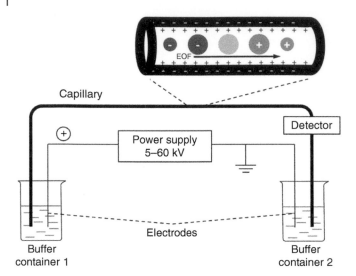

Figure 5.17 Schematic of electrophoresis capillary system.

stationary layer (called the Stern layer), due to interactions with the static negative charge present on the fused silica surface. Another layer formed, the external layer, forms a diffusion layer. The Hermann von Helmholtz model was the first to describe the electric double layer. The ions present in the diffusion layer will move toward the electrode having an opposite charge, resulting in the effect that the buffer and all dissolved compounds (sample) flow in one direction. In summary, electrophoretic mobility is responsible for the separation of the ions present in the solution, while the EOF causes the entire solution in the capillary to move toward the end of the capillary [1].

CE proved to be an ideal tool for biological analyses, where the amount of available sample is considerably limited, due to the possibility of analyzing samples present in femtoliter volumes. Such a small amount of analytical material requires application of highly sensitive detectors. Because of this, MS is becoming an ideal tool during such analyses and has many advantages over standard ultraviolet–visible (UV/Vis) detectors. The main limitation in coupling CE with MS is the need to use buffers containing volatile substances only. This excludes the possibility of using salts such as phosphates and borates, commonly used in combination of CE with spectrophotometric detection. The most commonly used substances for buffers preparation are formates, acetates, and ammonium salts at concentrations between 10 and 200 mM (pH range of 2–11) [2].

5.3.2 Types of Electrophoretic Techniques

The most popular type of CE is CZE. Separation of the substances is carried out in a fused silica capillary filled with a solution of constant pH along the entire capillary length.

Another type is capillary gel electrophoresis, in which the capillary is additionally filled with a crosslinked gel. Slightly different is the capillary isoelectric focusing. During such separation, the pH of the capillary buffer varies from the highest value at the cathode side to the lowest value at the anode side of the capillary. A pH gradient along the capillary can be achieved by an addition of suitable compounds called ampholytes. Ampholytes are a mixture of chemical compounds that gives a complete range of isoelectric points for reproducible linear pH gradients. During such electrophoresis, the analyzed ions move toward the opposite-charge electrode, encountering zones of different pH values. Proteins and peptides that represent amphoteric compounds will move until they reach a buffer zone where the pH of the solution equals their isoelectric point. At this place, they lose their electric charge; thus no longer can move in the electric field.

The last type of CE is capillary isotachophoresis. Here, the sample is placed between two solutions, significantly differing in electrophoretic mobility of the ions present in both electrolytes. The electrolyte placed before the sample contains ions with high electrophoretic mobility that exceed mobility of all analyzed ions in the sample, while the electrolyte placed behind the sample has low electrophoretic mobility, much lower than the ions present in the sample. During electrophoresis, the ions present in the sample are distributed between the electrolytes containing high and low electrophoretic mobility. Separated ions focus in this area in decreasing order of electrophoretic mobility. Ions originating from a sample of different mobility lay right next to each other, forming zones containing one type of species. As a result of this separation, a stepwise isotachogram is obtained, and each step corresponds to a zone containing ions of equal electrophoretic mobility [3].

5.3.3 Capillary Electrophoresis Linked to ESI

5.3.3.1 Introduction

Since the invention of electrospray ionization (ESI), which is one of the most popular ion sources for analyzing liquid samples, direct coupling of CE with mass spectrometer has been greatly simplified and can be applied under atmospheric pressure. Additional advantages of ESI are the ability to analyze high molecular weight molecules, even over 10 000 Da, creating multiple charged ions allowing mass spectrometers with analyzers working in limited m/z ranges to analyze molecules with such high molecular weights (see Section 4.1.3.2). This feature and ESI nature of "soft" ionization (no spontaneous fragmentation in the ion source) contribute to the popularity of this type of ionization in analyses of biological samples, such as nucleic acids, proteins, and peptides. Several types of interfaces were constructed to combine ESI sources with CE. The most popular are the sheath flow interface, the sheathless direct interface, and the liquid junction interface. The design and basic principle of CE-ESI-MS connection types are briefly described in the following section [4].

5.3.3.2 Liquid Sheath Connection

The most commonly used type of CE connection with MS is the sheath flow interface, which is shown in Figure 5.18. The end of the fused silica capillary, in which the electrophoretic separation is performed, is placed in the metal needle, through which an additional liquid is pumped. At the outlet of the fused silica capillary and the stainless steel capillary, the sheath liquid is mixed with a solution leaving the fused silica capillary of CE system. Mixing both solutions provides electrical contact between the metal needle and the solution inside the fused silica capillary. In some systems, the nebulizing gas is also applied to stabilize spraying conditions in the ion source.

The flow rate of the used sheath solution usually ranges from few to several dozen microliters per minute ($\mu l\,min^{-1}$), while the flow of solution through the fused silica capillary induced by EOF is on the order of nanoliters per minute ($nl\,min^{-1}$). Significant dominance of the sheath liquid contributes to lower sensitivity during analysis but has a number of advantages. The high proportion of sheath liquid in the electrospray ion source allows for partial elimination of problems associated with the composition of the electrophoresis buffer (presence of salts) and increases spray stability due to the high content of organic phase in the sheath liquid (usually 60–80% of methanol or acetonitrile). Addition of formic or acetic acids also facilitates the ionization process in the case of positive ionization, which is prevalent in the analysis of proteins and peptides.

5.3.3.3 Sheath-Free Connection

Another type of CE-MS connection is the sheathless interface, where the tip is covered with a thin layer of metal providing electrical contact (see Figure 5.19). The tip of the capillary must be suitably tapered and can be manufactured by various techniques, such as hydrofluoric acid digestion, hot-tapered in the flame or mechanically tapered. In addition, the tip of the capillary must be covered with a thin layer of metal, usually gold, platinum, or silver. This can be typically achieved using vacuum sputter, which is a common device in the electron microscopy facilities. The conductive layer may be also made of graphite dispersed in polyimide or polyaniline.

The direct connection of electrophoretic capillary with an electrospray ion source generates some restrictions on CE itself. It is necessary to use buffers containing only volatile substances in the electrolyte used for electrophoresis.

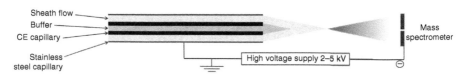

Figure 5.18 Capillary electrophoresis–mass spectrometry (CE-MS) sheath flow interface.

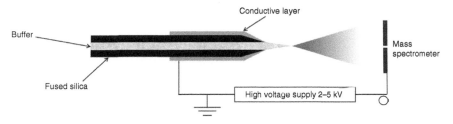

Figure 5.19 CE-MS sheathless interface.

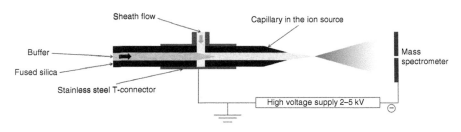

Figure 5.20 CE-MS liquid junction interface.

Widely applied electrolytes are formates, acetates, bicarbonates, and ammonium salts as well as volatile organic acids, such as formic acid or acetic acid. An additional limitation is an appropriate preparation of the sample, which should be free from salts and other nonvolatile substances, even in trace amounts, which may contribute to the ion suppression during the ionization process. The main disadvantage of a sheath-free connection is the short time of a fused silica tip, which needs to be replaced quite often, and the low repeatability of the analysis between various connections. The most common problems are due to the difficulty of obtaining the proper electrical contact during all analyses and reproducibility. Despite all the abovementioned disadvantages, sheath-free connection is characterized by the highest sensitivity, which results from the fact that the analyte leaving the capillary during electrophoresis is not diluted prior to MS analysis.

5.3.3.4 Liquid Junction

The last type of CE linked to MS is an electrical contact applied through a liquid junction interface. An example of such a connection is shown in Figure 5.20. The fused silica capillary, in which the electrophoresis is carried out, is connected to the capillary in the ESI ion source through a stainless steel connector, through which an auxiliary solution is introduced. The distance between the electrophoretic capillary and the emitter capillary in the T-tie is typically 10–20 μm. The auxiliary solution is mixed with the analyte leaving the

CE capillary inside the stainless steel fitting, which provides electrical contact required to perform CE and ESI.

This type of CE-ESI-MS connection exhibits higher sensitivity than the standard sheath liquid connection and requires less optimization due to fewer variable capillary parameters but has similar limitations. An additional difficulty may be the problem of crystallization of the salts inside T-tie, which is responsible for electrical contact with the solution.

5.3.4 Capillary Electrophoresis Linked to Matrix-Assisted Laser Desorption/Ionization

Another type of the soft ionization techniques used in MS is matrix-assisted laser desorption/ionization (MALDI). It is an excellent ionization technique for compounds such as proteins and peptides (see Section 5.1.5). In addition, it exhibits a far greater tolerance to salts, lipids, and other contaminants present in biological samples that significantly disturb ESI. The main problem in connecting CE with MALDI ion source is that ionization occurs in high vacuum. This is the main reason why solid samples are analyzed by this technique. There are some examples of CE-MALDI-MS connections using a time-of-flight (TOF) analyzer, used in conjunction with this ionization technique, due to the pulsed nature of the ion source operation (see below).

5.3.4.1 *Offline* CE-MALDI-TOF
The simplest and historically first MALDI-type CE connected with MS was the *offline* collection of fractions during electrophoresis and subsequent MALDI-MS analyses. Fractions can be collected in chromatography vessels or directly on the MALDI target plate. This kind of approach is based on the independent operation of the electrophoretic system and the mass spectrometer, which makes it possible to optimize each instrument individually. This simplifies the entire procedure and eliminates some of the limitations of CE-MS interfaces.

However, such analyses may also be troublesome. Collection of fractions having very small volumes may cause a technical problem. Because of the high resolution of CE, a high number of fractions need to be collected and analyzed, which is difficult to handle and consequently forces to automate the entire analytical process. There are number of ways to collect fractions on a MALDI target. After applying all fractions onto the surface of the plate, the solution containing matrix is applied. After drying and co-crystallization, the MALDI-MS measurements are performed. It might be helpful to use a sheath flow like in EC-ESI-MS, which leaves little residue of the eluent on the capillary tip, which reduces sample mixing.

5.3.4.2 Direct CE-MALDI-TOF
A slightly more complicated way of combining CE with MS is direct CE-MALDI-TOF, which collects fractions leaving the electrophoretic capillary directly on the surface of a continuous cellulose membrane (a sort of

continuous belt of cellulose). Cellulose membrane can be matrix coated. After deposition and drying the collected fractions, cellulose strip can be placed in a MALDI-type ion source. Analysis is carried out by measuring the mass spectra along the entire length of the belt. Laser shots are typically performed at 250 μm distances.

5.3.4.3   *Online* CE-MALDI-TOF

It is much more difficult to combine *online* CE with MALDI-MS. Commercial MALDI-type ion sources are only suitable for solid-state analysis. Such a connection must therefore be specially developed and encounters some difficulties, mainly due to the high vacuum in the ion source.

One combination of this type involves introduction of an analyte leaving CE capillary mixed with matrix solution at the end of the probe placed in the ion source. The laser beam hits directly at the tip of the probe placed in a high vacuum, where the desorption and ionization processes occur. The main disadvantage of this method is the ability to use specific matrices only, which greatly limits the practical application of such method.

Another method of CE-MALDI-MS uses aerosol. The sample and matrix mixture, when introduced into the ion source, is sprayed by a nebulizer, and then the solvent evaporates directly in the ion source. After drying of the aerosol, the created particles are ionized by a laser beam. The disadvantage of this technique is the deterioration of resolution of the mass spectra due to the formation of ions in the large aerosol area and not on the surface of the target plate as in typical MALDI ion source.

The last of the discussed methods of *online* connection of CE with MALDI-MS is based on the concept of the earlier description of direct CE-MALDI-MS connection. The difference is that fractions leaving the CE capillary are directly deposited on the moving elements in the ion source. Depending on the design of an ion source, rotating balls or rollers and rotating wheels or rewinding bands can be used. After evaporation of the solvent, the sample mixed with matrix appears as a trace deposited on the moving surface. One of the major difficulties are problems with the purity of the moving surface. A common problem is the so-called memory effect due to contaminated moving element in the ion source by the substances from previous analyses.

5.3.5   Summary

Connections of CE with MS detection can be applied to investigate diverse groups of compounds. In addition, it is possible to carry out fast analysis for many highly complex samples. Because of the low detection limit, this technique was mainly applied for biological samples. Combination of CE with MS comprises a truly universal detection of many different substances. At the same time, the results obtained are very selective and allow the analyses of each substance present in the mixture and identification of unknown compounds.

It should be highlighted that peaks obtained in CE are quite narrow; as a result, to get the appropriate number of mass spectra, instruments with high rates of data acquisition should be used [5].

Questions

- What is the key parameter affecting the electrophoretic mobility?
- What is the electroosmotic flow and how is it generated?
- Which are the most commonly used buffer in CE when coupled to MS?
- What types of electrophoretic techniques are commonly used?
- Describe how isotachophoresis works.
- What kind of interfaces are used to combine ESI sources with capillary electrophoresis?
- What are the advantages to use a dominance of the sheath liquid in the liquid sheath connection?
- What are the advantages to use the sheath-free connection? What are the disadvantages?
- Explain how the offline and direct CE-MALDI-TOF are carried out.

References

1 Baker, D.R. (1995). *Capillary Electrophoresis*, 19–51. New York: Wiley.
2 Kraj, A. and Silberring, J. (2008). *Proteomics. Introduction to Methods and Applications*, 55–63. Hoboken, NJ: Wiley.
3 Altria, K.D. *Capillary Electrophoresis Guidebook: Principles, Operation, and Applications*. Totowa: Humana Press.
4 Silberring, J. and Ekman, R. (2002). *Mass Spectrometry and Hyphenated Techniques in Neuropeptide Research*, 135–154. New York: Wiley.
5 Landers, J.P. (1997). *Handbook of Capillary Electrophoresis*, 2e. Boca Raton, FL: CRC Press.

6

Mass Spectrometry Imaging

Anna Bodzon-Kulakowska and Anna Antolak

Department of Biochemistry and Neurobiology, Faculty of Materials Science and Ceramics, AGH University of Science and Technology, Kraków, Poland

6.1 Introduction

Mass spectrometry (MS) imaging (MSI) is a method for surface analysis that was introduced at the end of the twentieth century, and since then a huge development of this technique can be observed. Currently, it is widely used in biological, biochemical, and medical research.

MSI technique combines two elements. The first is the ability to obtain the mass spectrum of a given point on the surface, which provides information about the molecules present in this particular point. The intensity of the peak of a given *m/z* value also allows for semiquantitation. Moreover, complete identification of the compounds is possible with the use of tandem mass spectrometry (MS/MS) option. The second element is the design of the ion source, which enables movement of the examined surface in the *x*- and *y*-axis. Therefore, the analysis of the entire grid of points on a given plane may be performed. The combination of these two elements allows for visualization of the distribution of substances on the analyzed surface by plotting the corresponding peak intensity for each point on the plane (Figure 6.1) together with its *m/z*.

Other imaging techniques, such as immunofluorescence staining, also allow for imaging, for example, proteins distribution in a sample. However, we need to know in advance what particular protein we are looking for to detect it with specific antibody labeled with fluorescent tag. In contrast, the main advantage of imaging mass spectrometry (IMS) is no a priori labeling. What is more, there is no need to know sample composition, and the samples do not require tedious methods for sample preparation characteristic for other imaging techniques.

Mass Spectrometry: An Applied Approach, Second Edition. Edited by Marek Smoluch, Giuseppe Grasso, Piotr Suder, and Jerzy Silberring.
© 2019 John Wiley & Sons, Inc. Published 2019 by John Wiley & Sons, Inc.

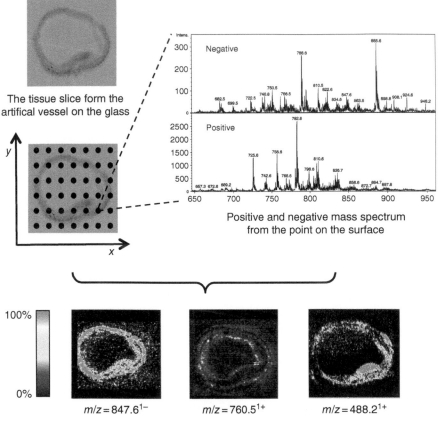

Figure 6.1 The idea of mass spectrometry (MS) imaging.

Currently, the most widespread MS techniques for imaging are secondary ion mass spectrometry (SIMS) (Section 4.1.7), matrix-assisted laser desorption/ionization (MALDI) (Section 4.1.5), and desorption electrospray ionization (DESI) (Section 4.1.3.4). Each of them allows for the analysis of slightly different types of molecules, which are characterized by different parameters (Table 6.1). Their advantages and disadvantages in the context of surface imaging as well as exemplary applications will be briefly discussed below (see also Fig. 6.2).

6.2 SIMS

SIMS (see Section 4.1.7) is the oldest MS technique used for surface analysis. During the analysis, a beam of single ions (e.g. Bi^+, Au^+) or ionic clusters (C_{60}^+, Bi_3^+, Au_n^+, Cs_n^+) transfers its kinetic energy through the so-called collision

Table 6.1 Comparison of the mass spectrometry (MS) imaging techniques.

Technique	Ion source	Ionization	Surface resolution (μm)	Mass range (Da)	Substances
Secondary ion mass spectrometry (SIMS)	Ion gun	Hard	<10	0–1000	Ions, small molecules, lipids
Desorption electrospray ionization (DESI)	Electrospray	Soft	100	0–1000	Small molecules, lipids
Matrix-assisted laser desorption/ionization (MALDI)	Laser	Soft	50	0–100 000	Lipids, peptides, proteins

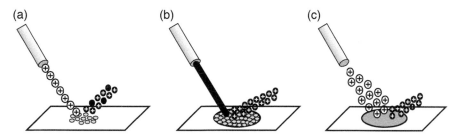

Figure 6.2 Ion sources used in mass spectrometry imaging techniques: (a) secondary ion MS (SIMS), (b) matrix-assisted laser desorption/ionization (MALDI), and (c) DESI.

cascade to the molecules located on the surface [1]. The energy is passed toward the surface of the sample, which causes sputtering of atoms and molecules. Because the transferred energy is relatively high, some of the molecules are being fragmented, and some remain intact. About 1% of the resulting molecules is charged and can be detected by the TOF analyzer, which is most commonly used with SIMS spectrometers. Thus, on the spectrum, we may observe molecular ions (radical cations), pseudomolecular ions (i.e. proton addition), and ions generated by the fragmentation of the molecule.

Compared with other MSI techniques, SIMS is characterized by very high resolution of the image, especially with monoatomic sources. This is because the primary ions can be precisely focused on the surface. In the case of ionic clusters, the achieved resolution is slightly lower due to the larger size of these species. Additionally, SIMS offers the possibility of imaging in three dimensions. The primary ions might be used to remove the subsequent layers of the sample, and thus, a spatial distribution of the molecules (deep into the surface) may be revealed. In the same manner, the layer of impurities may be discarded (e.g. the remaining of the cell culture medium), thereby improving the quality of obtained pictures of cell cultures. In particular, C_{60}^+ ion source is applicable [2]. Other techniques for removal of the surface comprise lasers or electron beams.

Of course, SIMS, like most analytical technique, has its own drawback. One of them is "hard" ionization, which causes fragmentation of molecules and therefore obscures detection of these with higher masses. In addition, commercially available devices do not have the possibility to perform MS/MS analysis of the obtained secondary ions. Thus, the identification of molecules based on the fragmentation spectrum is very difficult or impossible due to a mixture of various fragments belonging to numerous species. Nonetheless, with the aid of this technique, we can observe the distribution of individual ions, small molecules, and lipids on the analyzed surface [3].

6.3 MALDI-IMS

For the MALDI analysis, it is necessary to cover the sample surface with a matrix (Section 4.1.5.2), which is composed of small organic molecules that are able to efficiently absorb the energy of the laser. The laser pulse causes co-desorption of the matrix with the molecules present on the analyzed surface, which then undergo ionization. Matrix selection depends on the type of the analyzed compounds. Also, matrix deposition is critical. Its improper application (e.g. uneven distribution of the layer) may cause errors and hinder analysis. The resolution depends on the diameter of the laser beam used and is usually within 10–50 μm.

Ions generated in MALDI ion source are usually singly charged – proton adducts are formed. This ion source is currently the only imaging method that allows for protein analysis. For that purpose, two approaches are utilized: one is focused on the whole proteins, and the other is focused on peptides resulting from protein digestion on the analyzed surface.

In the first case, specific matrices that allow for analysis of proteins above 70 kDa are used (e.g. ferulic acid) [4]. In the case of protein digest analysis, it is necessary to remove lipids, salts, and all other impurities that could interfere with the signal from peptides (lipid masses usually overlap with the masses of peptides). This is usually obtained by immersing the analyzed tissue section in an appropriate concentration of ethanol prior to digestion of proteins. Array printers are used to place the small trypsin droplets (approx. 160 μl) on the analyzed surface. The proteins dissolve in such solution and undergo digestion. The resolution of the analysis depends on the droplet diameter (Figure 6.3). The operation is repeated several times, and the whole process is carried out at 37 °C and increased humidity. All the parameters, such as digestion time and trypsin concentration, must be optimized depending on the sample type. The peptides obtained in this manner are then analyzed in MS/MS mode, and fragmentation spectra, as well as peptide masses, are usually sufficient to identify the protein. However, the number of peptides obtained at a given point of

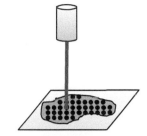
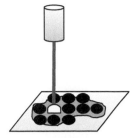

Classical MALDI analysis resolution
is limited by the laser diameter

Peptide analysis resolution
is limited by the droplet diameter

Figure 6.3 Limitation of resolution related to trypsin droplets applied. Substances on the surface are dissolved and may change their position in a droplet, which causes a loss of the resolution.

the surface may be a problem and may result in the complex spectra. Moreover, the masses of some peptides may overlap, which will make the analysis even more problematic [5].

6.4 DESI

In DESI ion source (Section 4.1.3.4), a stream of charged solvent molecules are generated analogously to the electrospray ion source. This stream is directed toward the analyzed surface. The resolution of DESI analysis is relatively low – typically around 100 μm – due to the nature of ionization. However, the significant advantage of DESI in comparison with SIMS and MALDI is practically in no need for sample preparation. Matrix application is not required, and the measurement can be carried out at atmospheric pressure.

DESI is a very good tool for lipid analysis. This method can also be used to analyze small molecules, e.g. drugs [6]. Except for a few experiments in which specific proteins have been observed (MS2 capsid from bacteriophage, blood albumin, and α- and β-globins), proteins have not been analyzed in biological samples by the use of DESI. This is because DESI does not ionize large particles as it utilizes low ionization energy. Moreover, during sample drying, the concentration of salts and impurities increases, which hinder further analysis.

6.5 Analysis of Tissue Sections Using MSI Techniques

Tissue sections are the most frequently analyzed samples using MSI techniques. During preparation of such samples, it is important to avoid thermal or enzymatic degradation of molecules in the sample. Additionally, the location of

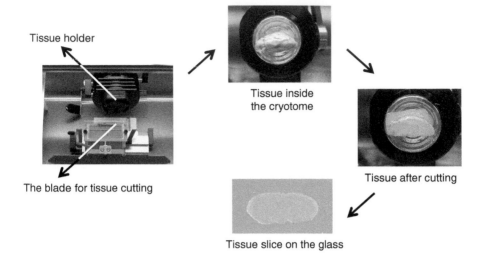

Tissue holder

The blade for tissue cutting

Tissue inside
the cryotome

Tissue after cutting

Tissue slice on the glass

Figure 6.4 Tissue cutting with a cryotome.

analyzed particles should remain intact. The simplest way to achieve this is an immediate freezing of the collected sample. Usually, samples are stored in liquid nitrogen, which has a temperature of −195 °C. That allows for retaining the intact structure of the tissue because during rapid freezing, the water does not form large crystals, which are responsible for the destruction of the cell membrane.

Prior to analysis, the biological material is cut into the slices of appropriate thickness with the use of cryotome. Cryotome (Figure 6.4) is a device used for cutting frozen tissue into very thin sections, e.g. 10 nm thick. To prevent thawing of the material, the cryotome is equipped with a chamber, where the negative temperature is maintained (within −15 to −25 °C, depending on the tissue). The resulting tissue slice is placed on a microscope slide (DESI) or on the indium tin oxide (ITO) glass covered with a thin layer of ITO to ensure electrical conductivity (SIMS, MALDI). Such samples can be stored at −80 °C prior to analysis.

Excellent research material from autopsies is available in hospitals. It is very valuable since it represents pathologically altered human tissue in many diseases. The problem is that in hospital practice, the collected tissue samples are usually fixed in formalin and hardened in paraffin before cutting. Such tissue slices cannot be directly measured by MSI techniques because paraffin suppresses ionization. In this case, appropriate procedures have been developed to enable analysis of such valuable material. At first, the paraffin is removed from the tissue by use of xylene or toluene. The tissue is then hydrated with ethanol solutions of increasing water concentration. Subsequently, the material is

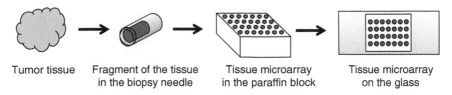

Figure 6.5 The formation of a tissue microarray matrix.

heated up to 95 °C in a Tris-HCl buffer at pH = 9. The aim of this procedure is to denature the crosslinked proteins that results in their increased susceptibility to proteolysis (see MALDI protein analysis; Section 6.3). All these treatments enable MSI analysis of tissue sections fixed in formaldehyde and embedded in paraffin.

Another type of samples used in the MSI analysis is tissue microarrays (TMAs). They are prepared from the needle biopsy material. In such procedure, the tissue is collected by puncturing the tumor with a needle. The obtained sample is in the form of a cylinder of 0.6–1.0 mm diameter and 4–6 mm height. During one biopsy, 4–8 samples can be taken. Then, the samples are fixed in formalin and spatially arranged in paraffin blocks (Figure 6.5), which are cut by the microtome. Following this preparation step, the slices are analyzed by MSI methods (see below) in order to indicate which biopsies contain tumor tissue [7].

6.6 Analysis of Individual Cells and Cell Cultures Using MSI Techniques

For the analysis of individual cells, SIMS and MALDI techniques are applicable due to their resolution. DESI can be used in this type of analysis for profiling only. It means that it is impossible to visualize a single cell with this technique; however, the chemical changes taking place simultaneously in many cells can be followed (Figure 6.6).

Prior to MSI analysis, cells should be appropriately prepared [8]. In particular, this concerns the analysis of material derived from cell culture. In this case, it is necessary to remove the culture medium, which is a mixture of salts, glucose, vitamins, and other substances necessary to keep the cells alive (e.g. proteins and pH markers). In order to remove residual medium, the cells should be washed with an isotonic ammonium acetate solution, which then quickly evaporates thanks to its volatility. In addition, evaporation can be accelerated by using a gentle stream of inert gas. The sample prepared in this way can be directly analyzed using MSI techniques or stored in liquid nitrogen until analysis.

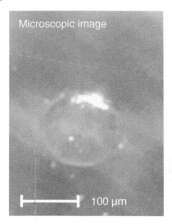

Figure 6.6 Microscopic image of the oocyte and the ion map for phosphocholine NaPC 36:2. MALDI analysis.

Also, cells can undergo the freeze-drying procedure before measurement. It involves lowering the temperature of the sample by liquid nitrogen and then lowering the pressure to cause sublimation of water from the cell. Then, the sample is gradually heated for several hours to reach room temperature. The technique of freeze-drying is most often used in the MALDI and SIMS methods, where measurements take place under high vacuum conditions. The water accumulated in the cells could interfere with the vacuum, and increased evaporation inside the device would lead to deformation of the sample.

There are also other procedures for preparing cells from *in vitro* culture for MSI analysis, and their selection depends on the type of analyzed cells and the expected results, e.g. cell fixation with formaldehyde or covering the sample with a gold layer.

6.7 Analysis with MSI Techniques: Examples

MSI techniques are very often used in cancer research because of the wide spectrum of molecules that can be analyzed using this technique. Transformation of a physiological cell into a pathological cancer is a complex phenomenon that involves changes in many pathways, affecting molecules such as proteins, lipids, and metabolites. The MSI methods can visualize a broad spectrum of species during single measurement. Besides, they are fast and do not require tagging. Comparison of spectra obtained from pathological and healthy tissues using statistical analysis (e.g. principal component analysis [PCA]) allows for indicating changes related to tumor development and may reveal mechanisms of tumorigenesis. In turn, by the use of artificial intelligence methods, the analysis of many samples allows for teaching the recognition system to distinguish between the images of healthy and pathological tissues. These techniques can be applied to tissue slices, as well as to the tissue matrices described above. The method is gradually developed and carries enormous potential to initiate a breakthrough in histopathological diagnostics [9, 10].

MSI techniques allow for indicating anatomical structures that cannot be determined using classical histochemical methods. Identification of these structures is based on the registered distribution of the profiles for various molecules. That was the case of the claustrum – a longitudinal structure made of gray matter, located in the ventral part of hemispheres, between the *nucleus lentiformis* and the insular cortex [11]. It is very difficult to indicate its boundaries using traditional histological methods due to its shape and proximity to other structures. The MSI analyses indicated that the ion of m/z 7725 occurs only within the boundaries of claustrum. In order to identify this molecule, a relevant part of the brain was cut and homogenized. The material was then fractionated using the reversed-phase (RP) HPLC method, and the appropriate fraction was separated by gel electrophoresis. Thanks to this procedure, it appeared that the signal at m/z 7725 corresponds to the G-protein gamma 2 subunit (Gng2). Determination of the boundaries of this structure provides a way to conduct further more detailed research on its functions.

Other discipline, which successfully utilizes MSI approach, is pharmacology. MSI allows for tracking the distribution of both the drug and its metabolites using their m/z values in various tissues, without labeling. For visualization of a drug with classical techniques like positron emission tomography (PET) or autoradiography, radiolabeling is required. This means that it is impossible to distinguish between a metabolite and a non-transformed drug, as both molecules will be characterized by the same signal. In contrast, MSI identification is based on its molecular weight and structure; thus a drug may be distinguished from its metabolites if only both molecules are ionizable. No label is necessary here. For pharmacological analysis, the sections of whole animals are prepared in such a manner that all major organs are exposed. This indicates that MSI techniques enable evaluation of drug penetration and accumulation in various tissues [12].

The disadvantage of this concerns a long time of analysis (it is necessary to scan a large area of tissue – e.g. a mice cross-sectional slice) and limited sensitivity related to the nature of the tissue and analyzed substance. The quantitative analysis may be hindered by the ions' suppression, which depends on the type of tissue and the presence of other molecules and salts. However, the internal standards may be introduced to overcome this issue.

6.8 Combinations of Different Imaging Techniques

The tissue analysis by MSI methods allows for indicating where the specific molecules (proteins, lipids, or small molecules such as drugs) are localized in the examined material. Such a map of the distribution of various molecules in the sample can be complemented with other techniques. That provides the opportunity to broaden our knowledge of the investigated material.

Histochemical staining with hematoxylin and eosin (H&E) allows for obtaining structural information about the analyzed tissue. Its combination with the data on molecular composition from IMS enables for the exact location of the substance in the tissue. This technique is very often used in the study of can-cerous tissues. The experienced pathologist is able to distinguish healthy tissue from the pathological one based on H&E staining. In contrast, MSI methods complete analysis with the information about molecules that differentiate the pathological state. Additionally, thanks to advanced statistical and IT methods, the computer can be trained in such a distinction, which may allow us to obtain even more accurate and fast diagnosis.

Immunofluorescence is a histological technique that indicates the location of protein expression in the tissue with a fluorescently labeled antibody. If particular protein is suspected to play a significant role in the pathological process, the image of its expression can be combined with the image of the distributions of various molecules obtained during tissue analysis with MSI techniques [13].

Another method used to detect specific protein expression is combining it with green fluorescent protein (GFP) that emits light. Expression of a particular protein may be the result of the occurrence of specific processes in the cell. Simultaneous application of GFP imaging technique and MSI methods provides more specific information about investigated process. For example, the single protein may be expressed under the influence of hypoxia. Labeling it with GFP allows for precise localization of the area where tissue hypoxia took place. The MSI analysis may, in this case, determine what changes at the molecular level are present in these specific regions under the influence of the process.

Other technique that can be combined with MSI analyses is magnetic resonance imaging (MRI), which provides additional morphological and spatial information about the investigated region [14]. The distribution of various compounds in the tissue may also be confronted with information stored in databases, such as Allen Brain Atlas [15]. Thanks to that, the gene expression in individual brain structures can be linked to the molecular analysis from MSI technique.

Each of the imaging techniques is characterized by slightly distinct parameters and allows for the analysis of different types of molecules; thus, it is of great importance to combine information obtained during the measurements with different methods (Figure 6.7). Besides localization of a given substance in two dimensions, we can also get a three-dimensional image of its distribution. The preparation of a slice series of same organ and the appropriate assembly of the resulting data allow for obtaining a spatial picture of the substance distribution in the tissue [16].

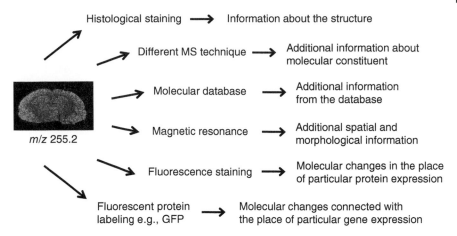

Figure 6.7 Combining different techniques with MSI approach.

6.9 Summary

Over the last two decades, the dynamic development of MS surface imaging has been observed not only in basic research but also in diagnostic and medical studies. There are many aspects of these applications that need constant improvements, such as resolution or the ability to analyze compounds of higher masses. The simplicity of material preparation and a small amount of sample sufficient to obtain reliable results definitely support utilization of these techniques. It seems that in the near future, a thriving development of these methods will be observed.

Questions

- What are the main advantages of imaging MS over other imaging techniques?
- Describe the various different ionization techniques used in MS imaging.
- What are the two different options to analyze proteins by MALDI imaging MS?
- How is it possible to remove lipids, salts, and impurities from the proteins on the surface to be analyzed by MALDI imaging MS?
- What is the main factor responsible for the limitation in resolution in MALDI imaging analysis carried out by protein digestion?
- What kind of molecules is mostly detected in the DESI imaging analysis?
- Why is it advised to freeze the sample tissue before the imaging analysis?
- What is the cryotome and what is its use?

- How is it possible to remove paraffin from biological tissue prior to the MS imaging analysis?
- How is it possible to analyze individual cells by MS imaging?
- Describe examples of application of MSI techniques on real samples.
- What are the advantages and disadvantages of applying MS imaging in pharmacology?
- In what cases it can be particularly advantageous to combine MS imaging with other analytical techniques?

References

1 Winograd, N. (2005). The magic of cluster SIMS. *Anal Chem* 77 (7): 1–7.
2 Piwowar, A.M., Keskin, S., Delgado, M.O. et al. (2013). C60-ToF SIMS imaging of frozen hydrated HeLa cells. *Surf Interface Anal* 45 (1): 302–304.
3 Passarelli, M.K. and Winograd, N. (2011). Lipid imaging with time-of-flight secondary ion mass spectrometry (ToF-SIMS). *Biochim Biophys Acta Mol Cell Biol Lipids* 1811 (11): 976–990.
4 Mainini, V., Bovo, G., Chinello, C. et al. (2013). Detection of high molecular weight proteins by MALDI imaging mass spectrometry. *Mol BioSyst* 9 (6): 1101–1107.
5 Cillero-Pastor, B. and Heeren, R.M.A. (2014). Matrix-assisted laser desorption ionization mass spectrometry imaging for peptide and protein analyses: a critical review of on-tissue digestion. *J Proteome Res* 13 (2): 325–335.
6 Vismeh, R., Waldon, D.J., Teffera, Y., and Zhao, Z. (2012). Localization and quantification of drugs in animal tissues by use of desorption electrospray ionization mass spectrometry imaging. *Anal Chem* 84 (12): 5439–5445.
7 Kallioniemi, O.-P., Wagner, U., Kononen, J., and Sauter, G. (2001). Tissue microarray technology for high-throughput molecular profiling of cancer. *Hum Mol Genet* 10 (7): 657–662.
8 Berman, E.S.F., Fortson, S.L., Checchi, K.D. et al. (2008). Preparation of single cells for imaging/profiling mass spectrometry. *J Am Soc Mass Spectrom* 19 (8): 1230–1236.
9 Eberlin, L.S., Norton, I., Orringer, D. et al. (2013). Ambient mass spectrometry for the intraoperative molecular diagnosis of human brain tumors. *Proc Natl Acad Sci U S A* 110 (5): 1611–1616.
10 Heeren, R.M.A. and Walch, A. (2014). MS imaging targets the clinic. *Anal Scientist* 12: 25–34.
11 Mathur, B.N. (2014). The claustrum in review. *Front Syst Neurosci* 48 (8): 1–11.
12 Prideaux, B. and Stoeckli, M. (2012). Mass spectrometry imaging for drug distribution studies. *J Proteome* 75 (16): 4999–5013.

13 Chughtai, K., Jiang, L., Greenwood, T.R. et al. (2012). Fiducial markers for combined 3-dimensional mass spectrometric and optical tissue imaging. *Anal Chem* 84 (4): 1817–1823.

14 Oetjen, J., Aichler, M., Trede, D. et al. (2013). MRI-compatible pipeline for three-dimensional MALDI imaging mass spectrometry using PAXgene fixation. *J Proteome* 90 (2): 52–60.

15 Abdelmoula, W.M., Carreira, R.J., Shyti, R. et al. (2014). Automatic registration of mass spectrometry imaging data sets to the Allen Brain Atlas. *Anal Chem* 86 (8): 3947–3954.

16 Eberlin, L.S., Ifa, D.R., Wu, C., and Cooks, R.G. (2009). Three-dimensional visualization of mouse brain by lipid analysis using ambient ionization mass spectrometry. *Angew Chem Int Ed Eng* 49 (5): 873–876.

7

Tandem Mass Spectrometry

Piotr Suder

Department of Biochemistry and Neurobiology, Faculty of Materials Science and Ceramics, AGH University of Science and Technology, Kraków, Poland

7.1   Introduction

One of the most important and effective analytical techniques among mass spectrometric methodologies is tandem mass spectrometry (MS/MS). During MS/MS procedure, the ions are fragmented (some of their covalent bonds are broken). Each fragment has its own mass-to-charge ratio (m/z) related to the m/z of the parent ion. By analyzing information received from the fragment ion mass spectrum, it is possible to find the structural information about fragmented substances. Exemplarily, it is possible to find an information about chemical substituents building a molecule and their connections and finally to reveal the structure of a molecule. This technique is commonly used for confirmation and identification of the chemical structure in scientific and industrial applications, forensic research, environmental protection, and doping control. MS/MS is also applied in the area of life sciences for sequencing in all "omics" strategies.

Tandem mass spectrometry in the literature is usually denoted as MS/MS, MS^2, or MS^n where "n" means a number of the following fragmentation steps (fragmentation levels). The technique is called multiple or sequential MS/MS [1].

7.2   Principles

Fragmentation procedure is composed of three main steps. All of them are visualized in Figure 7.1.

Mass Spectrometry: An Applied Approach, Second Edition. Edited by Marek Smoluch, Giuseppe Grasso, Piotr Suder, and Jerzy Silberring.
© 2019 John Wiley & Sons, Inc. Published 2019 by John Wiley & Sons, Inc.

Figure 7.1 Scheme of the tandem mass spectrometry (MS/MS) procedure.

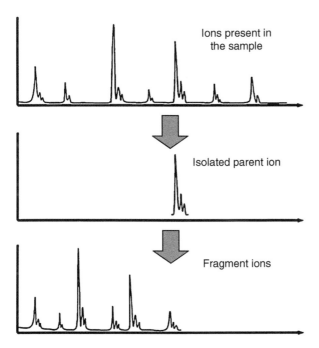

Figure 7.2 Ions detected by mass spectrometer in the consecutive stages of MS/MS experiment. Ion isolation can be considered as "purification" of the single molecule from the mixture. Fragments (daughter ions) derive exclusively from this specific parent ion, thus making possible its, at least partial, identification.

The first step can be understood as "purification" of the selected ion from other ions, visible on the mass spectrum. Before the next step, it is necessary to remove all other ions from the spectrum before fragmentation may take place. This assures that fragments generated in the collision cell derive exclusively from this particular ion, which makes a specific fingerprint of the molecule.

The second step is realized by adding the external energy to the isolated, parent ion, which is necessary for dissociation of covalent bonds in the ionized molecule. It may be done using collisions with inert gas molecules, laser irradiation, electron capture, etc. (see Section 7.4). The procedure, including steps 1 and 2, is shown in Figure 7.2.

Data from the MS/MS experiment include information on the parent ion and all daughter ions derived from it. Such information can be used to reveal amino acid sequence of a given protein/peptide or chemical structure of small organic compounds and their modifications or attached groups.

7.3 Strategies for MS/MS Experiments

There are two options that allow for separation of the parent ion from the remaining ions on mass spectrometry (MS) spectrum:

a) Separation in space (tandem in space)
b) Separation in time (tandem in time)

The first one (a) uses two (or more) analyzers combined in a single mass spectrometer (e.g. triple quadrupole time-of-flight [Q-TOF] system), while the second one (b) utilizes capabilities of other analyzers that can "store" ion of interest, removing all others, and then do its fragmentation [2]. Typical analyzers for such experiments are ion traps (IT) (linear or quadrupole IT, ion cyclotron resonance [ICR] cells).

7.3.1 Tandem in Space

The easiest and most intuitive way to perform MS/MS experiment is a combination of two analyzers in a single mass spectrometer. There is a collision chamber placed in between them, where fragmentation of a selected ion occurs (Figure 7.3). The first analyzer is responsible for isolation of the ion of interest from other ions observed on the spectrum. Also, background ions will be

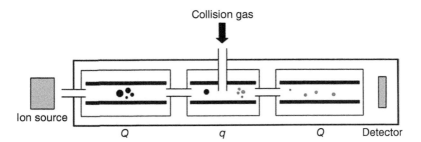

Figure 7.3 Scheme of the triple quadrupole (QqQ) mass spectrometer. In such construction, it is possible to perform "tandem in space" experiment. First analyzer (Q) separates ion of interest, which is guided to the collision chamber (q). In q, in the presence of collision gas, the ion decays into fragments, which are acquired in the second quadrupole (Q). Its main role is to scan fragment ions to generate MS/MS spectrum.

removed. Then, the selected ion is guided to the collision chamber, where it undergoes fragmentation due to the collisions with the inert gas (usually He or Ar). Daughter ions, being a characteristic pattern of a given molecule, are explored in the second analyzer.

Because fragmentation spectrum shows only ions belonging to the parent ion, the resulting information carries a unique information about structure of the analyzed substance.

An example of the mass spectrometer able to execute fragmentation separated in space is a triple quadrupole (QqQ) (triple quad). It consists of two quadrupole analyzers (Q_1, Q_2) separated by the collision chamber (q) (Figure 7.3). Ions, formed in the ion source, are transferred to the first quadrupole (Q_1). Then, the ion, which will be fragmented, is selected. The selection process can be done manually or automatically after programming of an algorithm consisting of basic criteria for selection (scan range, ion intensity, order of isolation, data-dependent isolation, etc.). The latter is particularly helpful during liquid chromatography (LC)-MS/MS experiments. Only the selected ion can pass first quadrupole and reach collision chamber (q). In this area, an inert gas, usually helium or argon, is present (collision gas). During the flight through the q, isolated ions collide with the gas and undergo spontaneous decay. Dependent on kinetic energy of parent ions and a pressure of collision gas, fragmentation occurs with frequency dependent on the internal energy of parent ion. Operator can increase kinetic energy of the parent ions in the first quadrupole, which intensifies collisions and in effect raises quantity of fragment ions leaving collision chamber. Finally, fragment ions travel to Q_3, where their separation and analysis occur. Alternatively, the pressure of collision gas can be adjusted appropriately. Elimination of the background noise during MS/MS experiment also improves sensitivity, i.e. signal-to-noise ratio is significantly better, which is frequently utilized during analyses [2].

The mass spectrometer described above can also be used for a typical MS analysis (full scan). In such case, collision chamber is switched off, and the first quadrupole is turned into the maximal transmittance mode. Then, Q_3 is the only working analyzer.

7.3.2 Tandem in Time

Second strategy of ion isolation/fragmentation process utilizes the analyzer capable of keeping and sequentially releasing ions at a particular strictly controlled moment. The analyzer fulfilling such requirements is an IT. Detailed description of this analyzer can be found in Section 4.2.4 (see also Figure 7.4). Here, only advantages of this device in the fragmentation process will be discussed.

The first stage of MS/MS experiment based on an IT is filling the trap with the ions delivered by the ion source. Then, by manipulating potentials on

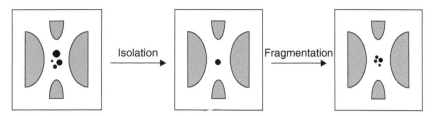

Figure 7.4 Tandem in time. The first stage is removal of all ions from the trap interior, except parent ion (isolation step). Then, parent ion receives additional energy and collides with the collision gas molecules in the same space that causes fragmentation. Transfer of fragment ions to the detector provides MS/MS spectrum.

trap electrodes, it is possible to remove all ions except those with predefined m/z value. This process corresponds to ion selection using Q_1 in QqQ mass spectrometer. The main difference between quadrupole and IT is that ions isolated for fragmentation are kept inside the IT until fragmentation, while quadrupole cannot accumulate ions – they must be constantly transferred to the collision chamber. This setup provides higher sensitivity of an IT as compared with the quadrupole-based fragmentation system. IT works also as a collision chamber, so the entire process involving isolation, fragmentation, and transfer to detector occurs in the same internal space of this analyzer.

One of the features of the IT is a relatively high pressure inside this analyzer during normal working cycle (c. 1×10^{-3} mbar of helium). Low-energy collisions between ions and helium molecules provide ion "cooling," which prevents ions from their escape from the IT. This allows to keep the ions inside the trap for a long time, reaching hundreds of milliseconds. Theoretically, trapped ions could stay on stable orbits for undefined time, but in practice, too long time of ion trapping is not desirable.

As the inert gas molecules are in the trap all the time, it is enough to start fragmentation by elevating internal energy of the ions. Using potential changes on the trapping electrodes, the ions are accelerated. This results in an increase of collision events, and, as ions are traveling faster (having higher kinetic energies), collisions generate sufficient energy to break some of the covalent bonds in the colliding molecule.

It is worth to remember that IT can also work as an analyzer, thus separating and collecting ions for the following fragmentation (operates like Q_1 in triple quads but increases ion quantity). For example, connecting trap with the time-of-flight (TOF) or Orbitrap analyzers leads to higher resolution of fragmentation spectra. In this approach, IT is responsible for high sensitivity (parent ion collection and efficient fragmentation), and the second analyzer for high resolution (m/z analysis only of daughter ions with high resolution).

7.3.3 Multiple Fragmentation

Inside IT, after isolation and fragmentation steps, there are still many fragment ions. At this moment, it is possible to repeat the entire process by selecting one of the fragment ions for the next isolation step and further fragmentation. After such selection, one of the fragment ions becomes parent ion for the next generation of fragments. Removal of all other ions from the trap allows for fragmentation of the previously selected one. Following this way, it is possible to receive MS/MS/MS spectrum (also called MS^3 spectrum). The process of consecutive isolation/fragmentation can be repeated almost infinitely, but due to various limitations and for practical reasons (decreasing sensitivity and elongated time), selection of parent ions in modern IT can be done up to 11 generations. In practice, it is not necessary to go deeper than to 3–4 generations, as fragmented molecule will decompose to very small ions. Additionally, with increasing fragmentation steps, the analyzer requires high initial concentration of the sample. Considering at least a few percent of an analyte lost during every step (as fragmentation is not 100% effective), after 4–5 generations of MS/MS, the initial concentration of the analyte should reach nanomolar concentration. Of course, it is possible to identify reaction products in organic chemistry, but in biological samples such quantities of analytes are rare and often impossible to obtain. Also, time of such multistep analysis is too long for LC-MS experiments. Overloading of the IT may lead to the space charge effect and loss of resolution.

Summing up, IT allows for complex, multistage ion trapping and multiple fragmentation (MS^n) where n reaches 10 or more (theoretically!). In practice, the level of n, which is still useful for molecular structure identification, is up to 5. Anyway, IT in the context of fragmentation analysis seem to be one of the best and cheapest analyzers available up to date. On the other hand, even IT have some limitations, which will be discussed in Section 7.4.

7.4 Fragmentation Techniques

7.4.1 Introduction

For the fragmentation process, the ion needs to gain additional energy, allowing for dissociation of some (or all) covalent bonds within the molecule. There are many ways how to increase the internal energy level of a molecule. Some of them, used currently, are listed below:

1) (Low-energy) Collision-induced dissociation (CID) or collisionally activated dissociation (CAD)
2) High-energy collisional dissociation (HCD)
3) Pulsed Q collision-induced dissociation (PQD)
4) Electron capture dissociation (ECD)

5) Electron transfer dissociation (ETD)
6) Electron detachment dissociation (EDD)
7) Negative electron transfer dissociation (NETD)
8) Infrared multiphoton dissociation (IRMPD)
9) Blackbody infrared radiative dissociation (BIRD)
10) Post-source decay (PSD)/laser-induced dissociation (LID)
11) Surface-induced dissociation (SID)
12) Charge remote fragmentation
13) Chemically activated fragmentation (CAF)
14) Correlated with ETD: proton transfer reaction (PTR)

It should be remembered here that at the current level of MS development, not every fragmentation technique listed above can be applied in every possible instrumental setup. A common strategy is to adjust particular fragmentation technique for particular mass construction of the spectrometer (like LIFT™ for TOF/TOF systems).

7.4.2 (Low-Energy) Collision-Induced Dissociation (CID)

Energy necessary for ion dissociation is delivered as a result of collisions of the analyte molecules with neutral gas present in the collision chamber. Collisional energy is usually below 100 eV, which is often insufficient for ion decay after a single collision. An ion designated for fragmentation must collide at least few times to accumulate adequate internal energy for covalent bond dissociation. Collisional energy and collision frequency are controlled by two parameters: pressure of the inert (collision) gas in the fragmentation chamber and the speed/ internal energy of ions. Fragment ions regain stable state as excess of energy was dissipated during fragmentation process. Such low-energy collisions are useful during peptide fragmentation, where peptide bond is usually the weakest one. Their selective breakage leads to sequencing of the peptides and, as a result, to their unequivocal identification. Fragment mass spectra of peptides carry vital information, which is easy to interpret. On the other hand, such mild fragmentation (delivering small quantities of energy to parent ions over a long time) can be disadvantageous due to inefficient dissociation. An ion possessing covalent bond at clearly lower dissociation energy will be fragmented only at this site. Thus, fragmentation spectrum will provide no useful data, showing only less important information (e.g. dissociation of NH_2^{1-} or amino radical: $NH_2^{\bullet}$ from peptide ion). In such cases MS^n fragmentation, where $n \geq 3$, might be valuable.

7.4.3 High-Energy Collisional Dissociation (HCD)

This type of fragmentation is mainly used in mass spectrometers equipped with the Orbitrap analyzer. In the Orbitrap-based MS, there is also a so-called

C-trap (modified linear IT) installed or a combination of C-trap and octapole. Fragmentation in such systems is usually performed in C-traps. Theoretical principles of HCD fragmentation are very similar to CID: ions in the linear IT are accelerated, which causes fast ions' movements back and forth along the trap. Inert gas is also introduced to the trap, so ions traveling inside are exposed to frequent collisions. A difference between CID and HCD lies in the amount of energy delivered to the ions. For CID, the energy is usually not higher than a few dozens of eV, while for HCD this value can reach keV levels. Thus, kinetic energies for fragmentation are much higher than in CID process. Under such conditions, even a single collision is enough for effective dissociation. Additionally, this reaction is more stochastic as dissociation energies of covalent bonds have minor influence on the order of bond fragmentation. Stochastic fragmentation causes an increase of fragment ions on MS/MS spectra, which expands structural information possible to read from HCD spectra. On the other hand, keeping fragment ions, still possessing significant excess of energy, in the IT is challenging for current constructions of IT. Up to date, the most popular solution to the problem of escaping ions is a joint action of two subsystems: C-trap (or linear IT) and octapole. After fragmentation process, daughter ions at low m/z have high kinetic energies, still sufficient to escape from the trap and cannot be retained in the instrument. Such behavior is observed with peptides labeled with the isobaric tags for relative and absolute quantitation (iTRAQ). Fragment ions after HCD fragmentations, including iTRAQ isobaric tags, are not trapped in the octapole and are overcoming the potential barrier, losing partially their energy. Highly energetic ions are intercepted by C-trap localized just after the exit from octapole. In C-trap they are additionally cooled by interactions with inert gas molecules. At a first glance, it looks odd, as in some cases inert gas contributes to the fragmentation of ions, while in other cases it stabilizes ions' movement. The difference lies in acceleration of the ions by manipulating external potentials: if the energy is constantly added to the ions to increase their kinetic energies, inert gas works as a medium allowing for fragmentation. But when internal energy of fragment ions is insufficient for further fragmentation (and they do not receive additional energy), inert gas works as a cooling agent, allowing ions for safe dissipation of kinetic energy.

During fragmentation process, the octapole consecutively releases remaining ions to C-trap to equalize internal energies of all fragment ions. After the time necessary for the equalization, all ions travel to the last analyzer, which is usually an orbitrap chamber.

7.4.4   Pulsed Q Collision-Induced Dissociation (PQD)

This fragmentation technique, in its theoretical principles, is relatively similar to CID fragmentation. The main difference is a possibility of efficient detection of fragment ions at much lower m/z ratios than the parent ion. This technique is useful during quantitative analyses of substances modified by iTRAQ iso-baric tags, similarly to HCD technique. Parent ions are isolated inside an IT or octapole, set on stable trajectories, and for a very short time (e.g. $100\,\mu s$) are exposed to a brief electric pulse at strictly controlled parameters. Such addi-tional kinetic energy leads to frequent collisions with inert gas present in the trap. This, in turn, causes fast accumulation of internal energy enforcing cova-lent bonds to break down in a very stochastic manner, leading to formation of very diverse population of fragment ions.

7.4.5 Electron Capture Dissociation (ECD)

This fragmentation method is entirely distinct from those listed above. Instead of provoking ion dissociation through addition of energy excess, ions are exposed to the low-energy electron beam (10–20 eV). Collision chamber is built similarly to the electron impact ion source. Electrons are captured by the positively charged ions, which increases total ion energy to values equal to those carried by electron. Such fast and substantial increase of internal ion energy leads to immediate fragmentation. It is worth noting that the ion cap-turing electron must be charged higher than singly, as ion ionized by charge 1+ only, after addition of electron, loses its charge and cannot be detected by the mass spectrometer. For this reason, fragmentation by ECD is usually combined with electrospray ionization (ESI), as using this ion source, molecules, espe-cially at higher molecular weight (MW), are preferentially ionized by the charge higher than single.

One of the biggest disadvantages of ECD is very long time necessary for effective recombination between electron and ion (reaching even seconds). Due to this property, fragmentation by ECD could be only done in analyzers able to keep ions for a time long enough for recombination. Additionally, slow, low-energy electrons can effectively react with the appropriate ions only in a very high vacuum, which eliminates IT from the array of analyzers used for ECD reaction. That is why the method is used only in the systems equipped with the ICR analyzer (for description see Section 4.2.6).

7.4.6 Electron Transfer Dissociation (ETD)

Similarly to ECD, this fragmentation method uses electron transfer to the parent ion, but in this case electron is transferred by chemical reaction, not like in the case of ECD, which utilizes electron bombardment.

ETD fragmentation is a multistep process, shown in Figure 7.5. In the first stage it is necessary to generate anion radicals, which can easily transfer elec-tron to the fragmented ion. The most commonly used substance providing such radical during ionization is fluoranthene (Figure 7.5). Additionally, a special module containing fluoranthene must be installed in the mass spectrometer.

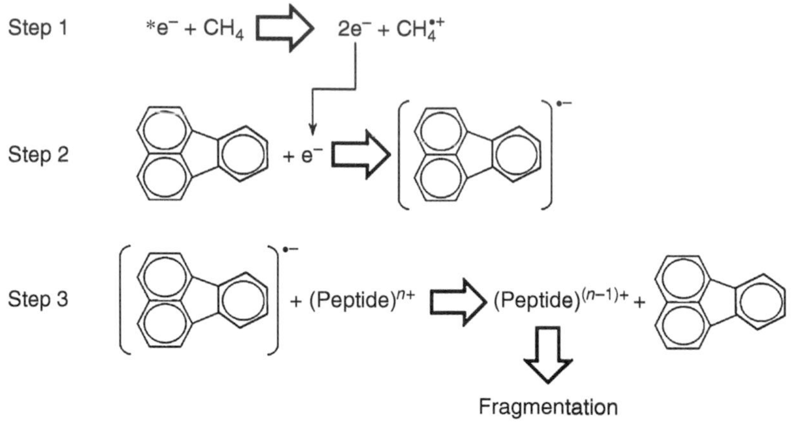

Step 1 $*e^- + CH_4 \Rightarrow 2e^- + CH_4^{\bullet+}$

Step 2

Step 3

Fragmentation

Figure 7.5 Electron transfer dissociation (ETD) fragmentation scheme. Step 1: Interaction between electron (60–80 eV, *e⁻) generated by EI ion source and methane molecule, leading to the formation of two low-energy electrons. Step 2: Reaction between fluoranthene and low-energy electron formed in step 1: fluoranthene anion radical is formed. Step 3: Transfer of the electron from fluoranthene anion radical to a molecule undergoing fragmentation. Such electron transfer lowers charge of the ion by a value of 1.

The first stage of fragmentation by ETD is introduction of small amounts of methane (or similar gas) to the electron ionization (EI)-type ion source being a part of the mass spectrometer. This ion source is only used to produce anion radicals of fluoranthene. Electrons at energies of c. 60–80 eV produced by the EI filament remove single electron from the methane molecule. As a result of such reaction, two low-energy electrons are formed. Such low energetic electrons can be safely intercepted by fluoranthene molecule without its dissociation. Finally, anion radical of fluoranthene can interact with the previously isolated parent ions, transferring excessive electron to the ion structure. This reaction, giving additional energy to the parent ion, causes immediate fragmentation of the covalent bond. Fragmentation is strongly stochastic and only weakly linked to the energy of the bonds. ETD allows for formation of a great variety of fragment ions, giving a great deal of information from generated MS/MS spectra.

ETD fragmentation has also some disadvantages. One of the most important is the inability to fragment positively singly charged ions (1^+). Similarly to ECD, in this type of fragmentation, interception of the single electron by 1^+ ion causes loss of charge and failure to detect neutral molecule. Only ions possessing n charges, where $n \geq 2$, are suitable for ETD fragmentation. Another problem is an increase in the fragment ions mass caused by the introduction of electron into the ion structure. The observed mass shift is small (equal to electron mass) but significant, especially during high-resolution analysis. This problem could be easily omitted by modification of the software, which should be taken into account during MS/MS spectra analysis.

7.4.7 Electron Detachment Dissociation (EDD)

Theoretical bases of this technique are similar to the fragmentation using electron capture. However, this strategy was designed for fragmentation of the negatively charged ions; therefore usage of electrons is limited. The main idea was to capture positron (e^+) by the multiply charged parent ion, which would cause an increase of its internal energy, followed by fragmentation as effect of this reaction. Unfortunately, the efficient source of positrons, which could be effectively used in mass spectrometers, is not available. This drawback has been solved by another system. Multiply charged anions are entrapped in the ICR analyzer and irradiated by electron beam at energy of 10–27 eV for 0.2–1.0 seconds. This process leads to the removal of electron, and, as a consequence, the local positive charge is created. Such anion radicals are very unstable and easily decompose due to movements of the "electron hole" along the ion structure.

Similarly to ECD technique, EDD is used in the ICR-based mass spectrometers as formation and delivery of low-energy electrons to the ion can be successfully realized in high vacuum only. Analyzers filled with low-pressure inert gases prevent interaction between electrons and ions.

7.4.8 Negative Electron Transfer Dissociation (NETD)

This technique allows for fragmentation of the multiply negatively charged ions, and this effect is similar to the EDD. Ions entrapped in the analyzer interact with the positively charged gas (e.g. cation radicals of argon, xenon, or krypton). After collisions, electron transfer from the ion to the charged noble gas is possible, thus neutralizing its charge. Fragmentation occurs in the same manner like in EDD technique, but the mechanism of electron transfer is less random in comparison to the removal of electron by electron beam. Fragmentation mechanism and ion generation are more predictable. Additionally, NETD can be performed in the analyzers that need small amount of gases for appropriate operation. That is why this type of fragmentation works well in quadrupoles and IT analyzers.

7.4.9 Infrared Multiphoton Dissociation (IRMPD)

In this method, additional delivery of energy to the parent ion is realized by irradiation of the sample ions by a laser beam. Thus, energy absorption does not occur as a result of collisions or charge transfer, as in the previously described methods. Ions undergoing fragmentation by IRMPD must be capable of absorbing energy from radiation at wavelengths equal to the applied laser. Laser beam passes the space, where ions circulate. ICR is the most often used analyzer here. Ions interacting with laser absorb its energy and transfer it into internal oscillation energy until the moment when its level is above the energy of the weakest covalent bonds. By reaching energy absorption limit, parent ion undergoes fragmentation.

7.4.10 Blackbody Infrared Radiative Dissociation (BIRD)

Here, the isolated parent ions receive additional energy through absorption of thermal energy (infrared radiation). According to one of the first description of this methodology, parent ions are introduced into the analyzer, and then this element is heated to c. 200 °C. It is worth to remember that pressure is very low inside the analyzer; thus energy accumulation is mainly realized by absorption of the infrared radiation. This fragmentation method is very simple; however parameter that seems to be the weakest part of the whole methodology (i.e. very long time of energy transfer to the ion) is one of the strongest points. Long time of energy acquisition by the ions (reaching even minutes) results in detailed analysis of fragmentation scheme of various chemical substances. This methodology is especially useful for fundamental investigations in organic and bioorganic chemistry.

7.4.11 Post-source Decay (PSD): Metastable Ion Dissociation

Ions leaving the ion source, depending on ion source parameters, have, in some cases, enough internal energy to spontaneously dissociate inside the following parts of mass spectrometer. This type of ions is called metastable ions. Dissociation of the metastable ions can be observed in sector instruments (e.g. doubly focusing instruments) equipped with EI/chemical ionization (CI) ion sources and in matrix-assisted laser desorption/ionization (MALDI)-TOF mass spectrometers. Metastable ions dissociate spontaneously having enough time during their flight through the analyzer. By manipulating the ions' energy gained in the ion source, it is possible to perform fragmentation process even in the spectrometers without components dedicated to MS/MS. Breaking down covalent bonds in metastable ions is called post-source decay, and in modern MS constructions, this effect is observed predominantly in MALDI-TOF systems; however the percent of ions leaving MALDI chamber and having sufficient energy for spontaneous decay is not very high. Additionally, construction of the TOF systems does not promote observation of ions fragmented using PSD mechanism, so the PSD spectra are insensitive. It is due to the observation that PSD fragments have significantly different m/z ratio from their parent ions, but their kinetic energy is almost the same; thus metastable and fragment ions will hit detector at the same time. The situation looks similar to a person with a luggage staying on the moving band. It is indifferent whether he/she will keep luggage in the hand or left it on the band: both will arrive at the end of moving band at the same time.

As PSD phenomenon is interesting from the analytical point of view, to enhance acquisition of fragment ions received after dissociation of metastable ones, it has been found out that such process is strongly supported by laser irradiation (LID). Such analysis is executed in TOF/TOF instruments. In this approach, metastable ions leaving first TOF analyzer are introduced to the collision chamber being also a second ion source. In this area, they can collide with neutral gas molecules or receive additional energy from laser irradiation, which significantly increases their internal energy, and, as a consequence, this leads to their decay. Second TOF analyzer perceives all the ions leaving "collision chamber" as fragment ions, which strongly improves intensities of the peaks on the MS/MS spectra. Fragment ions can be detected in linear or reflectron modes, which also improve resolution of the measurements. Detailed description of TOF/TOF analyzer can be found in Section 4.2.1.

7.4.12 Surface-Induced Dissociation (SID)

This technique is used for fragmentation of ions of relatively low molecular masses. Ions are accelerated to high kinetic energies (up to keV level) and collide with the appropriately prepared surface. Especially important in this technique is a need for optimization of many parameters influencing fragmentation process, including ion kinetic energy, surface type, angle between ion beam and collision surface, and others. On the other hand, it is possible to receive diverse information about the sample, depending on the parameters optimized from dissociation of non-covalent bonds to complete destruction of the molecular structure of the ion.

7.4.13 Charge Remote Fragmentation

This process results in dissociation of at least a single covalent bond and is induced by the charge localized remotely from the broken bond. This phenomenon was observed for several types of molecules: steroids, fatty acids, surfactants, porphyrins, complex lipids, etc. Ionization of these molecules gives them so much energy that decay of covalent bonds became theoretically possible. Such decay is dependent on charge localization on the molecule, but it is also important to provide oscillation energy sufficient to initiate bond dissociation. To make such fragmentation easier, there are also techniques to introduce the charge into the ionized molecule to a dedicated position with the aid of chemical reaction (see Section 4.1.3.1).

7.4.14 Chemically Activated Fragmentation (CAF)

Derivatization of molecules by various reagents able to control fragmentation is not a new idea. Adding charged ligand to a molecule may promote specific fragmentation patterns. Particularly interesting is fragmentation of peptides after reaction with CAF reagents. One of the first CAF chemicals was tris(2,4,6-trimethylphenyl)phosphonium (TMPP) derivative covalently bound to the peptide' N-terminal, providing stable positive charge to the peptide. During CID fragmentation in the mass spectrometer equipped with ESI source and IT analyzer, only b-type ions (see peptide fragmentation nomenclature in Section 7.6 and Figure 7.7) were observed, due to the stable positive charge present on the N-terminals of every labeled fragment ion. Such procedure makes MS/MS spectrum much easier for interpretation, as only one ion series is visible (signal is enhanced at least few times, which makes it visible over other peaks on the spectrum). Following the first idea, a whole set of substances derivatizing peptides was designed, which can be used for visualization of a single series of peptide fragment ions. One of the most popular agents was N-hydroxysuccinimidyl (NHS) ester of 3-sulfopropionic acid. In this case, N-terminal sulfonylation causes formation of the y-ions series mainly. Another similarly working substance was 4-sulfophenyl isothiocyanate (SPITC), widely used for peptide N-terminus derivatization. PSD in MALDI-TOF or CID in MALDI-TOF/TOF or Q-TOF systems can show clear y-ions series for such modified peptides. Currently, there are a vast number of derivatizing molecules used for peptide fragmentation – their spectra are definitely easier in interpretation.

7.4.15 Proton Transfer Reaction (PTR)

PTR does not cause fragmentation but is very useful during analysis of high MW peptides or small proteins in the low-resolution mass spectrometers equipped with IT. PTR allows for reduction of multiple charge (for positively charged ions from n to $n-1$) that, in some cases, is crucial for estimation of charge quantity per ion. Low-resolution mass spectrometers usually allow for clear separation of isotopic pattern for charges not exceeding $z = 7$ to 8 (in m/z ratio). Therefore, deconvolution of MW of small proteins becomes impossible, based on a single peak, when z value is higher than 10. In this case, isotopic peak having more complex isotopic envelope cannot be distinguished on the spectrum. As an effect, fragmentation of big peptides/small molecules generates fragment ions also having high MW. Such spectrum does not carry reasonable information and cannot be the basis for peptide/protein identification. Charge reduction, in some cases, may help in spectrum interpretation. Exemplarily, on the MS/MS spectrum, we observe all ions having charge 4+ or higher using mass spectrometer equipped with the IT. On such spectrum, charges 4+ and 5+ will be still estimated correctly. Uncertainty in ion deconvolution will increase with charge. For real spectra, charges equal to 9+ or higher are impossible to deconvolute due to insufficient resolution of a given instrument. If the quantity of ions with a charge higher than critical for successful calculation of the MW would be too high, MS/MS spectrum will be impossible to interpret. In such a case, PTR analysis could be an interesting alternative. PTR technique increases percent of ions for correct charge calculation. Sometimes, such simple procedure is sufficient for effective analysis and identification of whole proteins at MW = 20–25 kDa using fragmentation technique without any previous sample preparation (e.g. enzymatic degradation of protein).

PTR occurs as an effect of interaction in the gas phase of an ion having a multiple positive charge, with carbanions of the appropriate reagent (the most common is fluoranthene). During reaction, unpaired electron from carbanion is transferred to the ion that causes reduction of its total charge. Unfortunately, transfer of the electron to the singly positively charged ion results in a loss of charge and inability to detect it.

7.5 Practical Aspects of Fragmentation in Mass Spectrometers

Below, the most important parameters of mass spectrometers usually used in routine fragmentation of the ions are discussed.

7.5.1 In-Source Fragmentation

Almost every mass spectrometer is able to create fragment ions even if is not equipped with subsystems dedicated for fragmentation process. Fragmentation may occur in ion sources as a side effect of sample ionization. Depending on the ion source settings, this effect may be almost invisible or can be enhanced by the operator. Such type of fragmentation makes sense only if the sample is homogeneous and no other ions are observed. It is possible to trigger in-source fragmentation, keeping in mind that every fragment ion generated under these conditions is a derivative of pure sample molecule. Otherwise, we may end up with a mixture of fragments derived from several components. In such case, it is impossible to distinguish between fragment ions generated from the molecule of interest from fragment ions being products of other components. An interesting novel technique allowing for the acquisition of hundreds of compounds in complex biological samples has been recently introduced using fragmentation without isolation of the selected parent ion. This method, invented in 2012, is called Sequential Windowed Acquisition of All Theoretical Fragment Ion Mass Spectra (SWATH) and was described in Section 6.6.

For the sources operating under atmospheric pressure (like ESI or APCI), in-source fragmentation can be obtained by increasing voltage between parts of the ion source (e.g. between heated capillary and skimmer or funnel, depending on the ion source design). Acceleration of the ions traveling between these parts of MS provokes more frequent collisions with air components, as this segment of the ion source is under ambient conditions. Such experiments are realized by experimentally setting voltage parameters, adjusted during observation of MS/MS spectrum. Such settings are individual for each molecule; thus it is not recommended to use this technique for automatic runs.

7.5.2 Triple Quadrupole Fragmentation

QqQ discussed in the first part of this chapter is composed of two quadrupole analyzers separated by the collision chamber, as shown in Figure 7.6. This is one of the most popular arrangements in the MS. Previously, it was mentioned that setting second quadrupole (Q_3) in the focusing ion mode and turning off collision gas flow to the collision chamber (q_2), such instrument allows for a typical m/z scanning in the Q_1 analyzer. Of course, it is also possible to set Q_1 in the focusing ion mode and allow Q_3 to operate as the main analyzer.

After separation of a narrow range of m/z by Q_1 and introduction of a small quantity of collision gas to q_2, the quadrupole Q_3 generates MS/MS spectrum received after fragmentation of an ion introduced to q_2. This mode is referred

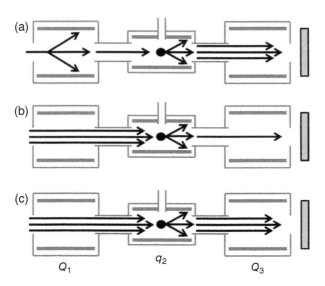

Figure 7.6 Synchronous modes of operation with two quadrupole analyzers (Q_1, Q_3) separated by the collision chamber (q_2). (a) Daughter ion scan: Q_1 releases only ions at defined m/z, and Q_3 separates all fragment ions. (b) Parent ion scan: ions from Q_1 are transferred to the collision chamber q_2, where they are fragmented, and Q_3 allows only a defined m/z ion to reach detector. This procedure allows for estimation whether the molecule of interest is present in the sample. (c) Neutral loss scan: both analyzers are working synchronously, but Q_3 transfers only those ions with previously defined m/z, i.e. neutral losses. All precursors that undergo the loss of a specified (predefined) common neutral will be monitored. These neutral losses comprise typical molecules or groups that dissociate from a molecule, e.g. fatty acids in phospholipids, H_2O, $-CH_3$, $-NH_2^-$, phosphate group, etc.

to as a daughter ion scan or product ion scan. It is easy to notice that we can simply change functions of Q_1 and Q_3: it is possible to allow Q_3 filtration of all m/z ratios except a selected a single m/z that will be guided to the detector. Then Q_1 should be set into the full scan mode. This setup will allow for the search for all parent ions derived from the identified fragment ion. Such type of experiment determines whether investigated sample contains the requested parent ion to assign it to specific molecule. The described approach is very useful during comparative analysis of substances having similar structures – it is possible to select fragment ion, which differentiates substances differing by location of a single ligand or modifying group.

Based on the synchronized activity of both quadrupole analyzers, we can also distinguish between few modes of fragmentation without full interpretation of MS/MS spectra. Techniques described below are used for fast monitoring of sample composition and are useful if the operator is mainly interested in confirmation or exclusion of a given compound in the sample. To distinguish whether the compound is present or absent in the sample, a knowledge about its fragmentation pathway is necessary. The presence of molecular/pseudomolecular ion is not sufficient to confirm the identity of a substance. Comparison between theoretical fragmentation pattern and that received experimentally is the base of such analysis. Fortunately, nowadays, access to the databases containing fragmentation data of thousands of substances is not a big challenge.

The last among the important and widely used modes of the QqQ operation is the so-called neutral loss scanning. Here, both analyzers (Q_1 and Q_3) are working synchronously, but scanned m/z ratios are differing of selected mass. Exemplarily, searching for the water loss from the investigated molecule, we may set Q_3 quadrupole to pass only the ions at a mass decreased by 18 Da compared with those delivered by Q_1. After such analysis we are able to observe only the ions, which lost 18 Da in the collision chamber or labile enough to lose this mass during transfer through Q_1. It is also possible to scan the ions characterized by a neutral mass gain, but from the practical reasons this mode of operation is of minor importance.

Due to the relatively low kinetic energies of ions present in the quadrupole-based mass spectrometers, fragmentation occurs due to the low-energy CID schema. To increase effectiveness and collision energy, ions are often accelerated inside the collision chamber. This enhances fragmentation in the described systems.

Also scanning with the neutral loss is very useful for detection of compounds containing typical chemical groups like posttranslational modifications (PTMs) of proteins and peptides. Exemplarily, scanning with 80 Da loss (MW of the phosphate group) allows for finding of phosphorylated peptides in the homogenate containing tryptic peptides. Similar strategy could also be applied for searching for metabolites containing particular functional group (like amino-, hydroxy-, etc.)

Selected reaction monitoring (SRM) or selected ion monitoring (SIM) modes are mainly used to confirm the presence of a specified substance in the sample. Identification usually occurs during LC-MS or GC-MS analyses, assuming that only the one, searched substance fulfills fragmentation criteria. To perform SRM/SIM procedure, a desired m/z ratio should be preselected, and this particular ion will be isolated in the analyzer (quadrupole Q_1) and detected. If this exact m/z value appears in the analyzer, it is guided to the collision chamber (q), where it undergoes fragmentation. After fragmentation, second analyzer (quadrupole Q_2) also passes the preselected fragment ions only (transitions from parent to daughters). This simplifies the procedure and saves much time as total ion current plot is not needed and the instrument scans only for selected m/z values. This also saves space on the hard disk. It should be clearly stated that before start of SRM/SIM analysis, it is necessary to preselect substances and their fragments we are searching for. It is also important to know fragmentation pattern(s). Thus, during the analysis, two criteria should be simultaneously filled:

a) Proper m/z of the parent ion introduced to the collision chamber.
b) At least one fragment ion of the searched substance.

Such type of analysis is strongly discriminative (it rejects a large amount of unwanted data, leaving only those of importance). It is widely accepted that fragment ion intensities observed during SRM analysis are proportional to the concentration of a compound present in the sample. Therefore, this type of analysis is often used for quantitative studies.

Another type of analysis used in QqQ systems is the multiple reaction monitoring (MRM). The idea behind the experiment is similar to SRM, but instead of monitoring single precursor ion and its single daughter ion, it is possible to simultaneously measure few or even a few dozens of the parent ions along with their corresponding fragments. This approach allows for monitoring the presence of many substances in a sample during single analytical run, and the entire procedure is still much faster than scanning over the entire m/z range. Scanning for more than a single fragment ion also allows for minimizing eventual false positives. This is a routine during forensic analyses. For example, in GC-MS systems, where registration of both parent and several fragment ions is strongly correlated with elution time from chromatographic column, such mode substantially improves analytical reliability. In SRM (like in every ion isolation technique), sensitivity of the system is higher than in the case of a broad scan range, the selected substance can be visualized even if this was not visible during routine MS analysis. Elimination of background plays crucial role in this case. Moreover, even for the coeluting compounds having similar structures, it is possible to separate them, finding predicted differences in their fragmentation patterns.

Both types of analysis are often used in the routine investigations searching for the defined substance in the vast number of complex samples (like derivatives of the drugs in the set of forensic samples or in doping control).

There are also some disadvantages of the SRM/MRM approaches. The most important is limitation of the data, which can be detected. This excludes other substances that might be present in the samples, except those predefined. A good example might be the recent improvement of anti-doping analyses using a combination of high-resolution MS, prolongation of the detection window, and broader search for metabolites, not limited to the few predefined [8]. Another problem is a minor but possible false-positive result under specific analytical conditions. During LC-MS/MS analysis, when the background suddenly rises, e.g. because of higher salt content, parent and its fragment ions will be detected as having higher concentration as they really have.

7.5.3 Ion Traps

IT is a typical example of the analyzer that can ensure fragmentation separated in time (not in space, like in the case of set of quadrupole analyzers). Although some basic information about IT was already delivered at the beginning of this chapter, this analyzer has few features that should be taken under consideration by the operator.

Fragmentation to occur requires delivery of sufficient energy to the ion to make collisions with neutral gas efficient. In IT, the energy is delivered by resonant excitation of ions by electric field at a high frequency provided by spherical and round electrodes. As every ion has its own oscillation frequency (dependent on its *m/z* ratio), it is possible to freely select particular ions for fragmentation without separation of the parent ion in space, as it was in the case of quadrupoles. Thus, in the IT, it is possible to select an ion among others present in the trap. Moreover, parent ion decay will cause that fragment ions will have distinct values of resonant frequencies and will not absorb additional energy, which is still delivered to the parent ion. During this process, it is possible to increase fragmentation energy slowly, causing gradual increase of oscillation energy of the ions. Such energy ramp allows for the fragmentation of the ion even if the exact energy for its fragmentation is unknown.

The second important factor associated with fragmentation inside the IT is the so-called low mass cutoff effect. To keep fragment ions inside the trap, it is necessary to decrease voltage on the trap circular electrode. Unfortunately, the electrode potential cannot be decreased too much, because this will be insufficient to generate fragment spectrum at reasonable quality. Additionally, when the potential applied to the circle electrode is too low, fragment ions have tendency to lose stable trajectories inside the trap and can escape from the trap space. Therefore, voltage level applied to circle electrode is a compromise between capabilities of the system to perform effective fragmentation and hold fragment ions inside. The effect of this compromise is that IT usually cannot retain fragment ions at MW lower than 30% of MW of the parent ion. Some manufacturers not only allow users to maintain the voltage of the circle electrode according to their requests, which results in lower-quality MS/MS spectra, but also offers a possibility to detect fragment ions at MW down to c. 15% of the precursor ion MW. Such option may be useful during some types of experiments (e.g. iTRAQ-based analysis; see Section 8.1.5.2).

There is also another method to avoid losing fragment ions due to the low mass cutoff effect by applying multiple fragmentation (MS^n, $n > 2$). Then, slow decrease in the observed MW of the fragment ions will be achieved after each fragmentation step. Cutoff effect is the reason why there is tendency to minimize collisional energy during fragmentation in the IT. In contrary, high-energy collisions result in generation of a vast number of daughters having low MW. This leads to a loss of significant amount of information from ions running out from the trap. So, multiple fragmentation in IT is used to gain important information from the low MW region rather than achieving other goals.

Comparing fragmentation process in IT and triple quadrupoles, it is worth to note that for quadrupoles, collisional energies of ions in CID technique are, in general, higher than in the traps. MS/MS spectra, derived from the quadrupole-based instruments, are usually richer in fragment ions than those received from traps. On the other hand, in spectrometers equipped with IT, it is possible to monitor SRM/MRM modes in more advanced combinations and also to investigate neutral losses. IT remain to be basic MS analyzers in instruments where there is no need for high-resolution spectra. High popularity of IT arises from the fact that these analyzers can accumulate ions for the long time; thus sensitivity of the system is higher than in the case of quadrupoles. Traps also have usually a slightly higher resolution achievable during scans over limited *m/z* range. However, selection between quadrupole and trap depends on specific applications. Quadrupoles are mainly used for quantitative analysis in detection of low MW compounds. IT are popular for higher MW molecules like peptides or proteins, and their multistage fragmentation delivers more information. This crucial feature (MS^n for $n > 2$) is especially useful where very mild fragmentation at the first stage does not provide sufficient information (like water or ammonia loss). Then MS^3 can be useful for successful fragmentation and confirmation of molecular structure.

7.5.4 Time-of-Flight Analyzers

Fragmentation in TOF analyzers needs the presence of a reflectron (see Section 4.2.1.6). Additionally, parent ions undergoing fragmentation must have enough internal energy to spontaneously decay during flight from the ion source to detector. The speed of the fragment ions arising in the TOF tube (as well as sum of their kinetic energies) is the same as speed of their parent ions. On the other hand, individual kinetic energies of fragment ions differ significantly. This allows for changes in their trajectories evoked by the reflectron's electrical field. The process allows for differentiation between traveled distances and a flying time of fragment ions. As *m/z* ratio is correlated with TOF, it is easy to calculate MWs of the fragment ions. Single TOF has significant disadvantage: the reflectron is able to effectively repulse fragment ions in a narrow mass range not less than 75% of the initial mass of the parent ion. To investigate lower *m/z* of fragment ions, it is necessary to repeat ionization process and change reflectron settings to repulse ions at lower *m/z*. Such segmental process must be repeated at least few times, which results in high sample consumption and a relatively long analysis time. Additionally, MS/MS spectra have low resolution in comparison with the typical reflectron-based MALDI-TOF MS spectrum.

To make a successful MS/MS analysis using TOF systems, it is necessary to precisely select parent ions from all other ions, generated in the ion source. In this case, electronic gate is used. Gate opens for a very short time to allow selected ions to travel along the TOF analyzer. Time length of gate opening is calculated according to the settings introduced by operator.

The most commonly used are MALDI ion sources combined with TOF analyzers. Additional energy necessary for PSD is acquired from the laser beam. In MS/MS experiments, laser power is significantly increased to optimize ratio of decays. Some spectrometers equipped with TOF analyzer also have IT. Such hybrid systems allow for convenient accumulation of the ions and fragmentation in IT, resulting in high-resolution analysis of arising fragments in TOF analyzer.

7.5.5 Combined Time-of-Flight Analyzers (TOF/TOF)

A very effective approach to overcome the above described problems with fragmentation in a single TOF analyzer is a combination of two TOF systems. In such constructions, resembling two quadrupoles working together, first TOF is responsible for parent ion selection, while the second one performs fragment ion analysis. During typical MS run, laser power is limited to avoid spontaneous fragmentation of the sample. First TOF analyzer is set to allow maximal transmission of the ions. During MS/MS analysis, laser power is elevated to the levels providing preferential metastable ion formation. As mentioned earlier, metastable ions traveling through the first analyzer decompose into a set of fragment ions, carrying together kinetic energy of the parent ion. Energy received from the parent ion is proportional to the masses of fragments, so traveling speed of fragments does not change. As a result, a group of fragment ions, derived from the same parent ion, flying through the analyzer, could be treated as a single object. If every ion formed in the ion source is metastable and their kinetic energies are the same, their derivatives also behave like a single "ion package." Then, first TOF separates them from each other, allowing for their travel to the second analyzer. All remaining metastable ions, except this of interest, are dissipated in the volume of the first analyzer. Next, a group of selected fragment ions is introduced into the second analyzer. Unfortunately, their flight trajectories, as well as their kinetic energies, are not optimal for MS analysis. Fragment ions traveling as a single package interact with others. Having the same charges, they are repelled by the Coulomb forces, which influence their trajectories. Therefore, there is additional device that can optimize fragmentation in the TOF-based systems. Construction of the device is patented by Bruker. Its name is LIFT. This device can add kinetic energy to the remaining parent and fragment ions. This process differentiates speeds of ions, which dramatically increases resolution of the second analyzer and finally allows for receiving better MS/MS spectra in the more convenient way. Such construction is sometimes referred to as a "second ion source" as additional kinetic energy supplied to the ions completely changes their flight parameters. It can be compared to the second acceleration process of the ions in the same way as in the MALDI ion source. All theoretical considerations of TOF/TOF instruments are in fact the same as for conventional TOF analyzer. The main difference is the second accelerating area ("second ion source"). Due to significant increase of the kinetic energy of fragment ions, reflectron of the second TOF can focus ions in the detector without changes in its own potential, so it is not necessary to repeat the experiment for different reflectron settings (for decreasing potentials of reflectron electrodes). Time of analysis is significantly shorter, as well as sample consumption in MALDI chamber. Such instruments are routinely used for fast fragmentation and routine identification of hundreds of samples, e.g. in proteomics.

Another MALDI-TOF/TOF setups use collision chamber for high-energy collision dissociations occurring between analyzers, but such construction seems to be more complicated.

7.5.6 Hybrid Instruments

As various analyzers have distinct features, in some instrumental constructions, there are at least two analyzers combined. Such instruments are called hybrid mass spectrometers. The most commonly used hybrids comprise quadrupoles combined with TOF analyzers Q-TOF or QqTOF. Such instrument resembles triple quad-based mass spectrometer, but instead of the second quadrupole, there is a TOF, usually in the orthogonal arrangement. The instrument provides all advantages of QqQ with additional, much higher resolution, precision of measurements, and faster scanning. Popularity of such mass spectrometers is also a result of solving some technical difficulties concerning fragmentation in TOF/TOF instruments (mainly LIFT calibration problems).

Another example of hybrid instrument is Q-Trap. Similarly to previous example, it is a modification of QqQ construction, where instead of the second quadrupole, an IT is mounted. This analyzer provides multiple fragmentation mode, as well as higher resolution in comparison with the instruments equipped with two quadrupoles. Such models are quite popular in combination with liquid chromatographs. LC-MS/MS systems are capable of acquisition of thousands of MS/MS spectra in a single chromatographic separation, which is not a big challenge for Q-Trap-based spectrometers.

Also, an interesting solution in development of hybrid instruments is connection between IT and ICR analyzers. ICR chamber has limited capacity; additionally, inside ICR, there is a very high vacuum kept constantly, so fragmentation in the ICR is difficult and time consuming. An IT located just before ICR inlet assures isolation, fragmentation, and mass measurement in ICR. This allows for a fast and convenient fragmentation, resulting in a very accurate fragment ion analysis.

7.5.7 Mass Spectrometers Equipped with Orbitrap Analyzer

For a few last years, MS with Orbitrap analyzer led to a breakthrough in biological sciences, due to its working parameters, capabilities, and ease of use; thus it definitely needs to be discussed separately.

Orbitrap has two features, which taken together distinguish it from other analyzers: high scanning speed (up to hundreds Hz) and high resolution (currently up to 1 000 000). Other advantages are trapping relatively high quantity of ions during single scan (especially comparing to Fourier transform ion cyclotron resonance [FT-ICR] systems) and lack of superconductive magnets in the analyzer construction and expensive cooling and maintenance.

A broad array of fragmentation options is available in the Orbitrap-based systems due to the modular construction of the instrument. Orbitrap is the last component of the whole analytical setup of MS. Fragmentation is performed in the earlier placed elements, like linear IT, quadrupoles, and other subassemblies, and the role of Orbitrap is focused on m/z ratio measurements of parent as well as fragment ions. One of the simplest constructions is a combination of quadrupole analyzer, linear IT, and Orbitrap. Quadrupole selects parent ion, which is guided to the linear IT, where fragmentation occurs (usually according to HCD mechanism). Finally, fragment ions travel to the Orbitrap chamber where they are analyzed. In the more sophisticated constructions, IT has its own ion detection system along with reagent delivery systems able to provide substrates for ETD fragmentation. In such cases, it is possible to perform CID, HCD, and ETD fragmentation. It is worth to mention that due to high resolution of the Orbitrap system, PTR reaction is not used in mass spectrometers equipped with this type of analyzer. Based on this device, it is easy to resolve multiply charged ions even with a tiny m/z difference.

7.6   Applications of Tandem Mass Spectrometry in Life Sciences

One of the most important applications of MS/MS in life sciences is peptide sequencing. Peptides, known as relatively big molecules, having numerous covalent bonds of the same energies, can dissociate during fragmentation process, generating complex MS/MS spectra. Due to their specific structure (amino acid residues of known MWs, linked by peptide bonds), it is possible to find their characteristic fragmentation patterns and codify fragmentation rules. This helps to interpret peptide MS/MS spectra and finally unambiguously determine peptide sequences. The most common nomenclature was proposed by Roepstorff and Fohlmann in 1984 [4], and after minor modifications introduced by Johnson in 1987, it is widely used up to date.

The idea staying behind this nomenclature is observation that peptide chain can be fragmented after dissociation of the CHR—CO, CO—NH (peptide bond), or NH—CHR covalent bonds. By observing N-terminal as well as C-terminal fragments, we can distinguish up to six series of fragment ions. N-terminal fragment ions are denoted a, b, or c series, while fragments consisting of C-terminal free carboxyl group are marked as x, y, or z series. Number associated with a letter represents number of amino acid residues present in the fragment ion. Exemplarily, an ion described as $b7$ arises after breakage of the seventh peptide bond, counting from the peptide N-terminal. Detailed description of the nomenclature is also shown in Figure 7.7. Different fragmentation techniques preferentially generate various fragments (series). For example, CID fragmentation technique is responsible for generation mainly of b and y ions, while ETD fragmentation forms usually c and z ion series. There are also rules of cyclic peptide fragmentation [7].

Knowing rules of peptide fragmentation, it is possible to use MS/MS for unambiguous protein sequencing. To achieve this goal, it is usually necessary to digest protein of interest by an enzyme of known specificity (like trypsin, pepsin, V8 protease, endoproteinases [Asp-N, Lys-C, Arg-C], etc.). The digest is introduced into the mass spectrometer, usually after preseparation by the LC

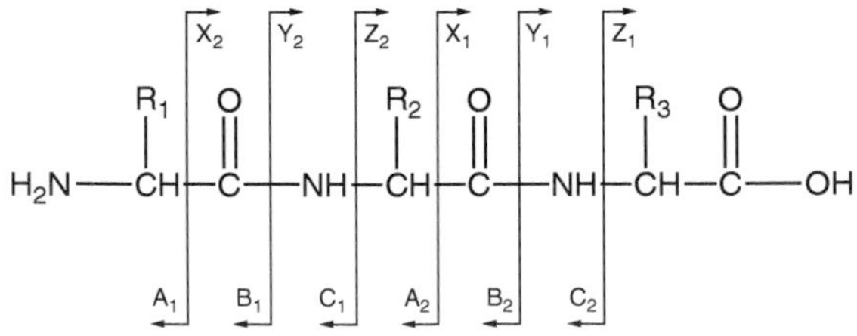

Figure 7.7 Roepstorff's and Fohlmann's nomenclature for fragment ions derived after peptide fragmentation.

system, in the case of the more complicated mixtures. MS/MS is performed, which generates fragment spectra for the peptides. Knowing enzyme truncation sites applied for protein digestion, at least N- or C-terminal amino acid residues are known for each peptide fragmented. Total protein sequence is reconstructed based on the algorithms supporting MS/MS spectra interpretation. Working with protein of a known sequence (deposited in the protein database; see Chapter 9), it is usually enough to receive 3–4 unique sequences from the set of fragmentation spectra. The situation is more difficult when protein sequence is unknown, is incorrectly uploaded into the database, or is not uploaded at all. In such cases, the result of database searching usually shows protein of the high sequence homology toward investigated protein. Bioinformatic algorithms also search for gene sequences that may generate a list of "hypothetical proteins" that do not exist but were read out from genome sequence. Such proteins should not be considered in data interpretation, and the only chance to reveal unknown protein sequence is to perform detailed studies on the isolated material. It is also important to state here that proteomic approach does not identify proteins directly. Information based on few tryptic peptides provides only a clue on protein identity. It does not provide any detailed information on eventual modifications (replacement of amino acids, protein truncation, all PTMs but few, etc.).

The single protein identification is a relatively simple process based on workflows described in detail in the available literature. However, with an increasing number of peptides in the sample, being a result of enzyme cleavage of high MW protein or proteomic analysis, identification of all components becomes more challenging. The problem is related to the limits of simultaneous analysis and fragmentation of peptides in a mass spectrometer. For instance, tryptic digest of a protein containing 200–300 amino acid residues usually generates 25–50 peptides [5]. The entire mixture will be easily separated using LC linked to ESI-MS and fragmented one after one. The problem arises when the number of peptides in a single sample is too high to be efficiently separated on the LC system. Peptides having similar hydrophobicity coelute from the reversed-phase column and must be simultaneously fragmented in mass spectrometer. Although MS algorithms are capable of selecting few parent ions and fragmenting them in an appropriate order, too many peptides in the ion source may exceed analytical capabilities of the instrument. In such case only part of them will be sequenced and identified, which leads to a loss of some vital information. Additionally, other unwanted effects like space charge (for FT-ICR, TOF, or IT) or ion suppression can occur – all of them may contribute to significant losses of information. Fortunately, in majority of cases, it is enough to achieve information on few peptides belonging to the same protein to identify it based on comparison with database records (with all reservations mentioned above). The most widely recognized databases collecting protein data are currently UniProtKB and NCBI (see Chapter 9). Thanks to their resources, it is currently possible to identify almost every protein found in biological samples. It should be, however, noted that database searching machines can detect homologous protein (e.g. from phylogenetically close species) instead of that predicted by the operator (e.g. mouse instead of human).

Applications of MS and hyphenated techniques are not limited to protein identification only. Proteins carry various modifications of their structure, which influence their biological function(s). The most common examples are disulfide bonds stabilizing 3D structure and PTMs providing significant changes in protein functionalities. All types of modifications are of strong interest of biological MS. Currently, various protocols allowing for the localization of the disulfide bonds in the protein structure can be proposed, mainly based on differential truncation of the same protein by proteases with and without prior reduction of disulfide bonds. There are many approaches for identification and localization of various PTMs along the polypeptide chain. Moreover, it is even possible to solve a structure of more complicated PTMs like glycosylation, mainly with the aid of multistep fragmentation.

7.7 SWATH Fragmentation

This special type of fragmentation was proposed by Ruedi Aebersold et al. in 2012 [3]. Up to date, this procedure of fragmentation includes isolation of the ion, its fragmentation, and, finally, identification of the molecule. SWATH technique omits parent ion isolation step. At the first glance, such strategy seems to make no sense, as, in every previously described fragmentation technique, isolation step was necessary to receive satisfactory results (daughter ions were observed after isolation of parent ion). But SWATH was designed for a special type of analyses; the technique is used for simultaneous sequence determination of hundreds of peptides derived from the proteomic samples (e.g. group of proteins derived from cell homogenate). Sample is prepared in the same manner like for *shutdown* proteomic approach (see Section 8.1.2.1). SWATH analysis is used in LC-MS/MS systems only. Additionally, high-resolution analyzer is necessary, such as FT-ICR, Orbitrap, or TOF. During the analysis, sample is separated on LC column, but the number of peptides eluted at the same time is much higher than capabilities of sequential fragmentation of mass spectrometer in typical approaches. Just to remind, for typical approaches, MS must isolate parent ion, fragment it, select next parent ion, fragment it, etc. The time necessary for a single cycle is relatively long; thus, if many peptides leave LC column at the same time, it is impossible to fragment them all. This obstacle causes substantial loss of important information coming from the analysis. Such loss is even higher with increasing number of peptides in the sample. The idea behind the SWATH was quite simple; the approach allows for selection of a wider range of *m/z* ratio for fragmentation, instead of selecting single parent. In fact, a difference is not so big, and mass spectrometers isolate peaks for fragmentation from a given m/z window set by the operator rather than selecting single peak with a window usually not wider than 3–5 Da. SWATH allows for setting the isolation window in much wider ranges. This method allows to set an isolation window width and the speed of sequential isolations of windows from the lowest to highest m/z. Mass spectrometer software isolates m/z window at a width of a few dozen units and makes fragmentation of all peaks visible within this range. Next cycle of fragmentation selects next m/z window, and the procedure is repeated. Mass spectrometer is capable of performing the entire fragmentation procedure over the entire m/z range in the time not exceeding the length of a single chromatographic peak for a typical high performance liquid chromatography (HPLC) analysis (10–30 seconds). For example, if the full scan range is 400–1200 m/z, the width of isolation window is set to 25 Da (from 400 to 425, 425 to 450 ⋯ to 1175 to 1200 m/z). In such case, mass spectrometer must repeat 32 isolation/fragmentation cycles to receive a full set of fragment spectra from the given scan ranges at a reasonable time, not longer than the mean time necessary for elution of a peak from the LC system. Assuming that HPLC or nanoHPLC (not ultra performance liquid chromatography [UPLC]) is connected to mass spectrometer, this should take no longer than 10–30 seconds. Judging from the speed of the currently available mass spectrometers, such setting is not a challenging task.

After SWATH analysis, the result file contains a mix of fragmentation data from a vast number of ions simultaneously fragmented. Such overwhelming pile of data looks hopelessly complex, but there are algorithms resolving peptide sequences from raw data. It is highly probable that single MS/MS spectrum represents fragment ions from few precursors. But the advantage is that all spectra are saved at high-resolution mode. Additionally, fragment ions can only be a combination of b- and y-ion series, which limits a possible number of combinations. This assumption allows for separation of populations belonging to the fragments derived from various parent ions on a single MS/MS spectrum. Advanced bioinformatic analysis of SWATH-based data allows for extraction of precise information about thousands of compounds in the proteomic samples. It seems that this approach provides the highest percent of identified proteins among currently available techniques.

Questions

- Describe the main principles at the base of tandem MS.
- How is it possible to separate the parent ion from the remaining ions on the mass spectrum?
- Which gases are normally used for ion fragmentation?
- What is the main difference between quadrupole and ion trap?

- How many times it is possible to conduct the fragmentation in MS^n experiments?
- How is it possible to increase the internal energy level of a molecule in order to obtain its fragmentation? What are the main parameters controlling collisional energy and collision frequency in CID?
- What is the main difference between CID and HCD?
- What is the main disadvantage of ECD?
- What is the most commonly used substance in ETD to generate anion radicals?
- What are the main disadvantages of ETD?
- What is EDD and how does it work?
- What kind of analyzers is mostly suitable in NETD?
- How is the energy delivered to the parent ion in IRMPD?
- How is the energy delivered to the parent ion in BIRD?
- What are metastable ions?
- What kind of ions is most suitable to be fragmented by SID?
- In what kind of molecules it is possible to observe the charge remote fragmentation?
- How is it possible to use charged ligand to a molecule in order to promote specific fragmentation patterns? Give some examples.
- In what cases it is useful to apply PTR?
- What is in-source fragmentation and how is it performed experimentally?
- How is triple quadrupole fragmentation carried out?
- Describe the different modes of operation in triple quadrupole fragmentation.
- What are the disadvantages of the SRM/MRM approaches?
- What is the low mass cutoff effect observed in ion traps?
- What is the main disadvantage of TOF analyzers and how can be overcome?
- Give examples of hybrid MS instruments.
- Give examples of applications of tandem mass spectrometry in life sciences.
- How does SWATH fragmentation work and in what kind of analysis is more convenient to be applied?

References

1 Gross, J.H. (ed.) (2004). *Mass Spectrometry*. Berlin: Springer.
2 Prasain, J. (ed.) (2012). *Tandem Mass Spectrometry: Applications and Principles*. Rijeka: IntechOpen.
3 Gillet, L.C., Navarro, P., Tate, S. et al. (2012). Targeted data extraction of the MS/MS spectra generated by data-independent acquisition: a new concept for consistent and accurate proteome analysis. *Mol. Cell. Proteomics* 11 (6): O111.016717.

4 Roepstorff, P. and Fohlmann, J. (1984). Proposal for a common nomenclature for sequence ions in mass spectra of peptides. *Biomed. Mass Spectrom.* 11 (11): 601.

5 Ekman, R. and Silberring, J. (eds.) (2002). *Mass Spectrometry and Hyphenated Techniques in Neuropeptide Research.* New York: Wiley.

6 Matrix Science. Peptide fragmentation. http://www.matrixscience.com/help/fragmentation_help.html (accessed 10 January 2019).

7 Ngoka, L.C.M. and Gross, M.L. (1999). A nomenclature system for labeling cyclic peptide fragments. *J. Am. Soc. Mass Spectrom.* 10: 360–363.

8 Geyer, H., Schanzer, W., Thevis, M. et al. (2014). Anabolic agents: recent strategies for their detection and protection from inadvertent doping. *Br. J. Sports Med.* 48: 820–826.

8

Mass Spectrometry Applications

8.1 Mass Spectrometry in Proteomics

8.1.1 Introduction

Vincenzo Cunsolo and Salvatore Foti

Dipartimento di Scienze Chimiche, Università di Catania, Catania, Italy

During the last decades, the rapid technological development in molecular biology, together with the advances in large-scale protein analysis, has contributed to move the attention of scientists from the characterization of individual proteins to a detailed investigation of complex protein mixtures and to the proteome of organelles, cells, or entire organisms. Proteomic studies have gained notable development principally as a consequence of the impressive increase of performance and versatility of the mass spectrometry (MS) instrumentation, which today represents an indispensable tool in proteomics [1]. As a general statement, the overall goal of proteomics is to understand the function of all proteins present in an organism. This aim implies proteins identification and systematic determination of their different properties, including quantification, characterization of primary structure and posttranslational modifications (PTMs), determination of protein interactions, and subcellular distribution. In particular, MS-based proteomics is well suited to the study of PTMs because such changes lead to characteristic shifts in mass and can be located with the resolution of a single amino acid through peptide fragment ion spectra. The most frequently studied types of PTMs are phosphorylation, ubiquitination, glycosylation, methylation, acetylation, and other types of acylation. The continuous and rapid evolution of proteomics has potential implications for the understanding of molecular mechanisms in clinical diseases, but it is already yielding important findings across a wide range of applications in numerous

Mass Spectrometry: An Applied Approach, Second Edition. Edited by Marek Smoluch, Giuseppe Grasso, Piotr Suder, and Jerzy Silberring.

fields, such as pharmaceutical, microbiological, agricultural, and food technologies. From an analytical point of view, proteomic investigations are based on the use of highly efficient separation techniques, such as reversed-phase high performance liquid chromatography (RP-HPLC) or two-dimensional polyacrylamide gel electrophoresis (2D-PAGE) combined with biochemical methods, enzymatic digestion, and MS, in order to obtain data suitable for searching, with the aid of a specific software or genomic and protein databases.

Given the wide range of proteomic applications and therefore the high variety and differences among proteins in terms of their chemical and physical properties, it is evident that a comprehensive proteomic exploration does not comprise the application of a single technique or a unique strategy for all purposes. Instead, proteomic studies are usually carried out using multiple technologies, and when possible, different approaches aimed to provide complementary results and therefore to improve proteome resolution and sequence coverage.

8.1.2   Bottom-Up Versus Top-Down Proteomics

Vincenzo Cunsolo and Salvatore Foti

Dipartimento di Scienze Chimiche, Università di Catania, Catania, Italy

Proteomic studies are usually performed by the use of multiple technologies aimed to provide high-throughput protein identification, protein expression, and protein–protein interactions. Proteomic experiments involve the simultaneous investigation of several hundreds or even thousands of protein components and therefore may require an efficient separation step (i.e. gel electrophoresis and liquid chromatography [LC]-based methods) to simplify the complex mixtures of peptides or proteins that are delivered to the mass spectrometer. Depending on the final goal of the research, the general strategy applied may vary. The currently existing MS-based approaches used in proteomics are the so-called bottom-up and top-down [2]. In the bottom-up proteomics, proteins are digested, using one or more proteases, into peptides for the subsequent MS and tandem mass spectrometry (MS/MS) analyses (peptide maps). Instead, the "top-down" strategy directly analyzes intact proteins, and their fragment ions are generated inside the mass spectrometer (mainly Fourier transform ion cyclotron resonance [FT-ICR] and Orbitrap instruments), without prior digestion. Below, we will briefly describe both approaches, together with their advantages and drawbacks.

8.1.2.1   Bottom-Up Proteomics

Bottom-up strategy represents the most widespread proteomic workflow and can be carried out by different ways; each has a specific purpose, performance profile, and a range of utility. In the so-called "break-then-sort" approach (Figure 8.1), the protein mixture is directly digested into a collection of

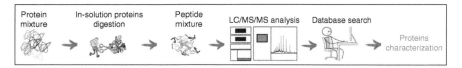

Figure 8.1 Workflow of a typical "break-then-sort" bottom-up approach.

peptides, which are separated by 1D or 2D chromatography (e.g. strong cation exchanger followed by RP-HPLC) coupled *online* to the electrospray ionization (ESI)-MS. The peptide ions are then transferred to the vacuum of a mass spectrometer, where they are fragmented in the gas phase to generate MS/MS (MS^2) spectra that contain information necessary to identify and quantify specific peptides [3]. Generally, peptides are fragmented by the collision-induced dissociation (CID) or higher-energy collision dissociation (HCD), but alternative methods, such as the electron transfer dissociation (ETD) or electron capture dissociation (ECD), are becoming widely available (see Section 7.4) because of their best performance in the fragmentation of large and modified peptides. This method carried out by means of MS data-dependent acquisition (DDA) is known as discovery (or shotgun) proteomics or multidimensional protein identification technology (MudPIT) [4]. In DDA-based methods, mass spectra of all ion species that coelute at a specific time on the chromatogram are recorded as a full scan (MS^1 level). The instrument alternates between the acquisition of full scan data and the acquisition of fragment ion spectra, in which as many precursors as possible are sequentially isolated and fragmented (at the MS^2 level). This approach is conceptually simple and aimed at achieving unbiased and complete coverage of the proteome but may result in greatly increased complexity of the generated peptide mixture. In fact, in this way, not all peptides present in the mixture may be observed or correctly identified by MS/MS analysis and database search. Moreover, only the peptides present at high relative abundances may be sampled, while information regarding the proteins represented as low abundant peptides in the complex mixture is usually lost. However, the MudPIT approach is now emerging in the proteomics field, thanks to the high resolving power of ultra-high performance liquid chromatography (UHPLC) systems, the high mass accuracy of MS instruments, and the high performance of ETD technique. Another bottom-up approach in which the protein mixture is directly digested into a collection of peptides is the so-called targeted proteomic strategy. This approach is aimed at the reproducible, sensitive, and streamlined acquisition of a subset of known peptides of interest by means of the selected reaction monitoring (SRM). Using preexisting information about known proteins of interest, peptide markers are selectively and recursively isolated and then fragmented over their chromatographic elution time. These data can be acquired by means of the ion trap,

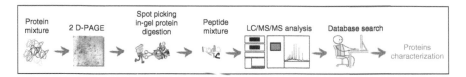

Figure 8.2 Workflow of a typical "sort-then-break" bottom-up approach.

triple quadrupole, or quadrupole/Orbitrap instruments. As an example, in a triple quadrupole configuration, SRM analysis is achieved by setting the first quadrupole to the expected precursor ion m/z ratio, the second one as a collision chamber, and the third quadrupole to the m/z ratio of an abundant fragment ion that is specific for the targeted peptide.

In the bottom-up approach called "sort-then-break" (Figure 8.2), the proteins are firstly separated by 2D electrophoresis; then the isolated components are subjected to the in-gel enzymatic digestion, followed by LC-MS/MS analysis. This gel-based approach, due to the unsurpassed ability of 2D electrophoresis to resolve complex protein mixtures, constitutes the long-established and popular strategy in proteomic studies. The popularity of 2D gel-based approaches in proteomics is mainly related to the reproducibility and robustness of the electrophoresis technique, as well as to the ease of performing quantitative analysis and interfacing with immunoblotting techniques. On the other hand, it is well documented that gel-based proteomics shows some limitations regarding the analysis of proteins with poor solubility (e.g. membrane proteins) [5]. Another drawback is related to a very high concentration range of proteins present in a biological system (e.g. up to 10–12 orders of magnitude in human sera). Consequently, only the most abundant proteins may be visualized in 2D gel electrophoresis (2DE) maps, while the low-abundant or trace components could remain undetected. In addition, it is important to highlight that in gel-based approaches, it is mandatory to minimize contamination with keratins, which are introduced by dust, chemicals, handling without gloves, etc., as the keratin peptides can easily dominate MS spectra. Moreover, bottom-up experiments require a lot of laborious steps in order to proceed from the protein to the peptide level. It is also important to underline that a crucial step in such experiments is the protein digestion, which is often the bottleneck in terms of time consumption. The use of trypsin is still considered the gold standard in bottom-up proteomics, but in specific cases, peculiar characteristics of the target proteins (e.g. amino acid composition or hydrophobicity) may require the selection of other more suitable proteolytic enzymes or multiple digestion strategies. For example, it is very reasonable that proteins focused in 2DE spots with extremely basic p*I* will probably contain many basic residues (i.e. arginines and lysines) and few acidic residues. Conversely, a highly acidic protein will be constituted of a lot of aspartic and glutamic amino acids and few arginines and lysines. Consequently, the uncritical use of trypsin may cleave a very basic protein into a high number of small-sized peptides and a highly acidic protein into a few large peptides. It is well known that too small and too big-sized peptides do not represent the best target for MS and MS/MS analyses performed in the most common MS instruments (i.e. with ion trap or quadrupole as analyzers). Therefore, alternative proteases such as Lys-C or Glu-C or combination of two different enzymes may be needed [6].

8.1.2.2 Top-Down Proteomics

If the "bottom-up" represents the traditional approach, the so-called top-down represents the emerging MS-based strategy in proteomic studies, providing information on both intact protein mass and its amino acid sequence. The top-down proteomics approach (Figure 8.3) utilizes molecular and fragment ions mass data obtained by ionizing and dissociating a protein in the mass spectrometer, without prior proteolytic digestion. In particular, the proteins present in a complex mixture are firstly fractionated and separated into pure single proteins or less complex protein mixtures. This step is followed by an *offline* static infusion of the sample into the mass spectrometer or by *online* LC/MS analysis for high-resolution mass measurement of intact protein ions. MS/MS of the mass-selected multi-charged ions of a protein then provides fragment mass values for its structural characterization [7]. The top-down strategy requires more complex instrumentation and methodology than the far more widely used bottom-up approach, but it allows to obtain far more specific data. It is important to note that the key to the success of this approach is an efficient fragmentation of intact proteins, as well as the exact selection of the multi-charged protein ions, and then the identification of the charge states of the resulting fragments ions. These objectives can be achieved by exploiting the high mass accuracy and high resolution of linear trap quadrupole (LTQ)-FT-ICR, quadrupole time-of-flight [Q-TOF], or Orbitrap mass spectrometers, further improved by the electron fragmentation methods (i.e. ETD or ECD; see Sections 7.4.5 and 7.4.6). In general, the top-down approach may provide structural information on proteins, which include an improved protein quantification and a more comprehensive characterization of PTMs, with a level of accuracy that is hard to achieve with classical bottom-up approaches [8]. Indeed, in the bottom-up experiments, information about a peptide can be lost during the LC-MS step for many reasons: because it may be too small or too

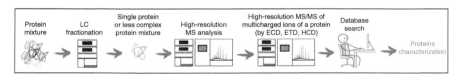

Figure 8.3 Workflow of a typical top-down approach.

polar and can be ignored in data-dependent experiments, because of its too weak MS signal, or because it can be too large to generate useful MS/MS spectra. Moreover, a single amino acid substitution or an unexpected PTM can prevent efficient database identification of the peptide. In addition, it should be evidenced that even when a protein is confidently identified in bottom-up experiments, no data about its molecular mass can be achieved. On the contrary, in top-down approaches, the experimental molecular weight of the intact protein, derived from its multiply charged ESI mass spectrum, represents the first information available. So, a rapid comparison between the calculated and the experimentally measured mass of the protein allows to ascertain the presence of PTMs or hypothesizes the existence of amino acid substitutions, deletions, or insertions. The subsequent MS/MS analysis of mass-selected multi-charged ions of the protein of interest provides, in principle, the characterization of its complete sequence. This has the advantage that all modifications (including PTMs) that occur on the same molecule can be measured together, enabling identification of the precise proteoform [9]. Another peculiar feature of the top-down approach is related to the MS/MS data, which are unequivocally assigned to "that particular" protein whose precursor ion was selected for fragmentation. On the contrary, in the bottom-up proteomics, in many cases, a set of MS/MS spectra does not permit the identification of a unique protein, but matches a group of proteins having similar sequences. Nonetheless, up-to-date top-down approach represents a relatively new field in MS and is not as widespread as bottom-up for some reasons. Indeed, although attractive in principle, top-down MS is experimentally and computationally challenging because of the greater difficulty in analyzing intact proteins in comparison with peptides. As mentioned above, top-down analysis requires a pre-fractionation step of protein samples in order to obtain more simple protein mixtures. Therefore, even if some studies have shown that large-scale profiling of intact proteins is feasible, most top-down applications aim at targeted studies of purified proteins. In this respect, separation of proteins represents one of the most important problems for top-down analysis. Traditional methods such as gel electrophoresis have limitations on the recovery of proteins in a form suitable for analysis by MS and frequently result in sample losses. On the other hand, chromatographic methods used to separate intact proteins may show poor resolving power for large proteins (>50 kDa). Another drawback to the natural evolution of top-down is the limited number of bioinformatic tools for data management and interpretation. Moreover, the indispensable use of high mass accuracy instruments may represent a restrictive requirement in a top-down approach. Indeed, these mass spectrometers are expensive to purchase and to maintain if compared with quadrupole and ion trap instruments, usually employed in bottom-up proteomics. Protein size may represent another important limitation in top-down proteomics. ECD and ETD methods provide efficient fragmentation for small- to medium-sized proteins [10]. Direct analysis of large proteins is more difficult because of an increased complexity of the gas-phase protein ion's tertiary structure with many non-covalent interactions. To address this limitation, the so-called "middle-down" strategy, involving limited digestion to produce larger peptides (>5 kDa), has been explored for very large (>200 kDa) proteins [11]. This approach joints the best features of top-down and bottom-up approaches. Middle-down proteomics takes its advantage by the high resolution and mass accuracy of MS instrumentation and by the high performance of electron-based (electron detachment [ED]) fragmentation methods. In ESI, medium-sized peptides (4000–10000 Da) carry out a higher number of charges in respect to smaller ones, thus enhancing the pattern fragmentation by ECD, ETD, and HCD methods and therefore improving protein sequence coverage and identification of PTMs. At the same time, the middle-down approach shows advantages related to the chromatographic analysis of peptides, which are much more efficiently separated than intact proteins, because of their narrower mass distribution, charge, and hydrophobicity.

On the light of their most important features (including advantages and drawbacks), it is obvious that top-down and bottom-up represent complementary approaches, as evidenced by a number of proteomic studies performed by coupling together these two strategies. In general, the bottom-up method is widely accepted for the routine identification of proteins in complex mixtures. As an example, in biomedical research, the identification of a protein, and therefore the indirect identification of its encoding gene, is usually more important than the full characterization (e.g. 100% sequence coverage) of the protein itself. In other cases, when a more extensive comprehension of the protein structure or specific data, such as polymorphisms or complete PTMs, is needed, top-down proteomics becomes an unavoidable tool. In any case, and independently from the adopted proteomics approach, it is important to highlight that the generation of the MS data is far from being the end of the experiments in proteomics, but rather represents an indispensable precondition for undertaking the second level of proteomic studies, the protein identification by the high-throughput bioinformatic methods, which will be in brief discussed in the next paragraph.

8.1.3 Database Search and Protein Identification

It is well known that the continuous improvement of the modern MS instrumentation allowed to reach an extraordinary level of protein characterization. As a result, the size of data produced in proteomic laboratories has increased by several orders of magnitude. So, handling and storing of the huge MS data sets has been possible by the contemporary availability of protein sequence databases and the improvement of computational analysis tools [12, 13]. In this respect, interpretation of MS data not only represents another fundamental step in proteomics, but it may also be considered a central and critical component of a successful proteomic experiment. In general, the characterization and identification of a protein may be obtained by comparing MS experimental data with calculated mass values obtained from a sequence database using a dedicated search engine (e.g. Mascot, MS-Fit, ProFound, Mass-Search, etc.) [14]. Taking into account that the enzymatic digestion of a protein generates a set of peptides, which represents its fingerprint, protein identification in bottom-up proteomics can be reached by two different methods, known as peptide mass fingerprinting (PMF) (or peptide mapping) and peptide fragmentation fingerprinting (PFF). In PMF approach, a list of enzymatic peptide fragment masses present in the mass spectra is matched against that calculated from the same computational proteolytic digestion of each entry in a sequence database. The proteins can be ranked according to the number of peptide matches. More sophisticated scoring algorithms take the mass accuracy and the percentage of the protein sequence covered into account and attempt to calculate a level of confidence for the match. It should be noted that the success in the identification of the protein depends on several factors. The number of experimental peaks observed in the mass spectrum and the mass accuracy of the mass spectrometer represent two important issues, but also the specificity of the protease used and the presence of the protein sequence in the database contribute to the successful identification. Generally, PMF is used for the rapid identification of a single protein component or very simple protein mixture. In the PFF method, database searching is performed using both the measured peptide mass and the list of peptide fragment ions produced by MS/MS. Theoretical MS/MS spectra are computed from the theoretical peptide sequences and correlated with the experimental MS/MS spectra in order to find the most similar (highest score) candidate peptide. Because the MS/MS spectrum of a peptide is strictly related to its amino acid sequence, protein identification is, in principle, more specific and discriminative than that of peptide mapping. MS/MS-based identification presents several advantages over PMF. In fact, because MS/MS data can also be used to search translated expressed sequence tags (ESTs) and other sequence databases containing incomplete sequences, the PFF approach is much more comprehensive than PMF. Moreover, MS/MS allows for direct analysis of protein mixtures. As an example, the bands from one-dimensional gels, when analyzed with the high sensitivity in the mass spectrometer, usually turn out to contain a lot of proteins. In such a case, the software reports a list of proteins, each matched by one or several peptides.

In the top-down proteomics, MS/MS data of the intact protein ions can be used in a similar way and, together with the experimentally high-resolution measured M_r, allow for protein identification. Irrespective of the approach adopted, the successful identification of the protein requires the presence of the corresponding protein sequence in the database. If the amino acid sequence of the protein under investigation is not present in the database, the best match will probably be the entry with the closest similarity and usually corresponds to a protein belonging to a strictly phylogenetic-related species.

A different approach, named *de novo* sequencing, consists in inferring knowledge about the peptide sequence directly from MS/MS spectrum, independently of any information extracted from a preexisting protein or DNA database. The complete or partial *de novo* sequences are then compared to theoretical sequences using specifically developed string similarity search algorithms (e.g. PEAKS, PepNovo, BLAST, etc.). *De novo* methods may overcome PFF methods when searching databases composed of homologous sequences in the case of cross-species identification or when analyzing a spectrum that originates from a mutated protein or a variant and may enable, in principle, identification of proteins also from species phylogenetically distant from organisms with completely sequenced genomes [15].

Finally, although MS-based approaches are mainly used to address qualitative aspects of proteomics (i.e. the identification and characterization of proteins), in order to understand the function of a protein, changes in gene expression often have to be determined. So quantitative protein profiling represents another essential part of proteomics. In this respect, quantitative information at the protein level may be very helpful to investigate and measure how the expression of proteins under different conditions (differential proteomics) changes. For this purpose, many strategies for a rapid, highly reproducible, and accurate quantification of proteins present in complex biological samples have been developed. Quantitative proteomics can be performed to obtain both absolute (using internal standards) and relative quantification of proteins using different techniques, which include gel-based, label-based, and label-free approaches (see Section 8.1.5). The label-based methods involve the labeling of peptides with stable isotopes introduced by either biosynthetic or chemical reactions. Quantification is based on the ratio of heavy/light peptide pairs. However, researchers are increasingly turning to the label-free shotgun proteomics techniques, which avoid the use of isotopes to label the samples under investigation, for faster, cleaner, and simpler results. MS-based label-free quantitative proteomics falls into two general categories of measurements. The first are the measurements of ion intensity changes, such as peptide peak areas or peak heights in chromatography. The second is based on the spectral counting of identified proteins after MS/MS analysis of their proteolytic peptides. A detailed description of quantitative proteomics approaches can be found in some review articles [16, 17].

8.1.4 In-Depth Structural Characterization of a Single Protein: An Example

Besides the capabilities of producing extensive identification of proteins in highly complex mixtures, it should not be neglected that the today available combination of UHPLC and high-resolution MS has also great potential for the in-depth structural characterization of a single protein in a complex mixture without previous isolation, including a fine characterization of unusual PTMs.

An example of such application is the recently reported characterization of the oxidation pattern of methionine and cysteine residues in rat liver mitochondria voltage-dependent anion-selective channel 3 (rVDAC3) [18]. VDACs, also known as mitochondrial porins, are the most abundant membrane proteins found in the mitochondrial outer membrane. Three isoforms of these pore-forming proteins (30–35 kDa), named VDAC1, VDAC2, and VDAC3, were originally identified in mitochondria. An increasing body of evidence indicates that VDACs play a major role in the metabolite flow in and out of mitochondria, resulting in the regulation of mitochondrial functions. All three isoforms show sequence homology and are expected to share several structural elements. The sequence of rVDAC3, the least abundant and least known isoform, includes seven cysteines in positions 2, 8, 36, 65, 122, 165, and 229 (UniProt Acc. N. Q9R1Z0). The three-dimensional structure of human VDAC3, which is expected to share structural similarity with rVDAC3, shows that VDAC3 cysteines protrude toward the mitochondrial intermembrane space, an oxidizing space of the cell, and that some of them are unlikely to be spatially proximal for disulfide bond formation. Therefore, the structural and, conversely, the functional role of cysteines remained unclear, and examination of their oxidation state was of interest. It is important to note that investigation of membrane proteins is generally difficult due to their low solubility. Moreover, analysis of rVDAC3 was complicated by the inability to separate the protein from other VDAC isoforms and other proteins present in the rat liver mitochondria extract. So, VDACs were enriched in the hydroxyapatite (HTP) eluate of Triton X-100 solubilized rat liver mitochondria. However, proteomic characterization showed that more than 300 proteins were still present in the enriched fraction.

Combined results of the nUHPLC/nESI-MS/MS analysis of the in-gel and in-solution tryptic and chymotryptic digests of the reduced and carboxyamidomethylated HTP-enriched fraction allowed to obtain a coverage of 96% of the rVDAC3 sequence and also showed that the N-terminal Met is absent in the mature protein and that N-terminal Cys is present in the acetylated form. Furthermore, it was ascertained that the mitochondrial VDAC3, in physiological state, contains cysteine residues 36, 65, and 165 oxidized to a remarkable extend to sulfonic acid. Cysteines 2 and 8 were observed exclusively in the carboxyamidomethylated form, suggesting that they may be involved in the formation of a disulfide bridge. Cys229 was detected exclusively in the oxidized form of sulfonic acid, whereas the oxidation state of Cys122 could not be determined because peptides containing this residue were not detected.

The oxidation state of the two methionines in positions 26 and 155 of the VDAC3 sequence was also investigated. It was found that in physiological state, both methionines are partially oxidized to methionine sulfoxide. A rough estimation of the relative abundance of the methionines oxidized to methionine sulfoxide and of the cysteines oxidized to sulfonic acid with respect to the cysteines detected in the carboxyamidomethylated form could be derived from the comparison of the absolute intensities of the multiply charged molecular ions of the respective peptides.

Questions

- What are the differences between the "bottom-up" and "top-down" approaches in proteomic?
- What are the differences between the peptide mass fingerprinting and peptide fragment fingerprinting approaches for protein identification?
- Describe the principles of a top-down method in proteomics.
- What are the major drawbacks of a 2D gel-based approach?
- What peculiar characteristics of the target proteins should be considered before the step of the protein digestion?
- In which way can stable isotopes be introduced in peptides for label-based quantitative determination?
- What does "*de novo* sequencing" in proteomics mean?

References

1 Yates, J.R., Ruse, C.I., and Nakorchevsky, A. (2009). Proteomics by mass spectrometry: approaches, advances, and applications. *Annual Review of Biomedical Engineering* 11: 49.

2 Kelleher, N.L., Lin, H.Y., Valaskovic, G.A. et al. (1999). Top down versus bottom up protein characterization by tandem high-resolution mass spectrometry. *Journal American Chemical Society* 121: 806.

3 Fournier, M.L., Gilmore, J.M., Martin-Brown, S.A., and Washburn, M.P. (2007). Multidimensional separations-based shotgun proteomics. *Chemical Review* 107: 3654.

4 Delahunty, C.M. and Yates, J.R. III (2007). MudPIT: multidimensional protein identification technology. *BioTechniques* 43: 563.

5 Rabilloud, T. (2009). Membrane proteins and proteomics: love is possible, but so difficult. *Electrophoresis* 30: S174.

6 Switzar, L., Giera, M., and Niessen, W.M. (2013). Protein digestion: an overview of the available techniques and recent developments. *Journal of Proteome Research* 12: 1067.

7 McLafferty, F.W., Breuker, K., Jin, M. et al. (2007). Top-down MS, a powerful complement to the high capabilities of proteolysis proteomics. *FEBS Journal* 274: 6256.

8 Wiesner, J., Premsler, T., and Sickmann, A. (2008). Application of electron transfer dissociation (ETD) for the analysis of posttranslational modifications. *Proteomics* 8: 4466.

9 Smith, L.M., Kelleher, N.L., and The Consortium for Top Down Proteomics (2013). Proteoform: a single term describing protein complexity. *Nature Methods* 10: 186.

10 Ge, Y., Lawhorn, B.G., ElNaggar, M. et al. (2002). Top down characterization of larger proteins (45 kDa) by electron capture dissociation mass spectrometry. *Journal of the American Chemical Society* 124: 672.

11 Han, X., Jin, M., Breuker, K., and McLafferty, F.W. (2006). Extending top-down mass spectrometry to proteins with masses greater than 200 kilodaltons. *Science* 314: 109.

12 Martens, L. (2011). Bioinformatics challenges in mass spectrometry-driven proteomics. *Methods in Molecular Biology* 753: 359.

13 Perez-Riverol, Y., Alpi, E., Wang, R. et al. (2015). Making proteomics data accessible and reusable: current state of proteomics databases and repositories. *Proteomics* 15: 930.

14 Eidhammer, I., Flikka, K., Martens, L., and Mikalsen, S.O. (eds.) (2007). Protein identification and characterization by MS. In: *Computational Methods for Mass Spectrometry Proteomics*. New York: Wiley, and reference therein.

15 Waridel, P., Frank, A., Thomas, H. et al. (2007). Sequence similarity-driven proteomics in organisms with unknown genomes by LC-MS/MS and automated de novo sequencing. *Proteomics* 7: 2318.

16 Panchaud, A., Affolter, M., Moreillon, P., and Kussmann, M. (2008). Experimental and computational approaches to quantitative proteomics: status quo and outlook. *Journal of Proteomics* 71: 19.

17 Elliott, M.H., Smith, D.S., Parker, C.E., and Borchers, C. (2009). Current trends in quantitative proteomics. *Journal of Mass Spectrometry* 44: 1637.

18 Saletti, R., Reina, S., Pittalà, M.G.G. et al. (2017). High resolution mass spectrometry characterization of the oxidation pattern of methionine and cysteine residues in rat liver mitochondria voltage-dependent anion selective Channel 3 (VDAC3). *BBA – Biomembranes* 1859: 301.

8.1.5 Quantitative Analysis in Proteomics

Joanna Ner-Kluza[1], Anna Drabik[1], and Jerzy Silberring[1,2]

[1] *Department of Biochemistry and Neurobiology, Faculty of Materials Science and Ceramics, AGH University of Science and Technology, Kraków, Poland*
[2] *Centre of Polymer and Carbon Materials, Polish Academy of Sciences, Zabrze, Poland*

8.1.5.1 Introduction

In quantitative research, absolute determination of individual proteins in a single sample may not be sufficient in many cases. A more specific and informative data can be obtained based on a comparison of the relative differences in their content between several samples, which reflect changed states of the organism or cells. However, to make such comparison possible with the aid of mass spectrometry (MS), proteins from the compared samples need to be appropriately labeled and simultaneously analyzed (multiplexed protein profiling). Otherwise, the results will be a subject to significant error related to the problems associated with nonidentical measurement conditions. In the following description, the terms "label" and "tag" refer to a low molecular weight substance that is covalently attached to the analyzed molecules (peptides or proteins) and capable of differentiation between parallel analyses.

Labeling methods using tags containing stable (nonradioactive) isotopes of some elements are commonly used (Figure 8.4). A "light" marker (containing,

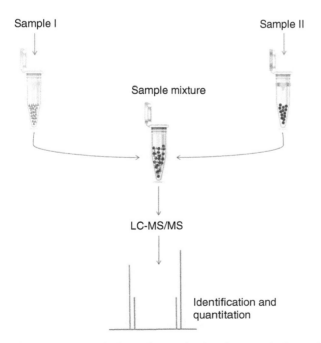

Figure 8.4 General scheme for conducting the quantitative analysis using tags.

for example, hydrogen atoms) is attached to the protein present in one of the samples, and components of the other sample are labeled with an analogous "heavy" marker (containing deuterium atoms). Stable isotope tags can also be prepared in the form of the appropriately modified amino acids, introduced into cells during the growth phase (see below). Regardless of the labeling method, the components of one of the samples act as an internal standard and therefore as a reference for the components of the other one. After labeling, the samples are pooled together and subjected to chromatographic separation, followed by mass spectra recording. Components of both samples are observed as pairs of peaks on the spectrum, whose difference in mass-to-charge (m/z) value corresponds to the difference in molecular mass between the tags. The intensities of these peaks correspond to the relative content of individual components in both samples. The labeled components differ in molecular weight, and most importantly, they are equally susceptible to ionization. For quantitative analysis, ^{2}H, ^{13}C, ^{15}N, or ^{18}O isotopes are usually applied.

8.1.5.2   Isobaric Tags for Relative and Absolute Quantitation (iTRAQ)

The quantitative determination of proteins in different samples can be made using isotopic labeling – isobaric tags for relative and absolute quantitation (iTRAQ), which relies on labeling peptides with isobaric labels (identical molecular weight). The use of iTRAQ (structures are shown in Figure 8.5) allows labeling of up to four or eight samples (4-plex or 8-plex) in one experiment.

The iTRAQ reagent consists of three parts (Figure 8.5): a reporter group, which is *N*-methylpiperazine; a carbonyl balance group; and succinimide ester that reacts with the N-terminal amino groups of peptides. There are eight types of iTRAQ reagents, each of them containing stable isotopes. Identical peptides in different samples can be labeled with various iTRAQ reagents, and the isobaric marker is attached to the peptides via amide linkage. Reporter fragments are characterized by diversified masses, depending on various isotopic combinations of ^{12}C/^{13}C, ^{14}N/^{15}N, and ^{16}O/^{18}O in individual reagents. The balancing groups also differ in mass in the range of 28–31 Da, which

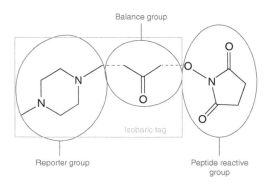

Figure 8.5 The structure of the isobaric tags for relative and absolute quantitation (iTRAQ) reagent.

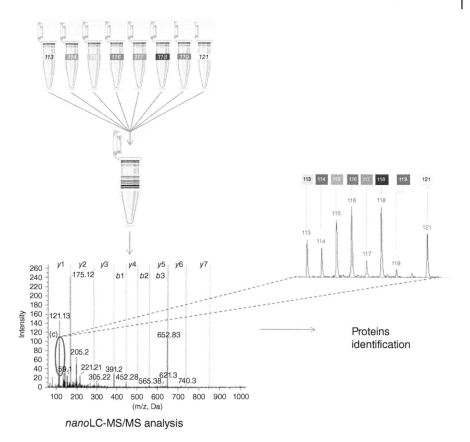

Figure 8.6 The iTRAQ experiment diagram using 8-plex.

guarantees a constant weight of the isobaric marker of 145 Da. Due to the constant mass of the label, the labeled peptides are eluted at the same retention times. Upon fragmentation of tagged peptides, the carbonyl group falls off as an inert fragment. At the same time, reporter ions with masses 113, 114, 115, 116, 117, 118, 119, and 121 Da for the 8-plex set (Figure 8.6) and 113–117 for the 4-plex set, allowing for quantification, are released. Characteristic peaks derived from the reporter groups are present in the fragmentation spectra and depict the relative content of each peptide in individual samples.

The MS^1 spectrum results in a high signal-to-noise ratio due to the biological nature of the sample, excess of derivatizing agents, etc.; therefore MS^2 spectrum is preferred, where the noise level is lower. Hence, for the iTRAQ method, modified peptides are analyzed using tandem mass spectrometry (MS/MS). The preparation protocol requires combining the content of each sample with particular iTRAQ reagent, which results in a mixture of samples containing the same peptides, which significantly enhances sensitivity of the peptide map.

Table 8.1 Advantages and disadvantages of the isobaric tags for relative and absolute quantitation (iTRAQ) method.

Advantages	Disadvantages
Comparative studies: 4- or 8-plex can be used for parallel experiments	Problems with quick search for peptides whose level is different in various samples; it is necessary to perform and analyze fragmentation spectra for all peptides
Ability to identify and quantify proteins in a short time in comparison with 2D electrophoresis	The need for tedious chromatographic separation of a very complex mixture before MS analysis
Large dynamic range, protein identification at high and low concentrations	The type of analyzer used may affect the quality of the obtained results
A very large group of peptides and proteins that can be identified and quantified during one analysis	

The ratio between peptides in different samples is expressed as the ratio of individual intensities of particular reporter groups. This implies that it is not a peptide to be quantitated, but the reporter group is specifically labeled. Initially, iTRAQ was designed to analyze two different biological samples using matrix-assisted laser desorption/ionization (MALDI) tandem time of flight (TOF/TOF). The possibility of simultaneous analysis of many samples using iTRAQ proved to be very useful and now can be applicable in various MS configurations: quadrupole TOF (Q-TOF), ion trap, and Fourier transform (Orbitrap) equipped with electrospray ion source. Application of iTRAQ in the ion trap was more difficult, but with the rapid technological development, most of the related problems (such as resolution, mass range, and cutoff range in the trap) have been solved (Table 8.1). The simultaneous analysis of complex peptide mixtures with their isobaric quantitative markers remains an obstacle in this particular case. For the ion trap, this problem results from the cutoff mass, which limits MS/MS spectrum in the range up to 30% of the m/z value of the precursor (1/3 cutoff rule). Several methods have been used to facilitate the quantitative analysis of the low mass reporter ions. Various fragmentation approaches were applied, including pulsed Q dissociation (PQD), higher-energy collisional dissociation (HCD), electron transfer dissociation (ETD), and MS^3 procedures (Section 7.4) [1–4].

8.1.5.3 Isotope-Coded Affinity Tagging (ICAT)
The isotope-coded affinity tagging (ICAT) method was introduced in 1999 by S.P. Gygi et al. as one of the first labeling methods and quickly became widely used in quantitative analysis. In this approach, the reagent is attached, via

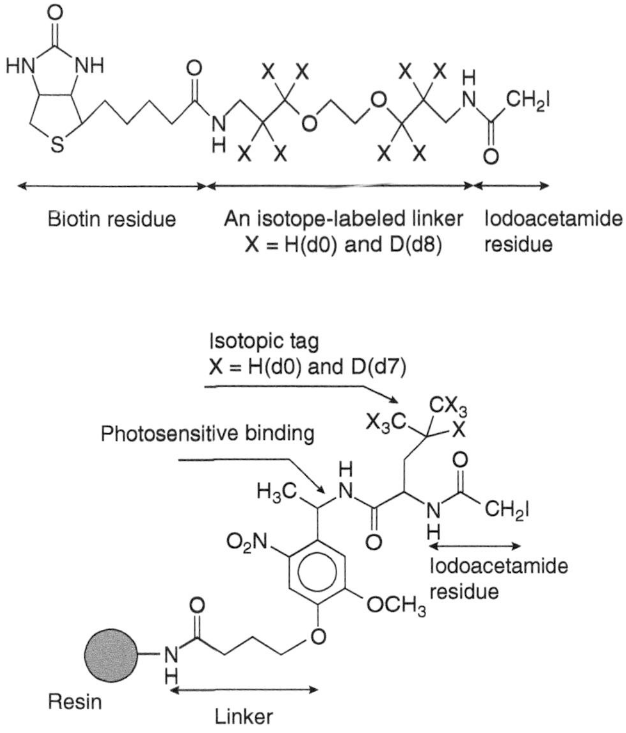

Figure 8.7 ICAT tags.

iodoacetamide residue, to amino acids containing thiol groups (cysteines). This tag occurs in two forms, containing eight atoms of hydrogen (d0) or deuterium (d8), which mark proteins in the two compared samples (Figure 8.7). Recently, ^{13}C isotope has been introduced due to the affected retention times in liquid chromatography (LC) for deuterium-labeled species. The samples are then combined and digested, usually with trypsin. The modified peptides are isolated from the reaction mixture by affinity chromatography due to the presence of biotin residue in the molecule (bound by immobilized streptavidin) and then analyzed by LC-MS. Peptides modified with "light" and "heavy" ICATs differ by 8 Da, and the intensities of the corresponding peaks are proportional to the content of peptides. The drawback of the method is related to the fact that as a result of a relatively large molecular difference in weight after attachment of the label, the measurement is subjected to an error due to the decreasing resolution. ICAT labels only proteins containing Cys residue, which may exclude components without this amino acid. Furthermore, peptide sequencing may be additionally complicated by fragmentation of the tag's molecule.

The procedure of conducting the ICAT experiment described above was developed by Aebersold et al., by creating a label that is immobilized on a solid support by a photosensitive binding. In this case, the proteins are first subjected to proteolytic digestion, and the resulting peptides are incubated with media containing an ICAT reagent with "light" (d0) or "heavy" atoms (d8) (Figure 8.8b). After labeling, impurities are washed out, and the samples are irradiated with UV. The labeled peptides are cleaved off the resin under UV light, released into solution, and then analyzed by LC-MS (Figure 8.6). Such way of conducting the experiment eliminates a need to isolate the modified peptides using chromatographic methods. The chemical structure of the marker makes it less susceptible to fragmentation during MS/MS spectra registration, which leads to easier sequence determination of the analyzed

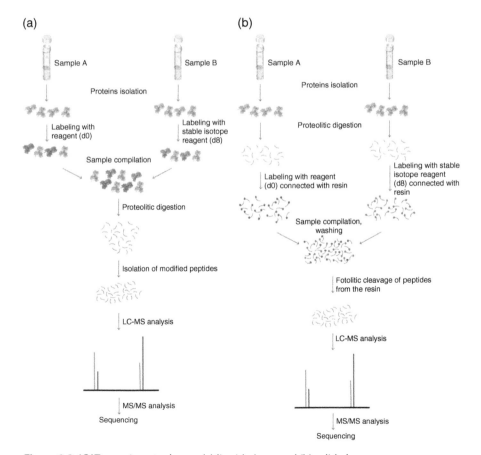

Figure 8.8 ICAT experiment scheme: (a) liquid phase and (b) solid phase.

peptides. The disadvantages of this and other methods using ^{1}H and ^{2}H atoms are the effect of deuterium label on the retention times in the reversed-phase high performance LC (RP-HPLC), because deuterium-labeled peptides leave the column at slightly lower organic phase concentration (i.e. shorter retention time). This makes interpretation of the results difficult, although the problem does not occur when using atoms of other elements as tags (e.g. ^{13}C).

The advantages of the ICAT method include sensitivity and specificity. Mass spectra are much simpler to analyze than the typical spectra of peptide maps of complex mixtures, because only those peptides containing cysteine residues, which are relatively rare in nature, are subjected to MS analysis. For the same reason, the ICAT method does not allow the assessment of the content of proteins lacking cysteine residues (such proteins consist of approx. 10% of all proteins). Due to the use of affinity chromatography, the isolated peptides can be concentrated, which facilitates the work with proteins at very low amounts [2, 5, 6].

8.1.5.4 Stable Isotope Labeling in Culture (SILAC)

Stable isotope labeling in culture (SILAC) uses the "light" or "heavy" forms of the amino acid in proteins (Figure 8.9). This method involves the addition of amino acids with substituted stable isotopes (e.g. deuterium, ^{13}C, ^{15}N) in various combinations. In the experiment, two populations/cell lines are cultured. Both are identical, except for the fact that one culture receives "light" and the other "heavy" form of the specified amino acid. The labeled analogue of an amino acid supplied to the cells in culture is included in all newly synthesized proteins. After many cell divisions, the presence of this particular amino acid in a sequence is replaced by its labeled analogue. Because there is no chemical difference between labeled and natural amino acids, labeled cell lines behave in the same way as the population of control cells cultured in the presence of a "normal" amino acid. Modified amino acids, such as ^{2}H$_4$-leucine, ^{2}H$_4$-lysine, ^{13}C$_6$-lysine, ^{13}C$_6$^{15}N$_2$-lysine, ^{13}C$_6$^{15}N$_4$-tyrosine, ^{13}C$_6$-arginine, and ^{13}C$_6$^{15}N$_4$-arginine, can be used in

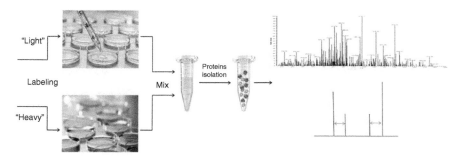

Figure 8.9 Scheme of the SILAC experiment.

this method. Frequent use of the Lys/Arg mixture for labeling opens the possibility of digesting the protein with trypsin in the next stage of sample preparation. As a result, one labeled isotope is usually obtained in each of the peptides, which allows for effective identification and quantitative analysis.

Populations of cells that are to be compared can be mixed and, at the same time, treated as a single sample, which allows preparation of the material without introducing additional quantitative errors. When analyzed by MS, the relative protein content can be calculated by comparing the intensity of "heavy" and "light" peptides. Numerous applications of this method in clinical and biological studies show that SILAC is a promising way of explaining cellular and pathophysiological mechanisms and identifying biomarkers. Initially, the use of the SILAC method was limited to the cell cultures or microorganisms. In order to adapt this technique to higher organisms, studies have been carried out to show that application of SILAC can be extended to mammalian model systems. Living mice can be labeled with SILAC, with no obvious effect of the method on their growth, behavior, or fertility. In mice, a non-lysine diet is used, which is supplied with $^{13}C_6$-lysine or $^{12}C_6$-lysine. Addition of lysine as a labeled amino acid resulted in the choice of Lys-C endoprotease as the most suitable proteolytic enzyme. Organs derived from the SILAC-labeled mice may serve as standards for subsequent comparative studies. Importantly, cell types, such as the intestinal epithelium, which are difficult to test *ex vivo*, can be tested *in vivo* in the SILAC approach. In addition, SILAC-labeled mice can serve as a reference model at any biological scale, from the entire organ, through specific cell types, to individual proteins of interest. In studies where a large number of laboratory animals are required, for example in toxicological experiments, labeled tissues and organs can be used as an internal standard for further testing [7, 8].

8.1.5.5 Stable Isotope Labeling of Mammals (SILAM)

Another approach in quantitative analysis is the method initiated by John R. Yates III, named stable isotope labeling of mammals (SILAM). The essence of the SILAM analysis is the labeling of all proteins in rodents (mice) using heavy ^{15}N nitrogen atoms. Supplementation of rodents with a ^{15}N nitrogen source is carried out by the addition of ^{15}N-rich (>99%) *Arthrospira platensis* (popularly known as spirulina) cyanobacteria. Such modification of the diet does not have a clear impact on the growth, behavior, or fertility. Metabolic inclusion of ^{15}N gives better results than ^{13}C labeling because there are a smaller number of nitrogen atoms than carbon atoms in the peptides. As a result, a better signal-to-noise ratio is obtained, and the chance of incorporating a "light" atom into a ^{15}N-labeled protein is lower than in the case of ^{13}C-labeled protein. As in the methods described above, the SILAM method can be applied to label tissues and organs used later on as an internal standard for quantitative analysis using a high-resolution mass spectrometer [2, 5].

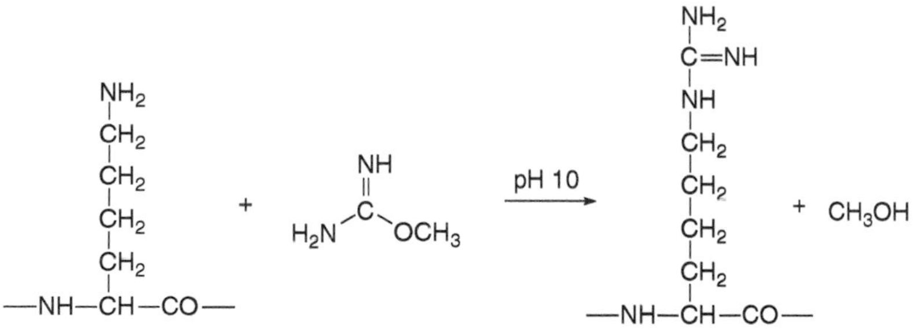

Figure 8.10 The mass-coded abundance tagging (MCAT) method – guanidylation of the lysine residue with *O*-methylisourea.

8.1.5.6 Mass-Coded Abundance Tagging (MCAT)

In contrast to the methods described above, the mass-coded abundance tagging (MCAT) method does not use stable isotopes commonly found in nature, but it relies on the selective guanidylation of ε-amino groups of lysine residues in peptide molecules, which arise during protein digestion. This reaction converts each of the lysine residues to homoarginine, which increases the molecular weight of the peptides by 42 Da for each modified residue (Figure 8.10). It should be emphasized that only peptides of one of the compared samples are modified, and the components of the second sample remain unchanged. Therefore, on the mass spectra, we observe pairs of peaks differing by 42 Da. On this basis, the relative content of individual components in both samples is compared. Analysis of MS/MS spectra of modified and native peptides also facilitates determination of their sequences. b-Series of ions without lysine residues are identical on peptide fragmentation spectra from both samples. The y-ions visible on the modified peptide spectra are shifted by 42 Da in relation to ions of the native peptide. This fact allows for a relatively easy distinction of peaks corresponding to y-ions and sequence determination. The guanidylation reaction, to some extent, changes the properties of peptides (especially those containing several lysine residues), including their susceptibility to ionization, which may lead to experimental errors [9].

8.1.5.7 Label-Free Techniques

In recent years, many methods of quantitative analysis have been published based on MS and using different labeling strategies. Most of these techniques involve samples labeling under different conditions using stable isotopes (^{2}H, ^{13}C, ^{15}N, ^{18}O). The label-free technique presents a completely different principle of operation, while two different approaches to this technique are known. The first approach is referred to as spectral counting and involves collecting and comparing the number of MS/MS fragmentation spectra obtained for the peptides of a given protein. Due to the fact that the number

Figure 8.11 Label-free scheme.

of tandem mass spectra of peptides increases with increasing amount of the corresponding protein (more protein – more identified peptides), it is possible to quantitatively determine the proteins between different samples. However, the use of this type of solution is controversial because the principle is based on quantitative calculation of the collected spectra rather than on the measurement of physical quantities. Nevertheless, the spectral counting method is widely used and being further developed. An alternative use of the label-free approach assumes measuring the area under the peaks of precursor peptide ions (Figure 8.11). Depending on the chromatographic method (e.g. reversed-phase LC), the peptides are separated according to their physical properties, then ionized in the ion source and finally analyzed. The intensity of the monoisotopic peak as a function of the retention time is shown on the extracted ion chromatogram (EIC) so that the area under the peak can be calculated [2, 10].

One of the disadvantages of quantifying peptides by comparing their signal intensities is that such analysis often encompasses experimental changes (retention times of individual peptides must be very reproducible between runs). The signal-to-noise ratio that can affect sensitivity is also an important aspect. Stable parameters are necessary to minimize changes during the elution time of the same peptide between subsequent analytical repeats. High quality and resolution as well as accuracy are also parameters that contribute to the overall success of quantitative proteomic experiments using the label-free technique. Increasing the number of repetitions, usually around three, can allow accurate comparisons if all analytical steps related to sample preparation and system stability are reproducible. Elevated background, noise coming from the samples and reagents, may further aggravate the result of protein identification and quantitative analysis. Another issue to be considered is coelution of peptides, which always accompanies the analysis of complex biological samples [1].

All the advantages and limitations of the label-free technique (Table 8.2) must be carefully verified before planning the optimal approach for the particular experiment. Currently, there are many statistical tools available to analyze data

Table 8.2 Advantages and disadvantages of the label-free technique.

Advantages	Disadvantages
The risk of introducing labeling that may cause changes in the sample has been eliminated	The relative amount of metabolic or chemical labels can be easily measured, while the label-free technique depends on many parameters
Sample preparation time is much shorter due to the reduction of several stages	For obtaining reliable quantitative results, the problem may be sequence coverage
Low-cost method in contrast to the methods where expensive chemical tags are used	The degree of fractionation of complex samples before analysis can affect its results

obtained from multiple LC-MS experiments, which allow to reduce the differences between successive repeats (runs), such as the following:

- Progenesis LC-MS (Nonlinear Dynamics, Durham, NC, USA, www.nonlinear.com)
- ProteinLynx (Waters, Milford, MA, USA, http://www.waters.com/waters/en_PL/ProteinLynx-Global-SERVER-%28PLGS%29/nav.htm?cid=513821&locale=en_PL)
- Elucidator (Rosetta Inpharmatics, Seattle, Washington, USA, www.rosettabio.com)
- Decider MS (GE Healthcare, Piscataway, NJ, USA)
- SIEVE (Thermo Fischer Scientific, San Jose, California, USA, https://www.portal.thermo-brims.com/index.php?ct=sub&id=69)

Questions

- How is it possible to tag a protein and why is this necessary for quantitative analysis?
- Describe the model for an iTRAQ reagent.
- How many iTRAQ reagents are currently available?
- What are the main advantages and disadvantages of the iTRAQ method?
- How does the ICAT method work?
- What effect the deuterium label has on the retention times in the reversed-phase chromatography? Can this be a problem in ICAT?
- What are the main advantages of the ICAT method?
- Describe the "modus operandi" of the SILAC experiment.
- Why ^{15}N is preferred to ^{13}C in SILAM?
- What is the tagging procedure used in MCAT?
- Describe advantages and disadvantages of the label-free quantitation techniques.

References

1 Latosinska, A., Vougas, K., Makridakis, M. et al. (2015). Comparative analysis of label-free and 8-plex iTRAQ approach for quantitative tissue proteomic analysis. *PLoS One* 10 (9): e0137048.

2 Chahrour, O., Cobice, D., and Malone, J. (2015). Stable isotope labelling methods in mass spectrometry-based quantitative proteomics. *Journal of Pharmaceutical and Biomedical Analysis* 113: 2–20.

3 Rauniyar, N. and Yates, J.R. 3rd (2014). Isobaric labeling-based relative quantification in shotgun proteomics. *Journal of Proteome Research* 13 (12): 5293–5309.

4 Liang, S., Xu, Z., Xu, X. et al. (2012). Quantitative proteomics for cancer biomarker discovery. *Combinatorial Chemistry and High Throughput Screening* 15 (3): 221–231.

5 Sethi, S., Chourasia, D., and Parhar, I.S. (2015). Approaches for targeted proteomics and its potential applications in neuroscience. *Journal of Biosciences* 40 (3): 607–627.

6 Aebersold, R., Griffin, T.J., and Donohoe, S. (2007). Isotope-coded affinity tagging of proteins. *CSH Protocols 2007*.

7 Chen, X., Wei, S., Ji, Y. et al. (2015). Quantitative proteomics using SILAC: principles, applications, and developments. *Proteomics* 15 (18): 3175–3192.

8 Ramberger, E. and Dittmar, G. (2017). Tissue specific labeling in proteomics. *Proteomes* 5 (3): 17.

9 Cagney, G. and Emili, A. (2002). De novo peptide sequencing and quantitative profiling of complex protein mixtures using mass-coded abundance tagging. *Nature Biotechnology* 20 (2): 163–170.

10 Sandin, M., Chawade, A., and Levander, F. (2015). Label-free LC-MS/MS ready for biomarker discovery? *Proteomics. Clinical Applications* 9 (3–4): 289–294.

8.2   Food Proteomics

Vera Muccilli and Rosaria Saletti

Dipartimento di Scienze Chimiche, Università di Catania, Catania, Italy

The constant and rapid evolution of the extremely efficient proteomic techniques has led to their increasing applications in the field of human and animal nutrition. Proteins are essential elements in the human diet, and the knowledge on their composition is very helpful for understanding the relationship between protein content and the nutritional and technological properties in foodstuffs, for developing new methods for food traceability, for assessing food quality and safety, or even for detecting genetically modified (GM) products and microbial contaminants. In addition, the detection and characterization of allergenic proteins present in some of the most widely eaten food (e. g. milk, eggs, cereals, etc.) represents a powerful tool for food control and may contribute to the production of hypoallergenic or nutraceutical foods (Pharma-Foods) [1].

Proteomic technologies applied to foodstuffs not only allow to obtain the identification of proteins expressed in each tissue type but also to accomplish investigation of differences in the protein composition before and after harvest/slaughter, cooking, or storage. It is worth noting that food proteins, prior to their consumption, are subjected to a wide range of postharvest/post-slaughter environmental and processing insults, which may impact structural modifications, including side-chain oxidation, crosslink formation, and backbone cleavage, which definitively influence some key properties, such as shelf life, nutritional value, digestibility, and health effects.

One of the main difficulties in the application of proteomics in the food analysis has been related to the incomplete knowledge of the genome of many plant species. This situation is now rapidly improving because the genomes of plants that are important for human and animal nutrition are now either sequenced or their sequencing is the topic of ongoing projects [2].

One of the most typical examples of the correlation between protein composition and technological properties is represented by the proteomic investigation of wheat proteins. It is well known that rheological properties of flour are mainly determined by its gliadin and glutenin composition. As an example, matrix-assisted laser desorption/ionization (MALDI) mass spectrometry (MS) (MALDI-MS) of crude wheat extract, which results in the complete profile pattern of high molecular weight glutenin subunits (HMW-GSs) [3], has been proposed as a routine method in breeding programs for rapid identification of lines containing subunits related to the high wheat quality and for cultivar identification [4]. Moreover, by coupling *online* reversed-phase high performance liquid chromatography/electrospray ionization MS (RP-HPLC/ESI-MS) and *offline* RP-HPLC MALDI-time-of-flight (TOF) MS approaches, it is possible to measure the molecular mass of intact low molecular weight glutenin subunits (LMW-GSs), and also by comparison of the molecular masses of the alkylated and non-alkylated subunits, it is possible to determine the number of cysteines present in their sequences, thus obtaining an information that is relevant for the characterization of this peculiar class of proteins.

MS approaches may be useful for a fast screening of milk protein composition and may provide suggestions for the destination of milk for drinking or cheesemaking because changes in protein composition, mainly due to the polymorphism of caseins (CN), deeply influence both the nutritional and technological properties of milk and milk-derived products. As an example, investigation of the CN fraction of an individual goat milk sample by means of LC-MS of the intact proteins and their tryptic digests allowed for the detection and sequence characterization of a truncated β-CN, associated with a "null" β-CN allele generated by a premature stop codon [5]. The absence of the full-length β-CN has direct consequences in cheese production, as milks lacking this protein show longer rennet coagulation time as compared with normal milks and their curd firmness is much poorer.

In addition to the studies on wheat and milk proteins, meat proteomics may also provide a relevant technological information. As an example, 2D gel comparison, followed by MS identification of muscle proteins, has been used to estimate the potential of tenderness from the live animal or carcass based on comparison between the two groups (very tender and not tender). This study has shown that the phosphorylation of muscle proteins plays an important role in the postmortem process and hence in meat quality, and this may elucidate the mechanisms behind meat storage and processing, which affects technological properties, tenderness, color, and its quality [6].

The manufacturing processes and storage conditions may induce significant changes in protein structure, including proteolytic degradation and many non-enzymatic posttranslational modifications (nePTMs), such as condensation, elimination, or hydrolysis of side chains, which render the proteome of stored food even more complex with respect to the raw and fresh materials [7]. Among the modifications induced by heat treatment, a complex series of reactions known as "Maillard reaction" [8] can impair the nutritional food value and may even show adverse health effects. As an example, in milk, the reaction between the amino groups of the whey proteins and lactose leads to the formation of lactosylated protein species, whose molecular mass increases by 324 Da per lactose unit. Using the most abundant whey proteins (i.e. α-lactalbumin (α-LA) and β-lactoglobulin (β-LG)) as molecular markers to monitor these reactions by MS, many studies have evidenced that the degree of protein lactosylation is strictly related to the storage conditions and to the thermal procedures used during industrial milk processing.

One of the main goals in food science is to provide accurate and precise methods for detecting harmful compounds, such as allergens that might be present in the food. The application of proteomic strategies to identify food allergens is referred to as "allergomics," and a typical workflow aimed to discover new allergens combines electrophoretic separation, electro-transferring onto nitrocellulose membrane, and IgE immunoblotting analysis with the sera of allergic patients, followed by a typical bottom-up approach to identify the proteins of interest [9].

Today, humans are exposed to a wide range of microorganisms and compounds that can be introduced into the body, mainly through food, water, air, and dermal contact. For these reasons, the governments of several countries have increased the amount of food safety-related legislations [10]. The major problem in food safety is food poisoning related to foodborne bacteria, which in many cases may represent a serious burden, especially in the light of increasing antibiotic resistance. MS-based proteomic methodologies have revealed to be a powerful tool for the identification and characterization of microbial food contaminants and represent a competitive alternative in their detection when commercial assays or specific immunological kits are not available [11]. In addition, because the assays may generate false-positive results, it is desirable to have a sensitive method to directly confirm the presence of toxin contamination, even if it is present in complex food matrices [12]. Proteomic studies using MALDI-TOF MS have also been established to obtain bacterial profiling, which provide a fingerprint specific for the analyzed microorganisms at a given time and physiological condition, in order to distinguish among different species and also strains [13, 14]. As a consequence, several commercial databases have been developed for bacterial identification by MALDI-TOF MS. Recently, a new public reference library, SpectraBank (www.spectrabank.org), which contains the mass spectral fingerprints of the main spoilage-related and pathogenic bacteria species from seafood, was created. This database includes 120 species of interest in the food sector [15].

Several bacteria including *Pseudomonas aeruginosa, Staphylococcus aureus, Staphylococcus epidermidis,* or *Escherichia coli* are the main causes of mastitis in dairy cows. This pathology leads to a decreased milk production and negatively interferes with its quality and usability for transformation processes. Proteomic investigations on mastitis have been performed either in milk and somatic cells, or in cow sera, allowing to reveal higher expression of some acute-phase proteins, that may represent possible biomarkers for subclinical mastitis and can be useful for the evaluation of the efficacy of specific therapies [16].

On the other hand, some bacterial species exert various health benefits in humans. In fact, by liberating lactic and acetic acids, they prevent the colonization of potential bacterial pathogens in the gastrointestinal tract, thereby maintaining a balance of intestinal flora. Bifidobacteria and lactobacilli are the most popular microorganisms that are added as live bacteria to produce probiotic dairy foods. MALDI-TOF MS was successfully applied for a rapid classification and identification of lactic acid bacterial strains from fermented foods [17].

Comprehensive characterization of the microbial machinery *in situ* is necessary to provide a relevant quality control of fermented foods. For this reason, proteomic investigation was aimed at the study of the microbial activity and interactions in complex samples, such as fermented food (metaproteomics). In particular, the proteome released from thermophilic lactic acid bacteria during the ripening of Emmental cheese was identified [18], and predominant bacterial and bovine proteins in experimental Swiss-type cheeses were also quantified [19].

Nowadays, there is a growing interest in promoting functional foods to gain healthy benefits beyond basic nutrition. Food peptidomics deals with the identification and quantification of nutritionally relevant peptides, usually referred to as "bioactive peptides," ranging from 2 to 100 amino acids, even if small peptides (2–6 amino acids) represent by far the largest category [20]. These peptides remain inactive within the sequence of native proteins but can become active by *in vivo* proteolysis (gastrointestinal digestion), during food processing (ripening, fermentation, and cooking), storage, or *in vitro* proteolysis, due to the presence of endogenous enzymes in the food matrix [21]. Their biological activities are numerous, including antihypertensive, antioxidant, hypocholesterolemic, immunomodulatory, anti-inflammatory, antimicrobial, antithrombotic, and anticancer properties. Some of the common sources of bioactive peptides are milk, meat, egg, fish, soybean, rice, sunflower, cereals, and maize proteins [22], but other organisms such as algae and organisms originated from the marine environment (e.g. sea urchins [23] and sea cucumber [24]) have been explored as a new source of bioactive peptides. In particular, antihypertensive activity generated during fermentation of porcine proteins by using lactic acid bacteria, leading to the release of angiotensin I-converting enzyme (ACE) inhibitory peptides, was demonstrated [25]. Moreover, novel sequences exhibiting *in vitro* ACE inhibitory activity, as well as *in vivo* antihypertensive activity, were identified from Spanish dry-cured ham [26]. Recently, extensive peptidomic study in commercial donkey milk samples was carried out by two different peptide purification strategies, resulting in a total of 1330 peptides identified [27].

Some databases collecting peptide sequences, activity, references, sources, and more – e.g. BIOPEP (http://www.uwm.edu.pl/biochemia/index.php/pl/biopep), PepBank (pepbank.mgh.harvard.edu), Antimicrobial Peptide Database (APB) (http://aps.unmc.edu/AP/main.php), and SwePep (www.swepep.org), which contain a repository of the MS/MS spectra relative to the identification of the peptides – are useful resources for bioactive peptides study.

Proteomic technologies are also used for the assessment of food authentication in different kinds of foodstuffs [28]. Milk adulteration represents one of the most common types of food frauds. In fact, the fraudulent addition of low-cost milk to the milk of higher costs and the undeclared use of admixtures of milk of different species to manufacture traditional products protected by denomination of origin (PDO) are the main problems in the dairy industry. As an example, by monitoring the profile of the most abundant whey proteins (i.e. α-LA and β-LG), the fraudulent presence of cow milk in ewe or buffalo products [29] and the adulteration of donkey milk have been demonstrated by means of a suitable, rapid, and sensitive MALDI-MS method, if compared with more laborious and time-consuming currently used techniques [30, 31]. Furthermore, for detecting the addition of powdered milk to fresh milk, several modified diagnostic peptides specific for powdered milk were identified by MALDI-MS and used as marker peptides in milk samples adulterated with powdered milk, subjected to in-solution digestion of the proteins [32].

Even cereals are a target for fraudulent actions. In particular, the molecular characterization of prolamins in local wheat varieties is an unambiguous way to trace the plant biodiversity that needs to be recognized, categorized, and preserved [33]. Proteomics has also been used as a complementary analytical tool to the existing safety assessment techniques for identifying side effects occurring in seeds from GM maize plants [34]. The substantial equivalence of GM crops applied, as an example, to tomatoes, maize, and potatoes, showed that the environment affects protein expression (in terms of the total number of protein spots) more than gene insertion and that the differential protein expression is often variety specific, supporting the idea that the GM does not influence global protein expression. In durum wheat, as an example, proteome analysis of kernels of transgenic and untransformed durum wheat lines, in which genes of the starch-branching enzymes were silenced by RNA interference, highlighted subtle differences, most of which were considered as "predictable unintended effects" due to the primary effect of the transgene within the starch biosynthetic pathway [35].

Recently, comparative proteomics of cotton leaves between the commercial transgenic Bt + CpTI cotton SGK321 (BT) clone and its non-transgenic parental counterpart SY321 wild type (WT) was performed [36]. Although some unintended variations of proteins were found between BT and WT cottons, no toxic proteins or allergens were detected, thus demonstrating that genetic modification did not sharply alter cotton leaf proteome.

Questions

- What is the role of bioactive peptides?
- How is it possible to perform the identification of microbial food contaminants?
- What is "allergomics?"
- What may be the limitations of proteomics in food analysis?
- How is it possible to detect food authentication or food frauds by proteomic technologies?

- Can the manufacturing processes and storage conditions induce significant changes in the structure of food proteins?
- Can proteomics be used for identifying side effects occurring in genetically modified (GM) or transgenic plants?

References

1 Cunsolo, V., Muccilli, V., Saletti, R., and Foti, S. (2014). Mass spectrometry in food proteomics: a tutorial. *Journal of Mass Spectrometry* 49 (9): 768–784.

2 Gašo-Sokǎ, D., Kovǎ, S., and Josić, D. (2010). Application of proteomics in food technology and food biotechnology: process development, quality control and product safety proteomics in food (bio)technology. *Food Technology and Biotechnology* 48 (3): 284–295.

3 Cunsolo, V., Muccilli, V., Saletti, R., and Foti, S. (2012). Mass spectrometry in the proteome analysis of mature cereal kernels. *Mass Spectrometry Reviews* 31 (4): 448–465.

4 Dworschak, R.G., Ens, W., Standing, K.G. et al. (1998). Analysis of wheat gluten proteins by matrix-assisted laser desorption/ionisation mass spectrometry. *Journal of Mass Spectrometry* 33 (5): 429–435.

5 Cunsolo, V., Galliano, F., Muccilli, V. et al. (2005). Detection and characterization by high-performance liquid chromatography and mass spectrometry of a goat β-casein associated with a CSN2 null allele. *Rapid Communications in Mass Spectrometry* 19 (20): 2943–2949.

6 D'Alessandro, A., Rinalducci, S., Marrocco, C. et al. (2012). Love me tender: an Omics window on the bovine meat tenderness network. *Journal of Proteomics* 75 (14): 4360–4380.

7 Pischetsrieder, M. and Baeuerlein, R. (2009). Proteome research in food science. *Chemical Society Reviews* 38 (9): 2600–2608.

8 Zhang, Q., Ames, J.M., Smith, R.D. et al. (2009). A perspective on the Maillard reaction and the analysis of protein glycation by mass spectrometry: probing the pathogenesis of chronic disease. *Journal of Proteome Research* 8 (2): 754–769.

9 Abdel Rahman, A.M., Kamath, S., Lopata, A.L., and Helleur, R.J. (2010). Analysis of the allergenic proteins in black tiger prawn (*Penaeus monodon*) and characterization of the major allergen tropomyosin using mass spectrometry. *Rapid Communications in Mass Spectrometry* 24 (16): 2462–2470.

10 Malik, A.K., Blasco, C., and Pico, Y. (2010). Liquid chromatography-mass spectrometry in food safety. *Journal of Chromatography A* 1217 (25): 4018–4040.

11 Venir, E., Del Torre, M., Cunsolo, V. et al. (2014). Involvement of alanine racemase in germination of *Bacillus cereus* spores lacking an intact exosporium. *Archives of Microbiology* 196 (2): 79–85.

12 Callahan, J.H., Shefcheck, K.J., Williams, T.L., and Musser, S.M. (2006). Detection, confirmation, and quantification of *Staphylococcal* enterotoxin B in food matrixes using liquid chromatography-mass spectrometry. *Analytical Chemistry* 78 (6): 1789–1800.

13 Fagerquist, C.K. (2017). Unlocking the proteomic information encoded in MALDI-TOF-MS data used for microbial identification and characterization. *Expert Review of Proteomics* 14 (1): 97–107.

14 Pavlovic, M., Huber, I., Konrad, R., and Bush, U. (2013). Application of MALDI – TOF MS for the identification of food borne bacteria. *The Open Microbiology Journal* 7: 135–141.

15 Bohme, K., Fernandez-No, I.C., Barros-Velazquez, J. et al. (2012). SpectraBank: an open access tool for rapid microbial identification by MALDI-TOF MS fingerprinting. *Electrophoresis* 33 (14): 2138–2142.

16 Piras, C., Roncada, P., Rodrigues, P.M. et al. (2016). Proteomics in food: quality, safety, microbes, and allergens. *Proteomics* 16 (5): 799–815.

17 Nguyen, D.T.L., Van Hoorde, K., Cnockaert, M. et al. (2013). A description of the lactic acid bacteria microbiota associated with the production of traditional fermented vegetables in Vietnam. *International Journal of Food Microbiology* 163 (1): 19–27.

18 Gagnaire, V., Piot, M., Camier, B. et al. (2004). Survey of bacterial proteins released in cheese: a proteomic approach. *International Journal of Food Microbiology* 94 (2): 185–201.

19 Jardin, J., Mollè, D., Piot, M. et al. (2012). Quantitative proteomic analysis of bacterial enzymes released in cheese during ripening. *International Journal of Food Microbiology* 155 (1–2): 19–28.

20 Lahrichi, S.L., Affolter, M., Zolezzi, I.S., and Panchaud, A. (2013). Food Peptidomics: large scale analysis of small bioactive peptides—a pilot study. *Journal of Proteomics* 88: 83–91.

21 Carrasco-Castilla, J., Hernández-Àlvarez, A.J., Jiménez-Martínez, C. et al. (2012). Use of proteomics and peptidomics methods in food bioactive peptide science and engineering. *Food Engineering Reviews* 4 (4): 224–243.

22 Korhonen, H. and Pihlanto, A. (2003). Food-derived bioactive peptides: opportunities for designing future foods. *Current Pharmaceutical Design* 9 (16): 1297–1308.

23 Schillaci, D., Arizza, V., Parrinello, N. et al. (2010). Antimicrobial and antistaphylococcal biofilm activity from the sea urchin *Paracentrotus lividus*. *Journal of Applied Microbiology* 108 (1): 17–24.

24 Schillaci, D., Cusimano, M.G., Cunsolo, V. et al. (2013). Immune mediators of sea-cucumber *Holothuria tubulosa* (Echinodermata) as source of novel antimicrobial and anti-staphylococcal biofilm agents. *AMB Express* 3 (35): 1–10.

25 Castellano, P., Aristoy, M.-C., Sentandreu, M.Á. et al. (2013). Peptides with angiotensin I converting enzyme (ACE) inhibitory activity generated from

porcine skeletal muscle proteins by the action of meat-borne Lactobacillus. *Journal of Proteomics* 89: 183–190.

26 Escudero, E., Mora, L., Fraser, P.D. et al. (2013). Purification and Identification of antihypertensive peptides in Spanish dry-cured ham. *Journal of Proteomics* 78: 499–507.

27 Piovesana, S., Capriotti, A.L., Cavaliere, C. et al. (2015). Peptidome characterization and bioactivity analysis of donkey milk. *Journal of Proteomics* 119: 21–29.

28 Ortea, I., O'Connor, G., and Maquet, A. (2016). Review on proteomics for food authentication. *Journal of Proteomics* 147: 212–225.

29 Cozzolino, R., Passalacqua, S., Salemi, S., and Garozzo, D. (2002). Identification of adulteration in water buffalo mozzarella and in ewe cheese by using whey proteins as biomarkers and matrix-assisted laser desorption/ionization mass spectrometry. *Journal of Mass Spectrometry* 37 (9): 985–991.

30 Cunsolo, V., Muccilli, V., Saletti, R., and Foti, S. (2013). MALDI-TOF mass spectrometry for the monitoring of she-donkey's milk contamination or adulteration. *Journal of Mass Spectrometry* 48 (2): 148–153.

31 Di Girolamo, F., Masotti, A., Salvatori, G. et al. (2014). A sensitive and effective proteomic approach to identify she-donkey's and goat's milk adulterations by MALDI-TOF MS fingerprinting. *International Journal of Molecular Sciences* 15 (8): 13697–13719.

32 Calvano, C.D., Monopoli, A., Loizzo, P. et al. (2013). Proteomic approach based on MALDI-TOF MS to detect powdered milk in fresh cow's milk. *Journal of Agricultural and Food Chemistry* 61 (8): 1609–1617.

33 Muccilli, V., Lo Bianco, M., Cunsolo, V. et al. (2011). High molecular weight glutenin subunits in some durum wheat cultivars investigated by means of mass spectrometric techniques. *Journal of Agricultural and Food Chemistry* 59 (22): 12226–12237.

34 Zolla, L., Rinalducci, S., Antonioli, P., and Righetti, P.G. (2008). Proteomics as a complementary tool for identifying unintended side effects occurring in transgenic maize seeds as a result of genetic modifications. *Journal of Proteome Research* 7 (5): 1850–1861.

35 Sestili, F., Paoletti, F., Botticella, E. et al. (2013). Comparative proteomic analysis of kernel proteins of two high amylose transgenic durum wheat lines obtained by biolistic and *Agrobacterium*-mediated transformations. *Journal of Cereal Science* 58 (1): 15–22.

36 Wang, L., Wang, X., Jin, X. et al. (2015). Comparative proteomics of Bt-transgenic and non-transgenic cotton leaves. *Proteome Science* 13: 1–15.

8.3   Challenges in Analysis of Omics Data Generated by Mass Spectrometry

Katarzyna Pawlak[1], Emma Harwood[2], Fang Yu[3], and Pawel Ciborowski[2]

[1] *Department of Analytical Chemistry, Faculty of Chemistry, Warsaw University of Technology, Warsaw, Poland*
[2] *Department of Pharmacology and Experimental Neuroscience, University of Nebraska Medical Center, Omaha, NE, USA*
[3] *Department of Biostatistics, University of Nebraska Medical Center, Omaha, NE, USA*

8.3.1   Introduction

In the last two decades, an exceptional technological boom has created an opportunity to generate unprecedentedly large, highly complex, and multidimensional data sets. At the same time, new challenges emerged such as data storage, transmission, and analyses. While the data storage and speed of data transmission problems have been successfully resolved through the offering of vast space in the cloud by many providers and the improved transmission between the cloud and local recipient, respectively, data analyses and presentation are lagging. The development of faster computing machines and user-friendly software for a variety of data analyses has significantly decreased the time between results coming from one experiment (study) and the initiation of subsequent experiments. Although these main segments of high-throughput-based (omics) studies are applicable to all areas of research, each space has a set of individual obstacles and challenges. In this chapter, we would like to focus on issues that are directly related to analysis of high-throughput proteomics and metabolomics data generated by studies employing mass spectrometry (MS).

One can argue that with the increasing size of data sets, including those in MS, there is an increasing opportunity to gain insight into how biological systems work. However, it must be noted that each data set generates important information as well as a proportional amount of background noise. Data analyses have to deal with this problem as well. At this point, a variety of methods have been used such as k-nearest neighbors (KNN), logistic regression, linear discriminant analysis (LDA), quadratic discriminant analysis (QDA), naive Bayes, and hidden Markov models (HMMs), etc. Each of these approaches has strengths and limitations, and the scope of this chapter does not allow for the detailing of each model. Thus, the second objective of this chapter is to give a general overview and guidance for subsequent exploration of analyses that best fit the message to be conveyed.

8.3.1.1 How Big Must Big Data Be?

Big Data is to some extent an elusive concept and quite often described with adjectives such as "so large or complex that traditional data processing application software is inadequate to deal with them" (Wikipedia). Some scientists remark that Big Data "is when the results of your experiments cannot be imported into an Excel file" [1]. Many other definitions can be found that target specific aspects of Big Data. In our view, the most important aspect is not volume, but complexity, since plurality is unstructured in nature. Some definitions place the center of Big Data in the volume measured in Gb and more. In our view, www.sas.com is close to the inherent nature of Big Data by saying, "...it's not the amount of data that's important. It's what organizations do with the data that matters."

From Big Data analysis we expect to see quantitative differences, regardless of the nature of questions asked, e.g. sales, climate change, response to drugs, or analytically acquired MS data. In each and every case, differences must be statistically significant to warrant further investigation. Therefore, statistical analysis is either a prerequisite of Big Data analysis or an inherent part. Summarizing our considerations, we conclude that omics-type experiments that are part of systems biology studies qualify as Big Data, even if they are relatively small in terms of number of bytes (Table 8.3).

8.3.1.2 Do Omics Experiments Generate Unstructured Data?

The objectives of working with large omics data sets are to perform wide-ranging analyses and to gain comprehensive interpretation. Again, it is to some extent arbitrary or elusive whether omics data analyses and interpretations are multidimensional or not. We consider omics data to be multidimensional when it is more than binary in nature, i.e. comparison of an experimental condition and a control, or when data sets consist of omics data from various platforms,

Table 8.3 Bytes as measure of data size.

Bytes value[a]	Metric (decimal)
10^3	kB (kilobyte)
10^6	MB (megabyte)
10^9	GB (gigabyte)
10^{12}	TB (terabyte)
10^{15}	PB (petabyte)
10^{18}	EB (exabyte)
10^{21}	ZB (zettabyte)
10^{24}	YB (yottabyte)

[a] The byte is a unit of digital information that most commonly consists of eight bits (https://en.wikipedia.org/wiki/Byte).

Omics input – data set

Statistical analysis

Bioinformatic analysis

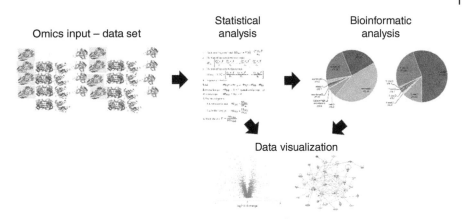

Data visualization

Figure 8.12 General workflow of OMICS data analysis. *Source*: www.google.com.

e.g. proteomic and transcriptomic. The importance of this aspect of data analysis is twofold: the increasing complexity in experimental design in pursuit of getting more information from one study and the growing importance of diagnostic and treatment strategies that are becoming more complex and more individualized and include information derived from different types of technology platforms and/or types of molecules (nucleic acids, proteins, carbohydrates, lipids, etc.). Subsequently, analysis of such data requires specialized algorithms that visualize the key points requested by the end user (Figure 8.12).

Based on many definitions of structured and unstructured data, we can say that MS itself generates structured data expressed by numeric values: mass to charge (m/z), peak intensity, area under the peak, etc. On the other hand, in conjunction with experimental design of omics experiments, we may be dealing with hybrid structured/unstructured data. For example, if a proteomics or metabolomics experiment is linked to the behavioral data of a model of a living organism under specific conditions, MS structured data will be linked to unstructured data and will pose a challenge in data interpretation. Inherent variability within the human population and individual responses of human subjects selected for any given control or experimental group will create a challenge in omics data interpretation. One approach to meet such a challenge is multidimensionality of data presentation and visualization, where structured and unstructured data are blended into one picture.

8.3.2   Targeted and Full Unbiased Omics Analysis Based on MS Technology

Mass spectrometers alone and/or in combination with other instruments constitute an astonishingly diverse and large family of analytical platforms, with a progressive list of applications in both qualitative and quantitative analyses [2]. Such a variety of instruments is obtained by unique combination/hyphenation of efficient ionization techniques with different mass analyzers, most frequently quadrupole (Q), time-of-flight (TOF), ion trap (IT), and Orbitrap, and in-line or offline fractionations. Each instrument or specific combination of any of them has specific limitations and strengths regarding accuracy, dynamic range, resolution, and speed of analysis. Nevertheless, the overarching challenge is quality of acquired data, which is discussed below.

8.3.2.1 Factors Affecting Data Quality

How can it be that two compounds registered with the same method in the same sample have a relative standard deviation (RSD) below 3% for one and above 20% for the other? What contributes to different variability of the data obtained for different compounds, and how can it be resolved? First, we have to examine the height of chromatographic peaks, which represent signal intensity, and compare them to the noise of the baseline in the chromatogram, which gives the signal-to-noise (S/N). If S/N is below 30, a higher standard deviation (SD) can be considered as a typical effect of randomly changed noise. If signals for both compounds are high, it is most probable that the signal with the higher RSD is interfering with other compounds present in the sample. For example, in metabolomics studies of urine, the preparation of a sample is often limited to dilution, centrifugation, and filtering. Such samples can be rich in different isobaric compounds, which will interfere with the signal of the analyzed compound. MS-based quantitative methods are based on multiple reaction monitoring (MRM), the transitions – pairs of fragmentation ions – are obtained from a specific precursor ion. Acquisition of signals of two different transitions, often called quantifier and qualifier, for each compound is a standard element of good practice in liquid chromatography–MS (LC-MS) quantitative methods. The second transition can also be used for the confirmation of the compound identity and for tracking the interferences. If this does not help, a highly performed separation by front-end in-line techniques (gas chromatography [GC], supercritical fluid chromatography [SFC or SCL], high performance liquid chromatography [HPLC], capillary electrophoresis [CE]) can help to separate interfering compounds from compounds of interest due to differences in volatility, polarity, hydrophobicity, or acidity [3]. Such sample preparations can be carried out offline, sequentially [4]. Resulting fractions will contain groups of separated, chemically similar compounds that can be analyzed separately. During extraction, a derivatization can be performed, which changes chemical properties of the compound, the molecular mass of the compound, and its fragmentation path. If available, a mass spectrometer with a higher resolving power can be used (Orbitrap or TOF), which offers better accuracy and can distinguish analytes from interfering compounds.

Combining selective sample preparation methods with high performance separation techniques and high resolution, MS increases the accuracy (identification and/or quantification) of the results and significantly decreases their

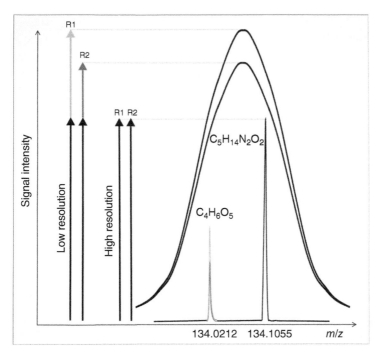

Figure 8.13 Resolution and repeatability of analytical results.

variability as long as the compound of interest is stable enough to survive all the steps of complex analytical protocol. On the same token, precision can be compromised by different "memory effects," which are registrations of peaks/signals corresponding to compounds that contaminated compartments of LC-MS in the past. For example, registration of a signal at m/z +115 is the most-known long-term memory effect caused by trifluoroacetic acid (TFA) in MS. It should also be pointed out that the instability of the compounds and memory effects are responsible for the drift of signals' intensities observed in time, which leads to the asymmetric distribution of data (Figure 8.13).

8.3.2.2 Speed of MS Data Acquisition: Why Does It Matter?

The first quadrupole (Q_1) is mostly used as an ion guide, sequentially focusing on selected ions before their separation or fragmentation in the collision cell and improving outcomes of the next compartments. The presence of Q_1 creates the possibility to (semi-)quantify target compounds by monitoring fragmentation products for selected reaction monitoring (SRM) and MRM. The number of monitored reactions is limited to the speed of mass analyzer and the time used for introduction of the sample to MS. Usually, with the exception of direct infusion, small volumes of samples are introduced as narrow peaks in the liquid mobile phase as they are eluted from the chromatographic columns. Changes in signal intensity, represented by a Gaussian curve, for different ions can be observed in the range of a few milliseconds (ms) to two minutes. Quadrupoles (Q_1 and Q_3) used in triple quadrupole, also known as tandem mass spectrometer, usually need 100 ms of dwell time for accurate acquisition of ions obtained for one specific measurement without significant loss of sensitivity. Thus, to obtain good precision, the shape of Gaussian's distribution of the sample's band needs to be defined by at least 10 points, acquired over 1 second. The same amount of time is required for each additional reaction, limiting significantly the number of transitions in the cycle. The number of observed analytes is usually increased by monitoring selected transitions in specified time windows (time scheduled MRM). However, such an approach is not satisfactory for ultra-performance liquid, supercritical fluid, and gas chromatography, which provide two seconds peaks. This problem was solved by connecting the quadrupole analyzer with other types of mass analyzers – for example, linear IT (QTRAP, Sciex) or other combinations of IT, i.e. Orbitrap (Thermo Scientific) with parallel reaction monitoring (PRM), which offers better sensitivity and decreases dwell time to 3 ms. As a result, high performance acquisition of 500 MRMs is possible in one analysis [5]. Nevertheless, it is limited to the monitoring of selected ions.

The problem of mass analyzer speed reaches a different level in the case of ions' scanning necessary for identifying compounds or looking for unknown compounds. Both a tandem mass spectrometer and a QTRAP mass spectrometer allow one to obtain a product ion mass spectrum within a few seconds, which is not satisfactory in the case of ultra-high performance separation techniques or high-frequency sampling (matrix-assisted laser desorption/ionization [MALDI] and desorption/ionization on silicon [DIOS]). TOF analyzers provide hundreds of mass spectra per second with high mass accuracy below 5 ppm. Applied in different configurations (tandem or hybrid with quadrupole), TOF provides millions of spectra in a single analysis, opening the window to high-throughput analysis. Following the technical development of mass spectrometers, three different approaches can be distinguished: MRM-specific analysis, multi-MRM selective analysis, and high-throughput scanning. Each offers a set of different advantages and limits that have to be faced when designing analytical strategy. One general rule should be always considered: the time proposed for 10 acquisition cycles in MS should not exceed time width of chromatographic peak. Although the last decade can also be characterized by rapid development of ultra-performance/ultrafast chromatographic techniques, the problem of faster acquisition of ions by MS remains [6] (Figure 8.14).

8.3.2.3 Analytical Strategies in Omics Studies

The main objectives of omics approach are quantification of biological molecules and characterization of how these changes influence function and dynamics of living cells, organs, and entire organisms. Following development of MS instrumentation, omics technology platforms can be divided into three

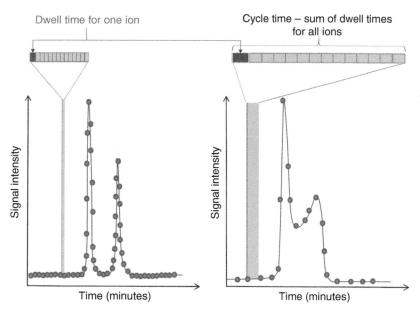

Figure 8.14 The influence of dwell time and cycle time on shape of chromatographic peak.

categories: untargeted or full unbiased, semi-targeted, and targeted. Each approach generates high-throughput data requiring statistical and bioinformatic tools for their analyses [7].

8.3.2.3.1 *Untargeted Approach*

This is an exploratory approach looking for any changes in molecular composition of the biological system being investigated. The hypothesis is very broad and usually includes discovery of changes resulting from biological system manipulation. Collection of experimental data is based on fast-scanning mass analyzers, most often TOF, in the wide mass range. Quite often molecules are not assigned to *m/z* values and are referred to as features. Scanning is performed in reference to a biological control group and differential analysis is performed to track features specific to each group. MS/MS data-dependent analysis (DDA) for each precursor ion requires much more time than MS scanning, resulting in significant loss of the sensitivity. To alleviate the limitations of the method, strategies have been developed that rely on neither detection nor knowledge of the precursor ions to trigger acquisition of fragment ion spectra. Those methods operate via unbiased data-independent acquisition (DIA) in the cyclic recording throughout the liquid chromatography (LC) time range. By using such scans, the link between the fragment ions and the precursors from which they originate is lost, complicating the analysis of the acquired data sets [8, 9]. Compounds can be identified by searching mass spectra obtained by DDA methods or by searching pseudo-DDA-MS/MS spectra reconstituted post-acquisition based on the coelution profiles of precursor and fragmentation ions. However, the latter approach is useful only in the case of rich databases like Sequential Window Acquisition of All Theoretical Fragment Ion Spectra (SWATH) in proteomics and mass profiler in metabolomics. The discovery approach delivers new data but offers the lowest precision. Therefore, identified compounds statistically indicated as potential markers should be analyzed by the semi-targeted or targeted method to confirm their reliability. To some extent, data acquired by the discovery method do not fully overlap or show full compatibility with data acquired by targeted methods. This is caused by significant differences in data variability and/or different ionization efficiencies and fragmentation levels in the ion source depending on the instrument design.

8.3.2.3.2 Semi-targeted Approach
This approach is based on selective analysis and relies on the state of knowledge represented by laboratory, databases, published data, and personal communications related to the biological system. In this approach, successful collection of experimental data will depend on several conditions such as maximizing the analytical potential of chosen instrumentation, knowledge derived from open sources, and specific analytical reagents (standards) that might not be readily available. All of this will contribute to the ability to correctly interpret the data. In this case, the data may not be Big Data per se, but will require advanced interpretation skills to derive biologically important conclusions. Typically, MRM types of techniques are used [5]. This can be applied to a single target or multiple targets. Importantly, the lower complexity of data collected using semi-targeted approach does not necessarily mean that data are easier to interpret, especially if data are placed in the broader context of how the biological system functions.

8.3.2.3.3 Targeted Approach
This approach starts with an assumption that selection of a group of compounds targets a specific aspect of a biological system, i.e. a specific metabolic pathway or process, so it is important to define how the biological system being investigated functions. This approach is also based on MRM technique but usually contains a limited number of analytes, allowing establishment of the most suitable sample preparation protocol(s), isolation/purification strategies, and MRM-based detection conditions. Similarly to the semi-targeted approach, this one may be seen as generating low complexity data, yet plugging in these data into a broader picture of how an organism functions remains a challenge [10].

8.3.3 Data Analysis and Visualization of Mass Spectrometry Omics Data

Systems biology data is no stranger to the Big Data boom. As technological and methodological approaches to high-throughput systems biology experiments expand and improve, so do the resulting data sets. With this growing complexity of data comes the need for clear and effective data visualization. Omics data generated by MS is no exception. In this section, we discuss the nature of MS omics data in general and introduce several strategies for its analysis and visualization. These methods might currently be considered to be the most applicable and widely used, but this list is a non-exhaustive introduction. Depending on the experimental goals, some or all may be incorporated into the analysis workflow.

8.3.3.1 A Brief Introduction to Data Visualization

Data visualization can be integrated at many stages of the data processing and analysis workflow. It may be used for exploratory purposes to gain an understanding of the data's underlying structure and patterns. It may help visually compare samples and identify outliers. It may show the effect of a statistical transformation or communicate the results of a statistical test. Thus, we introduce some background and foundational concepts of data visualization to keep in mind as you continue reading and implementing your own data analysis and visualization [11].

In general, data visualization is the presentation of data in a visual context to better communicate its contents. A major goal of data visualization is to present the data in an easy-to-understand manner to an audience that, quite often, has a wide range of expertise. A visualization should tell a story with the data in such a way that engages the audience and aids in discovery [12].

Human perception plays an important role in the realm of data visualization, since it is associated with human cognition. There are many elements of visualization that influence the viewer's interpretation of a graphical representation of data. Patterns, trends, and correlations predominate in the exploration and communication of a wide variety of data set types, and this is especially true of omics data generated by MS. This makes sense since humans inherently perceive objects as organized patterns. Elements of a figure can draw a viewer's attention to such patterns, making the visualization intuitive.

From a more psychological perspective, we may consider the Gestalt principles when relating human perception to data visualization. Gestalt principles describe the ways in which people organize visual information, recognizing that there are many parts that make up the whole of the visualization. These principles are relevant to the design of data visualizations because they aid in effectively encoding information. The *law of similarity* states that visually similar objects (those sharing the same mark or channel) form a group or pattern. The *law of proximity* states that objects close in space form a group, despite any differences in appearance. The *law of closure* states that objects separated in space tend to be perceived as a whole if enough of a shape is shown. The viewer's mind will essentially "fill in the blanks" to connect the parts of a visualization to form a complete figure. This idea is important to keep in mind when considering how to represent null or missing values, which may arise in MS data. The creator should avoid displaying information in such a way that implies a

Table 8.4 Gestalt principles.

Principle	Definition	Example
Law of similarity	Visually similar objects form a group or pattern	
Law of proximity	Objects close in space form a group	
Law of closure	Objects separated in space tend to be perceived as a whole if enough of a shape is shown	
Law of Prägnanz	Every stimulus is perceived in its simplest form	
Law of isomorphic correspondence	Interpretation of a visualization is informed by past experience	

connection that does not truly exist within the data. The *law of Prägnanz* posits that every stimulus is perceived in its simplest form (http://facweb.cs.depaul.edu/sgrais/gestalt_principles.htm). This has the effect of optimizing cognitive load or mental effort at a given moment. One application of this law is organizing data in decreasing order. Lastly, the *law of isomorphic correspondence* focuses on the idea that a viewer's response to and interpretation of a visualization is informed by past experience. It is essential to be aware of the conventions that may exist within a specific field [13, 14]. Table 8.4 gives brief descriptions of these principles and an example of each.

More concretely, marks are the most basic elements of a visualization and are one way of utilizing Gestalt principles. Examples include points, lines, links, areas, and containment. Table 8.5 shows several types of marks. Channels are the means of controlling the appearance of marks and include color, length,

Table 8.5 Marks.

Mark	Example
Points	
Lines	
Areas	
Containment	
Links	

Table 8.6 Channels.

Channels

Color		Sequence		Length	
Value/ gradation		Size/scale		Area	
Texture		Orientation		Proportion	
Symbol		Proximity/ density		Count	

texture, orientation, and symbol, among a few others. Table 8.6 depicts several channel types. Manipulating these elements gives the creator power to emphasize and organize data in a meaningful way [15].

These principles of perceptual organization and elements help guide the development of a visualization. They offer opportunities to "build context for information" and to add more layers of meaning [13]. This ultimately leads to more effective communication of what lies within increasingly complex data, expanding the boundaries of human understanding.

8.3.3.2   Exploration and Preparation of Data for Downstream Statistics and Visualization

In quantitative omics data generated by MS, the raw quantitative data will require processing to improve its quality prior to statistical analysis. Such processing is expected to prepare the data for further statistical analysis, leading to more accurate and reliable results. Examples of these suggested processing steps are discussed below.

8.3.3.2.1   Data Organization

When working with quantitative MS data, different software programs will produce different output. The output may need to be reorganized to make subsequent analysis easier. For example, it may be necessary to create separate columns for each sample's peak intensity values. This simple step makes it easier to compare samples, facilitating downstream visualization and analysis. The output may also include variables that are not needed in later analysis steps and as such can simply be deleted. Lastly, missing values in the data may have to be addressed by removing them from the data set or by finding a suitable imputation method. In MS data, many features may not be quantified in every experiment (sample). Usually, it is not advised to focus only on those data with complete quantification information because information will be lost in excluding the incomplete data. Thus, missing values may be imputed based on available data. Again, there are a number of methods ranging from straightforward to more sophisticated, and the decision depends on the experiment.

The need for these steps, like all parts of the workflow, depends on the raw output format, experimental design, and the overall purpose of the experiment. Regardless of the quantitation software used and how its initial output appears, data cleanup and reorganization will likely be a preliminary step in the workflow.

8.3.3.2.2   Principal Component Analysis (PCA)

Principal component analysis (PCA) is a useful exploratory tool for dimension reduction and visualization. It is a tool well suited to MS omics data, since mass spectrometers can measure levels of thousands of biomolecules in hundreds of samples. PCA offers a solution to the challenge of simple exploration and visualization of such high-dimensional data [16]. With proteomics data, for example, PCA will transform data into uncorrelated components, which are linear combinations of protein data. The components with large variation of the data will be called as principal components and will be used to represent the data. In this way, the data can be expressed using a small number of factors but retain a major amount of the data variation. We can use the principal components to draw plots with identified principal components as the axes. The samples located close to each other in the PCA will be expected to have similar distribution.

8.3.3.2.3 Transformation and Normalization

Processing steps will likely include some transformation and/or normalization methods to account for baseline noise and several sources of variation that are introduced in an experiment. The log transformation has been widely used in preprocessing high-throughput omics data including MS data. Because many statistical tests assume data normality, the transformation can reduce skewness and make the data more normally distributed. Transformation of the data allows the distributions of intensities of each sample to be visually compared by density plots, boxplots, violin plots, histograms, etc.

Additionally, the log transformation will be useful for reducing heteroskedasticity [17], converting the multiplicative relation into additive relation, and stabilizing variance. After log transformation, the fold change between conditions can be estimated using a difference in the log scaled expressions of the protein between the corresponding groups. A one-unit difference in the $\log_2$ scaled protein abundances equals a twofold change between conditions.

Omics experiments will usually have replicate data to shed light on variation stemming from the experimental setup and/or the biological system [18]. Normalization helps to reduce technical variation that might overshadow the effects due to biological conditions, which are what we seek. There are a variety of normalization methods used with omics data, and because the source of systematic bias is usually not known and every experiment is different, it is challenging to find the optimal method. One helpful open-source tool is "Normalyzer" [19], which generates a report using a number of different methods to guide the researcher in choosing the best fit. Normalyzer is an R package, but it also has an online implementation at http://quantitativeproteomics. org/normalyzerde.org/normalyzer. As with log transformation, distributions before and after normalization can also be visualized to aid in comparison. Figure 8.15 provides an example showing the distribution of MS data for one sample before the log transformation, after the log transformation, and after quantile normalization on transformed data in the form of histograms. Figure 8.16 shows boxplots before and after quantile normalization.

In general, the data processing choices depend on many factors, including initial data output structure, instrument type, and experimental setup. These important choices invariably affect downstream analysis results, where accuracy and reliability remain top priorities.

8.3.3.3 Differential Expression Analysis

One common goal of omics data analysis is to identify the features (e.g. proteins, metabolites, etc.) with differential expressions between different biological conditions. Specifically, we want to test against the null hypothesis of the log scaled abundance levels between conditions that are equivalent for each feature. Many statistical methods have been made available for differential expression analysis, depending on the experiment platform, the number of

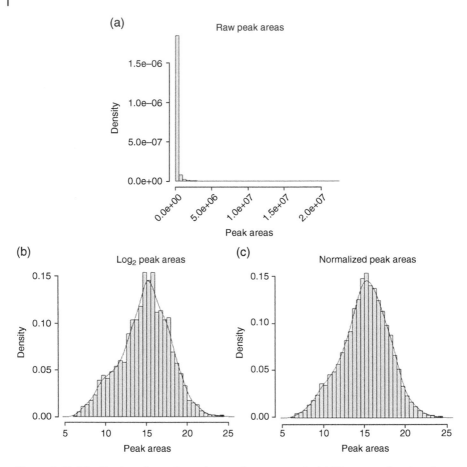

Figure 8.15 Distribution of protein peak areas for one sample. (a) Histogram showing the raw distribution of peak areas. (b) Histogram with density plot of the $\log_2$ transformed peak areas. (c) Histogram with density plot of $\log_2$ transformed and quantile normalized peak areas.

conditions under comparison, and the study design. When the study only contains two biological conditions, a simple two-sample t-test can be conducted on each feature. When the study has more than two biological conditions, analysis of variance or linear models can be used for data analysis. Recognizing that high-throughput omics experiments will quantify the expression data of genes, proteins, or other features in parallel, Bayesian methods have been developed for borrowing information across models for individual features.

In this chapter, we will use the linear models for microarray data (limma) [20] method to illustrate differential expression analysis. Limma was originally developed for modeling high-throughput genomics data. More recently, this

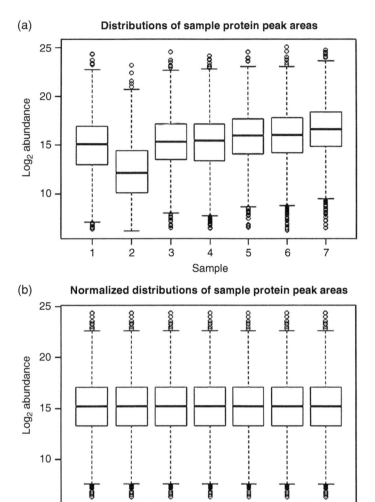

Figure 8.16 Distributions of protein peak areas for seven samples. (a) Boxplot of $\log_2$ transformed protein peak areas. (b) Boxplot of $\log_2$ transformed, quantile normalized protein peak areas.

method has been widely used for analysis of multiple types of omics data including proteomics and metabolomics data. When applying to proteomics data, limma fits linear models on the data along with empirical Bayes estimators to determine differential expression of proteins. Limma gained popularity because it fits a flexible model for high-throughput omics data. The use of linear models allows limma to handle studies with complex experimental designs. The limma models can test flexible hypotheses. Lastly, through empirical Bayes approach for estimation, limma leverages the parallel structure of the omics data and borrows strength between the protein-wise models and makes the statistical estimator more reliable when the number of samples is small. Limma can be implemented through R/Bioconductor packages.

Multiple comparisons with inflated type I errors arise when we conduct hypothesis testing on multiple features. False discovery rate (FDR) quantifies the risk of having false positivity among all significant tests. A control of the FDR for the large-scale studies like omics experiments will account for multiple comparisons. Specifically, the features are ordered with the mostly significantly differentially expressed features listed at the top. When certain features ranked at the top of the ordered list are selected as differentially expressed, the corresponding FDR can be estimated. The top features with sufficiently small FDR value will be claimed to be differentially expressed proteins. For example, the limma package in R includes the Benjamini–Hochberg method [21] to estimate and control FDR based on the resulting p value from the limma differential expression analysis.

8.3.3.3.1 *Visualizing Results*

The volcano plot [22] has been commonly used for visually identifying interesting features with the most meaningful change between conditions in comparison from high-throughput omics experiments. An example is shown in Figure 8.17. When the volcano plot is used for presenting proteomics data, for example, statistical tests will first be conducted for comparing the expressions of each protein between groups, and the negative log of the p values (usually on $\log_{10}$ scale) is calculated. The volcano plot is constructed as a scatter plot with negative log scale of the p values on the y-axis versus log fold changes on the x-axis. Because the x-axis is the log fold change, the changes with same magnitude but different directions appear equidistant from the center. In addition, we anticipate that the proteins with high magnitudes of fold changes will be located away from the zero center of x-axis and the proteins with low p values or high significance will appear toward the top of the plot. Therefore, the interesting proteins will be located in two regions toward the top of the plot and far to the left or right side of the plot away from the center of the x-axis. Overall the volcano plot incorporates information on both the statistical significance and the magnitude of change. This information allows identification of features expected to own both statistical significance and biological significance.

Another way to visualize significant results is with a heat map. Heat maps represent multidimensional data as a matrix of varying colors. Variables are placed in rows and columns of a table, and values are colored according to a defined scale. Thus, variance across multiple variables is displayed to reveal any similarities or differences. This technique is useful for MS data, where the rows might be defined as detected features and the columns as different

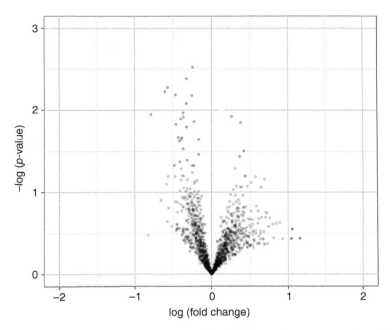

Figure 8.17 Volcano plot shows the results of differential expression analysis. Points with p value ≤ 0.05 are statistically significant. The p values are on the $-\log_{10}$ scale so that significant features are plotted at the top of the plot.

experimental conditions or samples. The relative abundances/peak intensities would correspond to some color scale, emphasizing which features are upregulated or downregulated across samples.

Cluster analysis pairs nicely with heat maps because it may help emphasize functional relationships among features and/or samples (https://docs.tibco.com/pub/spotfire/7.0.1/doc/html/heat/heat_what_is_a_heat_map.htm). It is a very useful data exploration tool for high-throughput omics data including MS data. With cluster analysis, the features or samples will be grouped into clusters so that the features or samples in the same cluster share a more similar pattern of expression profiles than those in different clusters. Hierarchical clustering is one commonly used cluster analysis method because it is computationally feasible for high-throughput omics data and its results are easy to interpret. In order to decide which features should be grouped together, we need to specify a distance metric to measure distance between features based on their expression levels. Additionally, we specify linkage criterion to specify the dissimilarity of sets as a function of the pair-wise distances of observations in the sets. Common choices of distance metrics and link functions include Euclidean distance, correlation distance, and average link. Specifically, the average link function will use the average distance between two features from two condition sets to evaluate their similarity. To better cluster features with similar profiles, we may normalize the expression values of the features before cluster analysis. An iterative procedure will be used to cluster features or samples into a hierarchy structure, in which the features or samples linked by the same branch are anticipated to share similarities in the expression level and the lengths of the branches between features quantify the strength of the similarity. Clustering is visualized as a dendrogram, a tree diagram showing the arrangement of the clusters [23]. In short, cluster analysis brings forth and emphasizes potentially important patterns and relationships.

One other suggestion to include with a heat map for more effective communication is a parallel coordinates plot, a technique for plotting multi-dimensional data [24]. An axis is drawn for each variable, and they are organized as parallel vertical or horizontal lines. The lines running from axis to axis connect series of values associated with each variable (axis). They can be useful for time series or dose–response studies, where the columns are inherently ordered [25]. Additionally, these plots can highlight small discrepancies between samples. The usefulness of these plots, however, diminishes as the number of profiles increases because information is layered. An example of heat map with a dendrogram and accompanying parallel coordinates plot is shown in Figure 8.18.

8.3.3.4 Strategies for Visualization Beyond Three Dimensions

As discussed above, data vary widely in complexity, and the way in which they are visualized influences utility and interpretation. Multidimensional data is challenging to both analyze and visualize. As it turns out, even the phrase "multidimensional data" is not well defined. The term "multidimensional" tends to be applied to data sets with three or more dimensions (https://en.wikipedia.org/wiki/Multidimensional_analysis). But what are the dimensions, exactly? In perhaps its most basic form, a data set can be thought of as an $m \times n$ matrix. In this way, dimensionality might refer to the number of attributes, or columns, present within the data. They may be categorical or numerical variables. For example, patient health data is oftentimes multidimensional, since multiple variables (sex, height, weight, blood pressure, temperature, etc.) may be recorded and compared among multiple patients. Hundreds to thousands of proteins or other features identified and quantified by MS for multiple technical replicates and biological samples could be considered multidimensional data. Data combining information from multiple omics platforms (e.g. proteomics, lipidomics, metabolomics, etc.) undoubtedly fall into the category of high-dimensional data. While it would be ideal to simply define dimensionality as the number of columns in a data set, this definition is challenged when one considers that oftentimes, some variables are interrelated or dependent on

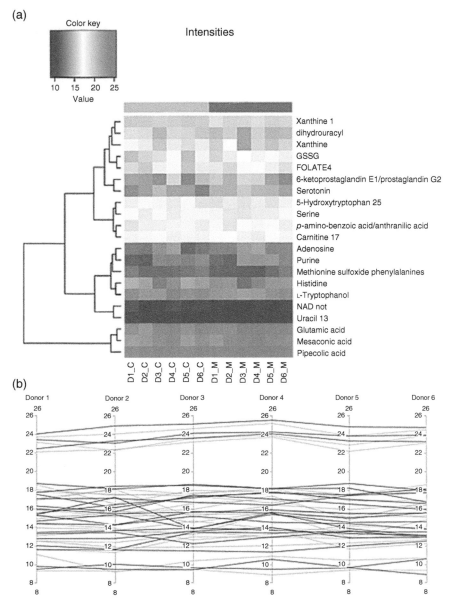

Figure 8.18 Displaying statistically significant features from differential expression analysis. (a) A heat map of the feature peak intensities (at log$_2$ scale) with a dendrogram from a hierarchical clustering algorithm. The six columns on the left (light gray) are control samples, and six columns on the right (dark gray) refer to drug-treated samples. (b) A parallel coordinates plot with the six samples as vertical axes measuring feature intensity, and each significant feature as a horizontal line in both cases: light gray for control and dark gray for treatment. The intensity changes (corresponding to the dark gray – white – light gray color key in (a)) can be tracked from sample to sample in each case, aiding in interpretation of the heat map in (a).

each other. Some dimensions can be reduced to summary statistics, such as the mean values. In other instances, some variables may even be useless in the context of certain applications or questions being asked (http://www.statisticshowto.com/dimensionality)(https://www.coursera.org/learn/big-data-machine-learning/lecture/PDGeA/dimensionality-reduction). Nevertheless, when the data has many dimensions, there are some strategies to effectively deal with the dimensionality.

One strategy to deal with high-dimensional data is to utilize dimension reduction techniques, such as PCA, mentioned above. Probably the simplest yet effective strategy to use when dealing with multidimensional data is to utilize some of the nonspatial graphical encodings (marks and channels) mentioned above. One could use shape, size, and/or color of data points to represent different dimensions plotted on a 2D plane. For example, color could represent patient gender, and the shape of a mark could represent different diseases. The possible combinations are many, so this is a plausible and easy way to incorporate many dimensions into one figure. It is important to keep in mind that 3D visualizations should only be used if one dimension of the data is inherently spatial. Visualizing the structure of biomolecules in space is one example, where the 3D location information of measurements translates to a representation of the 3D shape of the molecule [26]. Omics data, when quantitative, might consist of the feature name, peak intensity value, retention time, and m/z for multiple technical and/or biological replicates from different experimental conditions. Thus, more often than not, multidimensional data generated by MS can be represented in a 2D plane by taking advantage of the numerous visual encodings available.

8.3.3.5 Bioinformatics Tools

Bioinformatics is the "application of computational techniques to organize and analyze the information associated with biomolecules on a large-scale" [27]. Bioinformatics tools have the power to take data input and frame it in the context of a biological system. To use bioinformatics with omics data generated by MS, one can compile a list of features of interest or a list of differentially expressed features from a statistical test for input. Fold changes or p values might also be included, and it may be necessary to convert the input feature names to a certain ID type depending on the tool.

One example is enrichment analysis, which aims to categorize input (proteins, metabolites, etc.) into functional classes that may be over- or underrepresented in the data set, revealing biological patterns within the data. Any associated biological processes, cellular functions, disease phenotypes, or other biological meaning can then be assigned to the data. There are a wide variety of web-based and installed software-based tools for different types of omics data available (some may have a cost). In proteomics, a popular analysis tool is the PANTHER Classification System (www.pantherdb.org). The tool connects to the Gene Ontology (GO) project, which provides and maintains a "controlled vocabulary of terms describing gene product characteristics" based on peer-reviewed scientific publications (www.geneontology.org). For metabolomics data, Metabolite Set Enrichment Analysis (MSEA) analogously identifies biologically meaningful patterns in groups of metabolites by "investigating if a group of functionally related metabolites are significantly enriched" (http://www.metaboanalyst.ca/faces/docs/Faqs.xhtml#msea). MSEA is implemented through MetaboAnalyst, a complete set of online tools for metabolomics data processing, analysis, and interpretation (www.metaboanalyst.ca). Due to the challenging nature of lipidomics, there are not as many well-developed tools at this time. LIPID Metabolites And Pathways Strategy (www.lipidmaps.org) is a consortium with two goals being to advance the field of lipidomics and to provide resources to a broader range of the scientific research community.

Networks are another useful bioinformatics and visualization tool for some MS omics data to demonstrate how different data points interact. As with enrichment analysis, a list of significant features can be used as input to show relationships between them to help identify influential features. STRING is one database of experimentally verified and predicted protein–protein direct and indirect interactions for millions of proteins of many different organisms (string-db.org/cgi/about.pl). STRING generates a protein–protein interaction (PPI) network with default parameters, but the user may adjust them to generate another network. Figure 8.19 displays a PPI network from STRING data. The network can be exported for use in other platforms. ConsensusPathDB offers similar functionality for metabolite data in addition to other data types (cpdb.molgen.mpg.de). The network information generated by these tools can then be imported into a popular tool, Cytoscape, to manipulate, integrate, and further analyze the networks. Cytoscape can integrate networks with annotations, expression profiles, or other state data (http://www.cytoscape.org/what_is_cytoscape.html). For example, expression information can be encoded by coloring the nodes according to its intensity or whether it is up- or downregulated. From within the open-source software platform, Apps provide additional features to analyze networks and provide alternative layouts, among other things.

Although networks can be useful, a common challenge with some is their sheer size and complexity. Some nodes and/or paths might be obscured, ultimately camouflaging important patterns. These complex networks become uninterpretable and meaningless "hairballs." A more recently developed approach to network visualization is the hive plot. Its organization is based on meaningful properties of the network components, which can be selected to address specific questions. Examples and tutorials can be accessed at www.hiveplot.com. These properties might include measures such as connectivity, density, centrality, or other quantitative values (e.g. peak intensity). Nodes and edges can be manipulated using available data by changing color, size, or label

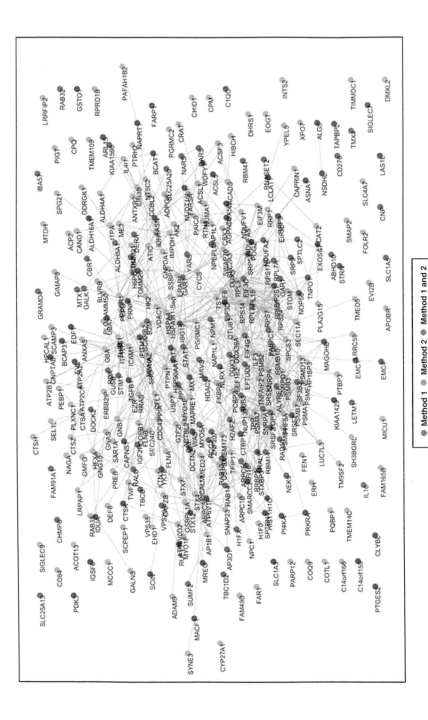

Figure 8.19 Protein–protein interaction network. A human protein interaction network generated using STRING. This particular network uses color to compare three different analysis methods that would represent three different PPIs.

● Method 1 ● Method 2 ● Method 1 and 2

to encode additional information. Hive plots offer the user great freedom to tailor a network to their data and to explore and compare alternative layouts, making this an invaluable tool for visualizing multidimensional data.

8.3.4 Databases and Search Algorithms

High-throughput omics experiments and fast-developing systems biology studies require high-quality multidimensional databases to be searched with data from unbiased as well as targeted experiments. An increasing number of investigators are becoming routinely involved in designing and executing all omics studies due to increasing knowledge. However, familiarity with systems biology approach does not mean in-depth knowledge with technology platforms, i.e. MS. At the same time, omics data sets are becoming larger and more complex, requiring not only more computer power for processing but also more in-depth understanding of the tools implemented in database searches. One avenue to target this challenge is to make the structure of database and search engines more user-friendly. In fact, much more comprehensive pipelines for proteomics data analyses have become available, meeting the broad demands of the research community. Although such development and improvement of databases and search algorithms have not eliminated manual interrogations, it allows limiting such interrogation to minimum.

The quality of information extracted from MS-based data sets depends on several factors: quality of data, quality of databases, and quality of algorithm(s) used for database searches. Steps following completion of MS-based profiling and transformation of all MS and MS/MS spectra into a desired format consist of choosing the database and search parameters. Although there are several databases available, a "non-redundant" type of database is preferential. Redundancy itself also has several dimensions. For example, one question is whether two alleles of the same locus or two isoenzymes from the same organism constitute redundancy. Another question is about the borderline between redundancy and completeness of information included in a database. There are not simple answers and individual investigators have to make their best educated decisions while dealing with the high complexity of biological data. Even curated Universal Protein Resource (UniProt) database provides a canonical sequence and other related sequences of isoforms and variants to address this issue. Therefore, we will experience ambiguity in data analyses and interpretation to some extent.

8.3.4.1 Databases for Proteomics
8.3.4.1.1 *Swiss-Prot*
Swiss-Prot was created in 1986. It is a biological database of manually curated or reviewed protein sequences, maintained by a number of experts. The database provides information in a uniform format that makes it easy to be used by

everybody, including those who have very limited expertise in proteins and proteomics. For that reason, Swiss-Prot provides reliable protein sequences associated with annotation with a minimal level of redundancy. An updated UniProt Knowledgebase (UniProtKB)/Swiss-Prot protein knowledgebase release and statistics is at http://web.expasy.org/docs/relnotes/relstat.html

8.3.4.1.2 *UniProt*

The UniProt is a collaborative providing a comprehensive, high-quality, and freely accessible resource of protein sequence and functional information. This is accomplished through various avenues, which include database curating, development of non-licensed software, and data annotation. The UniProt database includes the UniProtKB, the UniProt Reference Clusters (UniRef), and the UniProt Archive (UniParc). The UniProt Metagenomic and Environmental Sequences (UniMES) database is a repository specifically developed for metagenomic and environmental data. Moreover, UniProt database provides links to multiple related resources.

8.3.4.1.3 *NCBI nr*

The National Center for Biotechnology Information (NCBI) was formed in 1988 as a division of the National Library of Medicine (NLM) at the National Institutes of Health (NIH). Among other responsibilities, the NCBI facilitates the use of databases and software and performs research on advanced methods of computer-based information processing for analyzing the structure and function of biologically important molecules including proteins. The "nr" database is the largest database available through NCBI BLAST. The name "nr" is derived from "non-redundant," but this is historical only, because this database is no longer non-redundant.

8.3.4.1.3.1 Search Engines for Proteomics

After selection of database, the investigator needs to set up goals for the search output and, based on the employed strategy, needs to select proper search engines (tools). Multiple search engines are now at our disposal, and each of them has strengths and weaknesses. Therefore, it is very difficult to provide one way of such data analysis. Instead we would like to highlight important factors to be considered. They include search parameters and mass tolerance limits, number of miscleavages, and *in vivo* posttranslational modifications as well as those introduced purposely during sample preparation.

General databases can be compared to living organisms. They are constantly increasing in size, evolving, being modified and supplemented, and being integrated with other types of databases, i.e. cellomics, genomics, etc. Therefore, to accomplish the goals of effective searches, one must use multiple tools. Important to note is that organization of information is becoming more and more user-friendly, despite expansion and even if information included in protein databases is not complete. As the process of moving all data analyses into the cloud has already begun, soon it will not be necessary to move data from storage in the cloud to local computers.

8.3.4.1.3.2 Databases for Metabolomics Databases for metabolomics are currently under vigorous development. However, it is their diversity that is specific for the omics field. They can contain metabolic pathways, description of physical and chemical properties for metabolites, mass spectra, nuclear magnetic resonance (NMR) spectra, links, and hyperlinks to other databases and search engines. They can be divided into five main types:

A: Comprehensive Databases
- Human Metabolome Database (HMDB), http://www.hmdb.ca/spectra/ms_ms/search
- The Madison Metabolomics Consortium Database (MMCD), mmcd.nmrfam.wisc.edu
- NIH, Metabolomics WorkBench, www.metabolomicsworkbench.org
- BiGG Models, bigg.ucsd.edu

B: Biochemical Pathways Databases
- Kyoto Encyclopedia of Genes and Genome (KEGG), www.genome.jp
- MetaCyc, HumanCyc, BioCyc Compound, metacyc.org, humancyc.org, biocyc.org
- Sigma-Aldrich, http://www.sigmaaldrich.com/life-science/metabolomics.html
- Reactome: A Curated Pathway Database, reactome.org
- The Small Molecule Pathway Database (SMPDB), smpdb.ca
- WikiPathways, http://www.wikipathways.org/index.php/WikiPathways

C: Basic Physical and Chemical Information Databases
- MMCD Records, mmcd.nmrfam.wisc.edu
- PubChem, pubchem.ncbi.nlm.nih.gov
- ChemIDplus, https://chem.nlm.nih.gov/chemidplus
- ChEBI, http://www.ebi.ac.uk/chebi

D: Databases Limited to Specified Biological Object
- Yeast Metabolome Database, www.ymdb.ca
- *Escherichia coli* Metabolome Database, ecmdb.ca
- DrugBank, www.drugbank.ca
- Pharmacogenomics Database, www.pharmgkb.org
- Chemical-Protein Interaction Networks, stitch.embl.de

E: Platforms Supporting Interpretation of MS Data
- Golm Metabolome Database, gmd.mpimp-golm.mpg.de
- European Mass Bank, http://massbank.eu/MassBank
- Direct Infusion Metabolite Database, dimedb.ibers.aber.ac.uk

- XCMS, xcmsonline.scripps.edu
- METLIN, https://metlin.scripps.edu/landing_page.php?pgcontent=mainPage
- Fiehn Lab, http://fiehnlab.ucdavis.edu/19-projects/153-metabolite-library

Two databases were selected and described below with special emphasis on their unique character:

HMDB. HMDB is the biggest metabolomics database. It is a comprehensive, freely accessible database of metabolites found in the human body. It contains basic physical and chemical data, mass spectra, NMR, and clinical data, including that for food and pharmaceutical metabolites. It is one of the first databases dedicated to metabolomics. It started in 2007 by the Human Metabolome Project, founded by Genome Alberta, Genome British Columbia, and Genome Canada. The database is still expanding and can be presented as a good example of exponential growth of a data set (number of metabolites included 2180 in 2007, 6408 in 2009, 37170 in 2012, and 114101 in 2015). However, the number of metabolites with reference MS/MS mass spectra is about 2000.

XCMS. XCMS Online is a freely accessible MS processing platform integrated with the mass spectra library METLIN (number of analytes with reference MS/MS mass spectra 14000 (measured) and 220000 (obtained *in silico*)). The platform was released in 2004 and integrates univariate and multivariate data processing tools for metabolite feature annotation and metabolite identification. As a database dedicated to untargeted metabolomics, it requires high accuracy of the MS data (mass tolerance below 100 ppm). There are also two commercial platforms developed to facilitate data processing for metabolomics (Mass Profiler Pro (Agilent) and Metabolic Profiler (Bruker)).

It should be stressed that each platform has its exceptional capabilities. As all of them are under constant development, the number of metabolites with reference MS/MS spectra can be considered a universal critical parameter for their applicability.

8.3.5 Validation of High-Throughput Data: Current Challenges

Systematic evaluation and validation of high-throughput MS data (HTD) remains a challenge. The main goal of validation of a method or a combination of methods is to demonstrate that it/they is/are proper for the proposed experimental design. Widely accepted and applied principles for assessing the fit of analytical method(s) are specificity, accuracy, precision, detection limit (LOD), quantitation limit (LOQ), sensitivity, working range and linearity, robustness, and recovery. As much as we accept these criteria to validate analytical methods, HTD create a much more complicated environment for validation of multidimensional data consisting of different nature components. Thus, we expect that method validation will provide empirical evidence that methodology is reliable and consistent with accepted standards. If we consider that discovery of 0% of true positive differences is one (negative) end of spectrum and discovering 100% of true positive differences is the other side of spectrum, we have to ask whether it is possible to reach 100% discovery and, if not, what percentage of discovery would be acceptable as high-quality HTD. We envision three intertwining components of HTD validation: analytical, statistical, and bioinformatics. Each component requires different tools and provides part of information we seek.

8.3.5.1 Analytical Validation

Method validation is an important step in the analytical process. It is carried out to verify if the method under consideration has performance capabilities consistent with what the application requires. In other words, "fitness of the method to the purpose" needs to be justified (EURACHEM Guide). As the method validation can be very expensive, the definition of analytical requirements is crucial. First, method parameters critical for the performance of the method should be identified from the following list: specificity/selectivity, linearity/sensitivity, range, accuracy, precision, LOD, LOQ, and robustness. For example, if the method under development will be applied to establish amounts of the compounds in different biological objects or similar objects exposed to different conditions and statistical tests will be performed in order to find significant differences, the precision of the method is the most critical parameter. On the other hand, precision depends on S/N ratio, which comprises sensitivity, LOQ and LOD, and selectivity. As the analysis will be performed during a long period of time, it should be expected that some factors can vary in time (e.g. composition of solid phase in cartridges used for sample purification or modification of HPLC column or supplier of solutions). This means that system capacity to remain unaffected should be measured as well as robustness. When standards are available, the method is quantitative, and linearity, range, detection, and quantitation limits can be easily defined by performing calibration for the system using solutions with different standard concentrations. What can be done if standards are unavailable?

Linearity, an ability to obtain results proportional to concentration of the analyte, can be obtained by performing measurements of sample for different dilution levels. The response of the system should be proportional to the dilution factor (used to define the range) instead of the concentration of the analyte. Additionally, the presence of interfering matrix components can be found.

The dilution factor of the sample can also be applied to establish LOD and LOQ when critical values for S/N will be assigned as 3 and 10, respectively. Moreover, both of the parameters are strictly related to SD of the results obtained for a blank sample and can be estimated by multiplying SD by 3 and 10 times.

Selectivity is understood as the ability to assess unequivocally the analyte in the presence of components that may be expected to be present. It can be verified by changes of peak area or retention time when a different, selective to the analyte, transition is used in an MRM method. For example, if three chromatographic peaks are obtained for transition $128 \rightarrow 110$ and two for transition $128 \rightarrow 100$, peaks observed at the same retention time will correspond to the compound with precursor ion m/z 128 and fragmenting ions m/z 110 and 100.

Accuracy of the method can be verified only by comparing the results obtained by another method for the same sample. In the case of quantitative methods, results are compared following equations used to establish compound recovery. Qualitative methods are validated by establishing rates for false-positive and false-negative results.

The precision of an analytical procedure expresses the closeness of agreement (degree of scatter expressed as SD or RSD) between a series of measurements obtained from multiple sampling of the same homogeneous sample under the prescribed conditions. Precision may be considered at three levels: repeatability, established for the measurements of the same sample in the same conditions (RSD usually below 15% for trace components); intermediate precision, established for long-term measurements and randomly analyzed samples with different matrix composition (RSD usually below 30%); and reproducibility, established for the results obtained by the same method reconstructed in a different laboratory. All of these parameters should be reexamined after checking the system performance to ensure good quality of the data that is crucial for intermediate precision.

The reliability of the method is very important. It should be always kept in mind that if the results of the measurements cannot be trusted, then it has little value and measurements might as well have not been carried out at all.

8.3.5.2 Statistical Validation

Uncovering new biological mechanisms as well as broadly defined biomarkers vastly depends on effective statistical analysis of millions of data points within thousands, if not soon to be millions, of analyses. Demand for data analysis leads to the creation of new tools including statistics, which is expected to help us with introduced and inherent variability associated with empirical data. Variation in data can be introduced at numerous points during its course of acquisition. Scientists have to distinguish between variation that arises due to technical sources and variation that is a response of the biological system being studied. Technical variation obscures biological importance, so researchers face the challenge of reducing its effects when trying to make sense of collected data. Normalization is a process that is supposed to reduce the nonbiological variation, but with so many normalization methods available, choosing one is another challenge in and of itself. Comparing "similar" methods on the same data can potentially produce drastically different results. To explore this phenomenon, we used MS-generated metabolomics profiling data comparing control and illicit drug-exposed human primary macrophages. We borrowed normalization methods commonly used for microarray data that have also been deemed acceptable for MS data. In each case, we used the same set of raw data. First, we performed log_2 transformation on the data. Then we applied quantile normalization to the log_2 data to use as a second case or applied cyclic loess normalization as a third. Next, we performed limma for differential expression analysis on all three of these data sets. For the log_2 transformed data, limma identified 33 out of 156 features as differentially expressed between two conditions while 76 for the quantile normalization and 37 for loess normalization. Among these three separate analyses, 27 features were identified in all three sets. However, the question as to why the three normalization methods produce such different output remains. This analysis was executed using the "preprocessCore" and "limma" packages in R, and documentation for the "normalizeCyclicLoess" function describes cyclic loess normalization as "similar in effect and intention to quantile normalization." While these methods are widely accepted and used, and while the two produce some overlapping results, it is difficult to understand why quantile normalization gives around twice as many significant features as cyclic loess normalization. As much as answering this question would require further formal analyses, we postulate that an "ideal" workflow for statistically processing and analyzing MS-generated omics data requires comparison of multiple statistical methods, which we refer to as statistical validation. Otherwise bioinformatics analysis, a subsequent step, might be filled with too many false positives.

8.3.5.3 Bioinformatics Validation

Over the past two decades, development of omics approaches has led to a change in attention to data analysis of individual genes or proteins to large-scale networks that are able to capture the complexity of how a biological system works. As any other process, bioinformatics analyses also need to be validated. This can be seen either as direct validation of an output or as validation of bioinformatics tools, meaning the software. Validation of bioinformatics analyses is a challenge because as of now we have multiple tools but no "gold standard" for this process of inference from multidimensional data sets. In the majority of studies, GO (www.geneontology.org) is used; nevertheless some bioinformaticians create their specific ontologies shaped for a specific approach or even a specific project. For example, in UniProtKB entries, GO terms might be electronically or manually entered. Manually assigned GO terms found in UniProtKB/Swiss-Prot are associated with one of 15 GO evidence codes, while electronically assigned GO terms are found in UniProtKB/TrEMBL, but to some extent also in UniProtKB/Swiss-Prot, and are associated with the GO evidence code "IEA," which means "inferred from electronic annotation" (http://www.uniprot.org/help/gene_ontology). Summarizing, effective use of such resources requires good understanding of the inference of ontology association to avoid misinterpretation of the overall conclusions derived from any given data set.

Changes in proteins' expression measured by quantitative MS first have to pass statistical scrutiny to weed out those changes that, for many reasons, might be anecdotal. Changes that pass this step and represent reliable effects are then plugged into network analysis to reveal their role in various networks such as PPI networks, metabolic networks, cell signaling networks, etc. However, it has to be noted that while a network is characterized by the edges or connectivity of nodes, various types of data may generate different general network characteristics. Thus, the meaning of the nodes and edges used in what a network represents may depend on the type of data used to build the network.

In short, we do not present or propose solutions, yet we want to make readers aware of the importance and complexity of bioinformatics analysis, which is usually the final step in quantitative omics experiments. Nevertheless, we recommend that data analysis should be performed using more than one method and/or approach or with a set of bioinformatics tools before conclusions are drawn and new experiments are designed and executed.

8.3.6 Summary and Conclusions

Development of high-throughput analytical methods called omics brought with it a new set of challenges. While storage of Big Data has been successfully addressed by exporting them to the cloud, data analysis remains the bottleneck due to several reasons:

Firstly, lack of gold standards in analysis of specific data sets led to the use of a combination of various methods and approaches that might not be compatible; thus outputs might not be directly comparable.

Secondly, an ever-increasing number of resources, i.e. databases of which some are curated automatically or manually or using both methods which are not necessarily interconnected.

Thirdly, data normalization, in particular, generated by using different analytical platforms.

Fourth, validation to increase overall quality of data and output of data analyses.

It is not our intention to provide ready solutions or to promote one approach over the other. Our goal in this chapter is to make readers aware of the complexity as well as challenges they can face during high-throughput data analyses. A critical approach and constant self-education in this area are necessary to avoid mistakes in data interpretation prior to designing newer and bigger experiments.

Questions

- Is "Big Data" defined by size?
- What are main attributes of "Big Data?"
- Briefly characterize the term "unstructured data."
- What factors can influence the variability of a peak's area and the height of spectral peaks?
- Does sensitivity of LC-MS method influence data variability?
- How can selectivity of the LC-MS method be improved?
- Is the highest MS resolving power enough to obtain good quality data?
- What are "memory effects?" What are their sources?
- What can be a reason for asymmetric distribution of data?
- What factors can be responsible for higher noise and how it can be reduced?
- Is the parameter S/N related to data variability?
- What is dwell time and cycle time? How do they influence LC-MS data quality?
- Discuss and decide if chromatographic peak (width 20 seconds) should be defined/described by 5, 20, or 50 points (cycles of data acquisition) using quadrupole?
- What are the main approaches used in omics analyses? Discuss pros and cons.
- How does data throughput influence the precision of LC-MS method?
- What are the main goals of data visualization?
- What types of data sets can be easily visualized?
- How does human perception play a role in data visualization?
- What are the first steps to fulfill before data processing?
- What are the main outcomes of data transformation?
- What types of charts/plots visualize data distribution?
- Characterize the volcano plot and its function.
- Describe methods of visualization for many variables.
- When is data multidimensional?
- What exactly is "principal component analysis (PCA)" and how it is achieved?
- What is necessary to obtain biological meaning from omics results?
- What are the main tools for classification of data sets? What they are used for?
- Characterize enrichment analysis of data sets.
- What platforms can be used for data analysis and visualization?
- What are hive plots used for?
- What are three main factors influencing the quality of information extracted from MS data?
- Why are most databases redundant?
- What are main databases for proteomics?
- Name controllable parameters for proteomics search engines.

- What are the main databases for metabolomics?
- What is critical parameter limiting database applicability in metabolomics?
- Which parameters of analytical method influence precision?
- How accuracy of the method can be validated?
- What's the normalization of data and how it can be achieved?

References

1 Trifonova, O.P., Il'in, V.A., Kolker, E.V., and Lisitsa, A.V. (2013). Big data in biology and medicine: based on material from a joint workshop with representatives of the international Data-Enabled Life Science Alliance, July 4, 2013, Moscow, Russia. *Acta Naturae* 5: 13–16.

2 Putri, S. and Fukusaki, E. (2014). *Mass Spectrometry-Based Metabolomics: A Practical Guide*. London: CRC Press.

3 Norwood, D.L., Mullisa, J.O., and Feinberg, T.N. (2007). *Hyphenated Techniques*, vol. 8. San Diego: Elsevier.

4 Ivanov, A.R. and Lazarev, A.V. (2011). *Sample Preparation in Biological Mass Spectrometry*. the Netherlands: Springer.

5 Zhou, J., Liu, H., Liu, Y. et al. (2016). Development and evaluation of a parallel reaction monitoring strategy for large-scale targeted metabolomics quantification. *Analytical Chemistry* 88: 4478–4486.

6 Maher, S., Jjunju, F.P.M., and Taylor, S. (2015). Colloquium: 100 years of mass spectrometry: perspectives and future trends. *Reviews of Modern Physics* 87: 113.

7 Kaddurah-Daouk, R., Kristal, B.S., and Weinshilboum, R.M. (2008). Metabolomics: a global biochemical approach to drug response and disease. *Annual Review of Pharmacology and Toxicology* 48: 653–683.

8 Distler, U., Kuharev, J., Navarro, P. et al. (2014). Drift time-specific collision energies enable deep-coverage data-independent acquisition proteomics. *Nature Methods* 11: 167–170.

9 Rost, H.L., Rosenberger, G., Navarro, P. et al. (2014). OpenSWATH enables automated, targeted analysis of data-independent acquisition MS data. *Nature Biotechnology* 32: 219–223.

10 Chokkathukalam, A., Kim, D.H., Barrett, M.P. et al. (2014). Stable isotope-labeling studies in metabolomics: new insights into structure and dynamics of metabolic networks. *Bioanalysis* 6: 511–524.

11 Citation, W.P. *NIST/SEMATECH e-Handbook of Statistical Methods*. NIST Gaithersburg http://www.itl.nist.gov/div898/handbook (accessed 30 January 2019).

12 Krzywinski, M. and Cairo, A. (2013). Points of view: storytelling. *Nature Methods* 10: 687.

13 Wong, B. (2010). Points of view: Gestalt principles (Part 1). *Nature Methods* 7: 863.

14 Wong, B. (2010). Points of view: Gestalt principles (Part 2). *Nature Methods* 7: 941.

15 Krzywinski, M. (2013). Points of view: elements of visual style. *Nature Methods* 10: 371.

16 Ringner, M. (2008). What is principal component analysis? *Nature Biotechnology* 26 (3): 303–304.

17 Kvalheim, O.M., Brakstad, F., and Liang, Y. (1994). Preprocessing of analytical profiles in the presence of homoscedastic or heteroscedastic noise. *Analytical Chemistry* 66: 43–51.

18 Ciborowski, P. and Silberring, J. (2016). *Proteomic Profiling and Analytical Chemistry: The Crossroads*. the Netherlands: Elsevier.

19 Chawade, A., Linden, P., Brautigam, M. et al. (2012). Development of a model system to identify differences in spring and winter oat. *PLoS One* 7: e29792.

20 Smyth, G.K. (2004). Linear models and empirical bayes methods for assessing differential expression in microarray experiments. *Statistical Applications in Genetics and Molecular Biology* 3: 1027.

21 Benjamini, Y. and Hochberg, Y. (1995). Controlling the false discovery rate: a practical and powerful approach to multiple testing. *Journal of the Royal Statistical Society, Series B* 57: 289–300.

22 Li, W. (2012). Volcano plots in analyzing differential expressions with mRNA microarrays. *Journal of Bioinformatics and Computational Biology* 10: 1231003.

23 Sakai, R., Winand, R., Verbeiren, T. et al. (2014). dendsort: modular leaf ordering methods for dendrogram representations in R. *F1000Research* 3: 177.

24 Saatci, E. (2018). Correlation analysis of respiratory signals by using parallel coordinate plots. *Computer Methods and Programs in Biomedicine* 153: 41–51.

25 Gehlenborg, N. and Wong, B. (2012). Points of view: heat maps. *Nature Methods* 9: 213.

26 Gehlenborg, N. and Wong, B. (2012). Points of view: into the third dimension. *Nature Methods* 9: 851.

27 Luscombe, N.M., Greenbaum, D., and Gerstein, M. (2001). What is bioinformatics? An introduction and overview. *Yearbook of Medical Informatics* (1): 83–99.

8.4 Application of the Mass Spectrometric Techniques in the Earth Sciences

Robert Anczkiewicz

Institute of Geological Sciences, Polish Academy of Sciences, Krakow, Poland

8.4.1 Introduction

Mass spectrometric techniques have always been of central interest to the geoscientists. Since the discovery of the mass spectrometer in the early twentieth century, radioactive nuclides were immediately applied to constrain timing of processes shaping the Earth, as well as to the studies of extraterrestrial materials. The second truly revolutionary moment came in the 1990s when plasma source mass spectrometers and laser ablation (LA) were developed. This enabled analyzing the wealth of new isotopic systems, led to an improved spatial resolution of analyses, and opened a wide range of new research fields, which made geochemistry, and isotope geochemistry in particular, the dominant discipline in the earth sciences. The most successful and the most popular analytical methods involve isotope ratio measurements by thermal ionization mass spectrometry (TIMS), inductively coupled mass spectrometry (ICP-MS) with quadrupole mass filter, and high-resolution magnetic sector, multi-collector inductively coupled plasma mass spectrometry (MC ICP-MS). Detailed accounts on the vast number of applications of mass spectrometry in the earth sciences is beyond the scope of this contribution and can be found in the textbooks such as [1–5]. Below, we present only a brief summary of the major applications of the mass spectrometric measurements useful in the earth sciences which are grouped in three areas: (i) conventional geochronology, (ii) *in situ* geochronology, and (iii) geochemical and isotopic tracing.

8.4.2 Conventional Geochronology

The main goal of geochronology is to provide precise and accurate time constraints of geological or cosmogenic processes. Under the term "conventional geochronology," we understand wet chemical methods involving combined isotope dilution and isotope composition measurements by TIMS or MC ICP-MS. Due to technical advances, geological processes as young as several million years can be tackled even with radioactive nuclides like ^{147}Sm whose half-life is as long as 106 billion years, and thus, even on the geological time scale, very little daughter isotopes are available for analyses. Younger episodes, down to about 1 Myr (million years) or less, can be dated with the U–Th–Pb or K–Ar, and its modification, Ar–Ar, systems.

By far the most popular geochronological method is U–Pb dating of an accessory mineral zircon, commonly found in the Earth's crust. U–Pb single-crystal dating with precision of less than 0.1% (2 RSD) for as little as several pg of radiogenic Pb can be achieved by TIMS, which for decades has been the best tool for the most precise and accurate isotope ratio measurements. This technique remains unchallenged by any other analytical method and helped to precisely define timing of many key events in the Earth's history. Other geochronometers like ^{87}Rb–^{87}Sr, ^{147}Sm–^{143}Nd, or ^{187}Re–^{187}Os are also successfully applied to dating ore deposits and magmatic, metamorphic, or sedimentary processes. Although all isotopic systems mentioned above arguably are best approached by TIMS, they may also be applied using MC ICP-MS whose higher sensitivity and larger sample throughput make it often a method of choice, despite still a bit lower external precision of measurements. Considerably higher ionization temperature of Ar plasma than that of the TIMS source enabled development of new geochronometers based on the elements with high ionization energy. The best example is ^{176}Lu–^{177}Hf method whose early success achieved by TIMS was substantially enhanced with the appearance of MC ICP-MS that allowed high precision isotopic analyses of small amounts of Hf, which permitted dating of minerals like garnet, xenotime, lawsonite, or apatite (Figure 8.20).

The apparent drawback of conventional geochronology is the lack of spatial resolution offered by *in situ* techniques like secondary ion mass spectrometers (SIMS) or LA (MC) ICP-MS described below. However, thanks to the possibility of carrying out precise measurements even of the very small amount of an analyte, crystal fragments cut with the use of more or less sophisticated devices do provide a possibility of spatially resolved analyses (Figure 8.20). This is most feasible for large, typically rock-forming minerals like garnet, but fragments of very small crystals, such as zircon, have also been successfully analyzed. Furthermore, major and trace elements can be extracted during the ion exchange chromatography and subjected to ICP-MS measurements. In the same way, Hf isotope composition can be determined, which provides valuable petrogenetic information helping to link isotopic ages with specific conditions of mineral formation. Although this is a very tedious approach, it is the only solution for the geochronometers that cannot be applied *in situ*.

8.4.3 *In Situ* Geochronology

The possibility of coupling LA with plasma source mass spectrometer opened new possibilities in the high spatial resolution geochronology, which had earlier been reserved for SIMS. Nowadays, LA (MC) ICP-MS dating is practiced in numerous laboratories worldwide with similar to SIMS age precision (about 1% 2 RSD) but at the significantly lower costs and considerably larger sample throughput. Spatial resolution of LA (MC) ICP-MS analyses at the level of tens of microns (but for some applications even several microns) is routinely

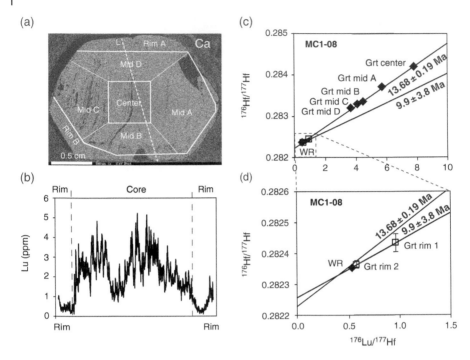

Figure 8.20 High-resolution Lu–Hf geochronology of a single garnet crystal conducted using multi-collector inductively coupled plasma mass spectrometer (MC ICP-MS) on three growth zones (center, mid, and rim) defined on the basis of internal mineral texture and Mg concentration X-ray map (a). Each domain represents 50–100 mg of a crystal fragment providing c. 5–10 ng of Hf for isotopic analyses (a). Lu–Hf dating results show high precision age obtained for the enriched in Lu crystal core and mid-zone and low precision age obtained for Lu-depleted rim (b–d) as documented by laser ablation ICP-MS Lu concentration profile (c) measured along dashed line L shown in (a). Grt, garnet; WR, representative whole rock powder. *Source*: Data from Ref. [6].

achieved. This method requires much less sample preparation in comparison with conventional geochronology but typically needs additional investigations of internal mineral structure with the help of cathodoluminescence (CL), X-ray element mapping, or backscattered electron (BSE) imaging (Figures 8.20a and 8.21a). This is necessary in order to identify homogeneous areas representing "equal time zones" that can subsequently be dated, avoiding mixing with other domains (Figure 8.21a).

Another gain of *in situ* analyses comes from the fact that they may easily be combined with complementary measurements within the same domain, which merge geochronology with petrogenesis. Such multisystem analyses are commonly applied to zircons whose U–Pb dating is frequently accompanied by LA (MC) ICP-MS determination of trace element abundance and Hf isotope composition and with SIMS measurements of oxygen isotopes (Figure 8.21). Similarly, U–Th–Pb monazite dating may be combined with trace elements and Nd isotope measurements. Another achievement of LA ICP-MS is a far more frequent dating of minerals containing relatively large amounts of initial common Pb like rutile, titanite, or apatite.

(a)

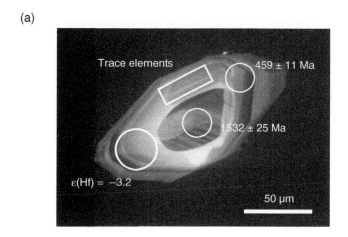

(b)

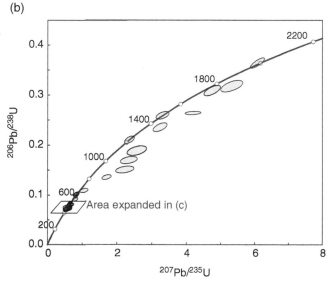

Figure 8.21 *In situ* sub-single-crystal LA ICP-MS zircon geochronology combined with trace elements and Hf isotope composition determination for a single age domain (a). Concordia plot shows multiple inherited age domains, typical of rocks of crustal origin (b) and dominant zircon population defining age of volcanic eruption (c). Rare earth element pattern for individual zircon analyses provides additional information on rock petrogenesis.

(c)

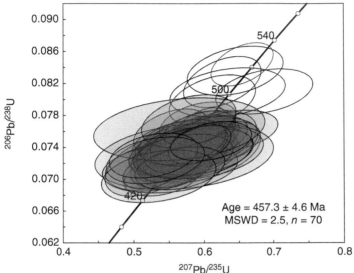

(d)

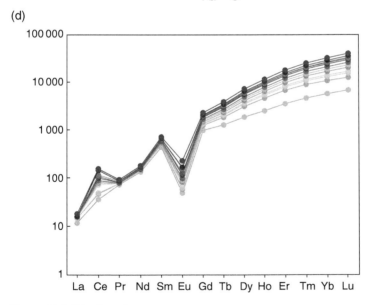

Figure 8.21 (*Continued*)

An interesting, and indeed very appealing, variant of multisystem analysis is offered by the so-called "laser ablation split stream" (LASS) technique. It is based on simultaneous use of two plasma source mass spectrometers connected to LA. This enables concurrent dating and trace elements or Hf (or Nd) isotope measurements from exactly the same mineral volume. This is time- and material-saving method with a great potential, but due to the reduced amount of analyte, which is split between the two instruments, precision of analyses is compromised. Certainly, where high precision is not required, *in situ* LASS (MC) ICP-MS dating is a very powerful tool.

A very exciting application of *in situ* measurements that must not be omitted here is element, isotope, and age mapping using LA (MC) ICP-MS. Interpolation of dense net of stationary (spot) or raster mode analyses allows to construct a map with spatial resolution of about 10 μm or even less. This is still much poorer than submicron resolution offered by modern electron microprobes (EMP) for element mapping but much wider range of elements analyzed by ICP-MS along with the possibility of isotopic measurements, and the detection limits in the low range of ppb (inaccessible for EMP) make this technique very valuable.

Although currently *in situ* geochronology is limited to U–Th–Pb dating, new generation of the triple quadrupole ICP-MS demonstrated potential for separating ^{87}Rb and ^{87}Sr isobars, which should soon result in a high-quality *in situ* Rb–Sr dating.

8.4.4 Geochemical and Isotopic Tracing

Isotopic and geochemical tracing aim at determining the source of a sample by referring to a reservoir of known composition. Radiogenic isotopes of Sr, Nd, Hf, Os, and Pb mentioned above are most commonly applied. Noteworthy, extinct short-lived nuclides can also be useful as isotopic tracers. A good example is ^{146}Sm decaying to ^{142}Nd with a half-life of 103 ky, which appears supportive in the studies of crustal fractionation in the early Earth evolution.

Besides radioisotopes, there is a large group of stable isotopes that undergo fractionation in response to chemical, biological, or geological processes. Isotopes of Li, B, Se, Cr, Mg, Ca, Fe, Zn, Cu, Cl, Hg, Tl, Si, or Mo are often termed "nontraditional stable isotopes," which distinguishes them from "traditionally" investigated stable isotopes of C, H, O, N, and S. Early advances in this field were made by TIMS and gas source mass spectrometric measurements, but the development of MC ICP-MS hastened the number of studies involving new stable isotopes that were applied to the studies of climate change, weathering, mantle–crust interaction, origin of life, rock petrogenesis, or planetary geology. One of the important new features brought by MC ICP-MS is a possibility of coupling it with LA, which allows for *in situ* analyses of stable isotopes. TIMS is still preferred in Ca isotope measurements, which are problematic by plasma source instruments due to isobaric interferences of ^{40}Ca and ^{40}Ar. However, addition of a collision/reaction cell to MC ICP-MS design may eliminate this obstacle as proved by the "extinct" MC ICP-MS *IsoProbe*.

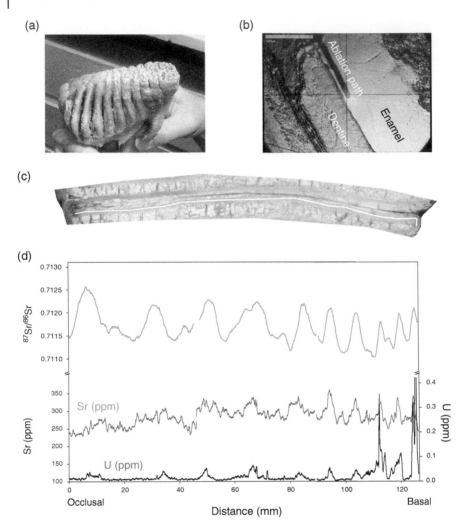

Figure 8.22 Laser ablation MC ICP-MS analyses of Sr isotope composition of the mammoth molar (a–b). Analyses are conducted within enamel along the enamel–dentine junction (b–c). Isotope $^{87}Sr/^{86}Sr$ ratios document cyclic changes caused by seasonal mammoth migration (d). Single plate (c) bears a record of approximately 10 years of mammoth's life and was determined with time resolution of about two weeks. Trace element measured by LA ICP-MS along the same path show positive correlation of U and Sr content with $^{87}Sr/^{86}Sr$ ratios and also point to two distinct migration regions. *Source*: Data from Ref. [7].

A very promising new field of application of *in situ* isotopic measurements is paleoecology of the extinct species. Element and Sr isotope analyses of tooth enamel, in combination with histological studies, not only offer a possibility of reconstructing migration paths of animals and humans but also may bring information about their diet. LA (MC) ICP-MS can be conducted with time resolution of about two weeks, which allows resolving even sub-seasonal changes. Variations in $^{87}Sr/^{86}Sr$ ratios recorded in tooth enamel in a single plate of a mammoth molar shown in Figure 8.22 document cyclic variations over a period of about 10 years. This pattern is interpreted as seasonal mammoth migration between two regions characterized by specific Sr isotopic composition. Strontium isotope composition analyses may be facilitated by element abundance measurements, which help to estimate the degree of postmortem alterations and provide additional characteristics of the migration paths (Figure 8.22).

Such methods are becoming the invaluable tools also in the studies of archaeological artifacts, for which virtually nondestructive LA (MC) ICP-MS analyses are particularly valuable. One of the biggest obstacles in applying isotopic tracing in geoarchaeology, at present, is insufficient characterization of potential source areas of the isotopic and geochemical signatures. However, this is still rather early time of geoarchaeology, and with the increasing number of studies, this gap will certainly be rapidly filled.

Questions

- What are the most successful and popular analytical methods in earth science?
- What is "conventional geochronology?"
- What is the most popular geochronological method?
- How is it possible to circumvent the lack of spatial resolution offered by *in situ* techniques like secondary ion mass spectrometers (SIMS) or LA (MC) ICP-MS in conventional geochronology?
- What is the "laser ablation split stream" (LASS) technique?

References

1 Faure, G. and Mensing, T.M. (2005). *Isotopes: Principles And Applications*. Hoboken, NJ: Wiley.
2 Dickin, A.P. (2018). *Radiogenic Isotope Geology*. Cambridge: Cambridge University Press.
3 Allègre, C.J. (2008). *Isotope Geology*, vol. 512. Cambridge: Cambridge University Press.

4 Teng, F.-Z., Dauphas, N., and Watkins, J.M. (2017). Non-traditional stable isotopes: retrospective and prospective. *Reviews in Mineralogy and Geochemistry* 82 (1): 1–26.

5 Kohn, M.J., Engi, M., and Lanari, P. (2018). Petrochronology: methods and applications. *Reviews in Mineralogy and Geochemistry* 83 (1): 1–12.

6 Anczkiewicz, R., Chakraborty, S., Dasgupta, S. et al. (2014). Timing, duration and inversion of prograde Barrovian metamorphism constrained by high resolution Lu-Hf garnet dating: a case study from the Sikkim Himalaya, NE India. *Earth and Planetary Science Letters* 407: 70–81.

7 Kowalik, N. Anczkiewicz, R., Müller, W., Wojtal, P., Wilczyński, J., Bondioli, L., Spötl, C. 2018. Sr isotope, trace elements and oxygen isotope record in molar enamel as an indicator of seasonal wooly mammoth migration in Central Europe. *Geophysical Research Abstracts* vol. 20. https://meetingorganizer.copernicus.org/EGU2018/EGU2018-16058.pdf (accessed 30 January 2019).

8.5 Mass Spectrometry in Space

Kathrin Altweg

Physikalisches Institut, Universität Bern, Bern, Switzerland

Mass spectrometry (MS) is a powerful tool for chemical analysis in the laboratory. But, since many decades, it has also been successfully applied in space research. For a long time, meteorites on Earth were analyzed using MS. This method proved very useful, and therefore already very early mass spectrometers were also carried into space, first on rockets and then on spacecraft. *In situ* instruments with very high sensitivities and/or high mass resolution have been flown successfully in the last decades, making elemental, isotopic, and molecular abundance measurements in a multitude of environments. Of course, *in situ* observations are confined to our solar system, and *in situ* exploration is expensive. However, in the cases where composition of atmospheres, exospheres, solar wind, plasma, and dust could be investigated by *in situ* MS, the scientific output was in most cases overwhelming.

The first mass instruments flown were Bennett-type RMS-1 cyclic radiofrequency mass spectrometers on rockets in 1957, sent by the USSR [1]. The goal of these experiments was the exploration of the terrestrial atmosphere at altitudes of 90–210 km. This research of our own atmosphere was followed up since then by many more rocket studies over more than 50 years. The mass spectrometers became more and more sophisticated using magnetic MS (e.g. Ref. [2]) and then time-of-flight (TOF) methods and developing more elaborate ionization methods (e.g. proton transfer) for the analysis of organics and aerosols. We concentrate here on mass spectrometers flown on spacecraft.

MS in space poses its own problems. There is first question of the weight of the instrument. While in the lab mass spectrometers are often very heavy, sometimes weighing easily more than a ton, especially magnetic analyzers containing big electromagnets, in space weight is very limited. This then needs special techniques. For example, big electromagnets have to be replaced by small permanent magnets, which have the disadvantage of relatively large fringe fields. But also physical dimensions are limited. To resize an instrument from large lab dimensions to dimensions compatible with a space mission can only be achieved by extremely precise mechanical dimensions of the ion optics. Figure 8.23 shows a drawing of the ion optics of the Rosetta Orbiter Spectrometer for Ion and Neutral Analysis (ROSINA) double-focusing mass spectrometer (DFMS) instrument flown on the European spacecraft Rosetta [3]. This instrument was resized from a lab instrument weighing 1500 kg to a space instrument of 16 kg. Accordingly, all dimensions of the ion optical elements had very tight requirements on their precision. For example, the toroidal electrostatic analyzer had a precision of better than 20 μm over the total length of

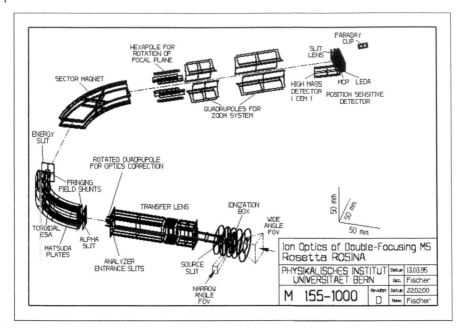

Figure 8.23 Drawing of the ion optics of the Rosetta Orbiter Spectrometer for Ion and Neutral Analysis double-focusing mass spectrometer (ROSINA-DFMS). This instrument was scaled from a 1500 kg laboratory instrument and weighted 15 kg at launch. *Source:* Courtesy University of Bern.

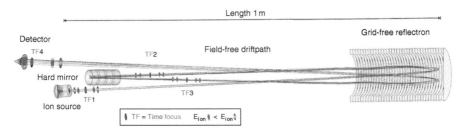

Figure 8.24 Ion optical design of Rosetta/ROSINA reflectron time of flight (RTOF) showing how the flight path of a time-of-flight mass spectrometer can be enlarged within the confined space of a spacecraft. *Source:* Courtesy University of Bern.

20 cm in order to achieve the mass resolution of 9000 for a mass 28 Da. But it also means that the precision and stability of the power supplies has to be extremely good. TOF instruments gain their mass resolution partly from the length of the flight path. In space, the flight path is limited by the size of the spacecraft, and this can only be compensated by reflecting the ion beam once or even multiple times. Figure 8.24 shows the ion optical principle of the

ROSINA reflectron time-of-flight (RTOF) instrument [3] where a flight path of almost 4 m was achieved by a reflectron and an additional ion mirror in a spacecraft with dimensions of less than 2.5 m. The short flight time also needs very fast electronics, which is associated with high power needs, again something which is not abundantly available on a spacecraft. The small dimensions pose another technical challenge connected to the problem of high voltages within a very confined space. All mass spectrometers use high voltages of the kV range, sometimes exceeding 30 kV (e.g. the Solar Wind Ion Composition Spectrometer [SWICS] instrument on Ulysses [4]) due to the energy range of the ions to be analyzed. Environmental challenges include high vibration levels during launch, large temperature gradients between the sunlit side of spacecraft and instrument and their anti-sunward side, and large temperature ranges encountered by the instrument along the mission with ranges frequently exceeding $-100\,°C$ to $+50\,°C$, all affecting precise ion optical alignments.

Another challenge is the variety of species to be analyzed. Solar wind compositional analysis requires a mass spectrometer with a high energy and angular acceptance combined with an excellent mass resolution for very low masses as the solar wind consists of highly charged atomic ions. To analyze plasma in space, mass spectrometers have to be able to analyze also the energy of the incoming ions and resolve their charge states. The mass/charge is below a few $Da\,e^{-1}$ for most species (e.g. He^{2+}, O^{8+}, Fe^{12+}) in the solar wind, and energies vary from thermal to 10 seconds of $keV\,e^{-1}$. On the other side, there are the very complex thermal organics in the coma of comets, requiring a very high mass resolution and a mass range from mass 12 to well above 100 Da and at the same time an excellent sensitivity and a very high dynamic range. For neutrals, most of the mass spectrometers in space use electron impact ionization from hot filaments as this presents the easiest and most reliable method. In order to overcome the sometimes very tenuous conditions with partial densities not exceeding 1 particle cm^{-3}, some instruments are equipped with the closed ion sources in order to concentrate the atmosphere, whereas in other cases very open ion sources are preferred to avoid chemical reactions in the source. Even more demanding are dust mass spectrometers in space, which are used to analyze interstellar, cometary, and planetary dust particles. Depending on the mission, dust is ionized by its impact onto a target plate if the relative velocity of the spacecraft and the dust is high enough (e.g. the PUMA instrument on the VEGA spacecraft at Halley's comet [5]) or secondary ion mass spectrometry (SIMS) can be used. This requires detection of the particle first by the optical means (microscope) in order to direct the ion beam onto the dust grain and to obtain secondary ions from sputtering (Cometary Secondary Ion Mass Analyzer [COSIMA] on Rosetta [6]).

The advantage of a mass spectrometer operating in space is, in most cases, the intrinsic ultra-high vacuum, which space itself presents. Neutral and ionized particle densities in free interplanetary space are approximately 0.01 and

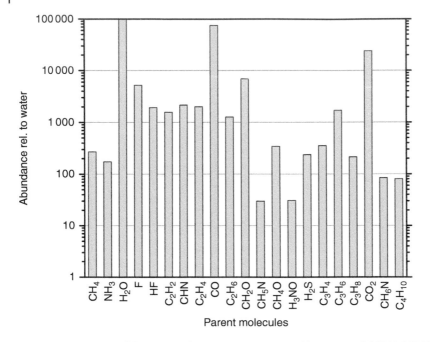

Figure 8.25 Spacecraft background composition measured by Rosetta/ROSINA-DFMS in March 2007, three years after launch, somewhere between Mars and Earth. Abundances are relative to water with water set arbitrarily at 10^5.

$1\,\mathrm{cm}^{-3}$, respectively, a vacuum by far not achieved in the ultra-high vacuum chambers on Earth. However, in the case of the very sensitive mass spectrometers like the ROSINA-DFMS, spacecraft outgassing can play an important role even after many years in space. Normally, spacecraft are built with a very high cleanliness level. This, however, does not prevent that organic material (e.g. from electronic boards, multilayer thermal blankets, glue, etc.) outgas due to desorption of water, diffusion from the interior of the spacecraft and decomposition by UV and cosmic rays. In addition, spacecraft carry fuel, mostly for attitude control, as well as for orbit maneuvers, which is another source of contamination. A study by the ROSINA instrument in 2008 revealed a broad mixture of organic compounds found in the vicinity of the Rosetta spacecraft (Figure 8.25).

And last but not least, mass spectrometers in space have to function over long timespans, without interaction. Voltage adjustment in space is cumbersome and dangerous due to the sometimes very long signal travel times. Broken filaments cannot be replaced; detector aging has to be taken into account. Electrical discharges can lead to the loss of instruments. This happened to the neutral mass spectrometer (NMS) and simultaneously to the high-energy range sensor of the ion mass spectrometer (HERS-IMS) on the spacecraft Giotto, when a cometary dust particle hits the spacecraft leading to a high pressure inside the instruments, destroying both detector systems.

Looking at the many parameters outlined above, mass spectrometers should be individually designed for each mission to cope with environmental challenges, as the instruments address the very different science goals. In the following text, we will point out some of the most important measurements done in space with MS. This list is by far not exhaustive.

8.5.1 Solar Wind and Plasma

The first MS experiments on the rockets inspired scientists to use this technology also on satellites for the exploration of the terrestrial plasmasphere and magnetosphere and the interaction with the solar wind. One of the first mass spectrometers was flown on the Orbiting Geophysical Observatory (OGO) five satellite in 1968 [7]. The ion spectrometer separated masses 1, 4, and 16 by a magnet and detected terrestrial hydrogen, helium, and oxygen ions. More sophisticated instruments followed soon after the Ion Composition Experiment (ICE) on the Geostationary Operational Environmental Satellite (GEOS) [8]. Using magnetic mass spectrometers allowed coupling with the energy analyzers, which in the case of plasma around planets helped to distinguish between solar plasma and planetary plasma due to their different energy ranges. Energy ranges extended soon into the keV range, whereas mass ranges were concentrated mostly at low masses. These measurements helped to characterize the ionized atmosphere of the Earth and its interaction with the solar wind. A novel type of mass spectrometers based on the TOF technique, combined with an energy analyzer (SWICS [4]), was developed for the Ulysses spacecraft measuring, for the first time, solar wind coming from the poles of the Sun and later on for Solar and Heliospheric Observatory (SOHO), which is studying the Sun over a long time. These measurements allow studying the dynamic behavior, the flow of ions, the interaction of our ionosphere with solar storms, and space weather. Similar instruments were flown on the spacecraft to other planets, e.g. Venus (Venus Express [9]) or Mars (Mars Express [9]) and many more. The last decades saw rich scientific achievements on plasma physics from such instruments, and this is still ongoing, even more so with the growing interest in space weather.

8.5.2 Atmospheres of Planets and Moons

The first atmosphere investigated by MS outside the Earth was the Mars atmosphere analyzed by the Viking 1 and 2 (Nier et al. [10]). Carbon dioxide was found to be the major constituent of the atmosphere below 180 km. The isotopic composition of carbon and oxygen in the Martian atmosphere was found to be similar to that of the Earth, while ^{15}N is enriched in Mars' atmosphere by a factor of 1.62 ± 0.16. Venus' atmosphere was investigated with a quadrupole mass spectrometer installed on the Pioneer Venus orbiter [11]. Major constituents have been found to be CO_2, CO, N_2, O, and He. CO_2 is the most abundant constituent below 155 km in the terminator region. Above this altitude atomic oxygen is dominating. Isotopic ratios of O and C were found to be close to terrestrial. The Galileo Probe entered the atmosphere of Jupiter on 7 December 1995. Measurements of the chemical and isotopic composition of the Jovian atmosphere were obtained by the mass spectrometer [12] during the descent over the 0.5–21 bar pressure region for a time period of approximately 1 hour. The mixing ratios of the major constituents of the atmosphere hydrogen and helium have been determined, as well as mixing ratios or upper limits for several less abundant species including methane, water, ammonia, ethane, ethylene, propane, hydrogen sulfide, neon, argon, krypton, and xenon. MS measurements in neutral atmospheres have been performed very successfully by Titan with the Cassini mission and Huygens. A quadrupole mass spectrometer coupled with a gas chromatograph was installed on the European Space Agency (ESA) lander Huygens measuring altitude profiles of the atmosphere [13, 14]. Another quadrupole instrument (ion and neutral mass spectrometer [INMS] [15]) was measuring very successfully, from Cassini for the last 13 years, the upper atmosphere of Titan, as well as the composition of the plumes during flyby of Enceladus. Highlights of these measurements are the detection of a dense nitrogen atmosphere with some hydrocarbons, mainly methane, ethane, and nitriles in Titan. The INMS on Cassini also detected water in the plumes of Enceladus as well as abundant ^{40}Ar.

8.5.3 Comets

Before 1986, the composition of comets was badly understood. The only species detected in the coma of comets by remote sensing were radicals like CN and OH. The goal of the Giotto mission to the Halley's comet was to confirm the presence of water and possibly CO/CO_2. Several mass spectrometers were on board: an NMS ([16]) with a mass range from 12 to 36 Da two IMS (HIS and HERS [17]) with mass ranges up to $56 \, \mathrm{Da \, e^{-1}}$ and the dust mass spectrometer Particle Impact Analyzer (PIA) [18]. The flyby was short and fast, just a few hours, but the scientific return was overwhelming. Not only was water, CO, and CO_2 confirmed, but also the mass spectrometers were able to measure, for the first time, HDO/H_2O and many organics. The dust mass spectrometer detected so-called CHON grains, consisting mostly of organics, apart from the determination of the elemental abundances of many refractories.

While in the years following Giotto, remote sensing observations were extended to different wavelength and the antenna got bigger, allowing many more detections of parent species and isotopes in cometary coma, the highlight was certainly the Rosetta mission, where again several mass spectrometers were on board. Two spectrometers, a magnetic and a TOF-MS (ROSINA [3]), were analyzing the coma over more than two years, and a dust mass spectrometer using SIMS (COSIMA [6]) was analyzing hundreds of dust grains. Two miniaturized mass spectrometers (Cometary Sampling and Composition [COSAC], a GC-MS [19], and Ptolemy [20], an ion trap MS) on the lander Philae explored the comet from up close. This suite was complemented by Rosetta Plasma Consortium Ion Composition Analyzer (RPC-ICA) measuring solar wind ions [21]. The most important results were the measurements of a very high D/H value in water [22] and an even higher D_2O/HDO [23] pointing to presolar water ice and xenon isotopic measurements [24], which can explain part of the terrestrial atmosphere by cometary impacts, the very volatile argon [25] and N_2 [26], a surprisingly high O_2 [27] abundance, and a multitude of complex organics [28]. Most of these results could not have been achieved by remote sensing due to observational bias or too low abundances.

8.5.4 Interstellar and Cometary Dust

To analyze dust in space is difficult, but worthwhile. This can be seen when comparing results from the Stardust mission and the Giotto and Rosetta mission. Particular care was taken to collect the dust in the coma of Wild 2 and to bring it back to Earth. Nevertheless, most dust grains, especially organics, were quite heavily altered by the impact and/or contaminated on Earth, which makes analysis very difficult. While dust analysis in space is difficult, the results from the Giotto-PIA [18], VEGA-PUMA [29], and Rosetta-COSIMA [19] instruments show agreement in the refractory abundances. PUMA detected organic grains at comet Halley [30], which probably correspond to the macromolecular refractory organics seen by COSIMA [31]. The analysis is by far not finished at the time of writing this chapter.

Many more missions carrying mass spectrometers are planned, like BepiColombo to Mercury, the Europa Clipper mission to Europa, and the Jupiter Icy Moon (Juice) mission to Europa, Callisto, and Ganymede. MS has proven to be an important part of planetary exploration.

Questions

- What were the first extraterrestrial objects investigated by MS?
- What are the problems encountered in the use of MS in space?
- What is the main advantage of a mass spectrometer operating in space?

- Explain what is the "spacecraft outgassing" and why it can represent a problem for MS investigation in space.
- Give examples of the most important measurements done in space with mass spectrometry.

References

1 Istomin, V.G. (1962). Investigation of the ion Composition of the earth's atmosphere on geophysical rockets 1957-1959. *Planetary and Space Science* 9: 179–193.
2 Hoffman, J.H. (1970). Studies of the composition of the ionosphere with a magnetic deflection mass spectrometer. *International Journal of Mass Spectrometry and Ion Physics* 4 (4): 315–322.
3 Balsiger, H., Altwegg, K., Bochsler, P. et al. (2007). Rosina–Rosetta orbiter spectrometer for ion and neutral analysis. *Space Science Reviews* 128 (1): 745–801.
4 Gloeckler, G., Geiss, J., Balsiger, H. et al. (1992). The solar wind ion composition spectrometer. *Astronomy and Astrophysics Supplement Series* 92: 267–289.
5 Kissel, J. (1986). The Giotto Particulate Impact Analyser (PIA). *European Space Agency Special Publication* 1077: 67–68.
6 Kissel, J., Altwegg, K., Clark, B.C. et al. (2007). COSIMA–high resolution time-of-flight secondary ion mass spectrometer for the analysis of cometary dust particles onboard Rosetta. *Space Science Reviews* 128 (1): 823–867.
7 Harris, K.K., Sharp, G.W., and Chappell, C.R. (1970). Observations of the plasmapause from OGO 5. *Journal of Geophysical Research* 75 (1): 219–224.
8 Balsiger, H., Eberhardt, P., Geiss, J. et al. (1976). A satellite-borne ion mass spectrometer for the energy range 0 to 16 keV. *Space Science Instrumentation* 2: 499–521.
9 Barabash, S., Lundin, R., Andersson, H. et al. (2006). The analyzer of space plasmas and energetic atoms (ASPERA-3) for the Mars Express mission. *Space Science Reviews* 126 (1): 113–164.
10 Nier, A.O., Hanson, W.B., Seiff, A. et al. (1976). Composition and structure of the Martian atmosphere: preliminary results from Viking 1. *Science* 193 (4255): 786–788.
11 Niemann, H.B., Hartle, R.E., Kasprzak, W.T. et al. (1979). Venus upper atmosphere neutral composition: preliminary results from the Pioneer Venus orbiter. *Science* 203 (4382): 770–772.
12 Niemann, H.B., Atreya, S.K., Carignan, G.R. et al. (1998). Chemical composition measurements of the atmosphere of Jupiter with the Galileo Probe mass spectrometer. *Advances in Space Research* 21 (11): 1455–1461.
13 Niemann, H.B., Atreya, S.K., Bauer, S.J. et al. (2005). The abundances of constituents of Titan's atmosphere from the GCMS instrument on the Huygens probe. *Nature* 438 (7069): 779–784.

14 Niemann, H.B., Atreya, S.K., Demick, J.E. et al. (2010). Composition of Titan's lower atmosphere and simple surface volatiles as measured by the Cassini-Huygens probe gas chromatograph mass spectrometer experiment. *Journal of Geophysical Research: Planets* 115 (E12).

15 Waite, J.H. Jr., Lewis, W.S., Kasprzak, W.T. et al. (2004). The Cassini ion and neutral mass spectrometer (INMS) investigation. *Space Science Reviews* 114 (1–4): 113–231.

16 Krankowsky, D., Lämmerzahl, P., Herrwerth, I. et al. (1986). In situ gas and ion measurements at comet Halley. *Nature* 326–329.

17 Balsiger, H., Altwegg, K., Buehler, F. et al. (1986). The Giotto ion mass spectrometer. In: *The Giotto Mission: Its Scientific Investigations* (ed. R. Reinhard and B. Battrick), 129–148. Noordwijk: ESA Publications Division, ESTEC.

18 Kissel, J., Brownlee, D.E., Büchler, K. et al. (1986). Composition of comet Halley dust particles from Giotto observations. *Nature* 321: 336.

19 Rosenbauer, H., Fuselier, S.A., Ghielmetti, A. et al. (1999). The COSAC experiment on the lander of the ROSETTA mission. *Advances in Space Research* 23 (2): 333–340.

20 Todd, J.F.J., Barber, S.J., Wright, I.P. et al. (2007). Ion trap mass spectrometry on a comet nucleus: the Ptolemy instrument and the Rosetta space mission. *Journal of Mass Spectrometry* 42 (1): 1–10.

21 Nilsson, H., Lundin, R., Lundin, K. et al. (2007). RPC-ICA: The ion composition analyzer of the Rosetta Plasma Consortium. *Space Science Reviews* 128 (1): 671–695.

22 Altwegg, K., Balsiger, H., Bar-Nun, A. et al. (2015). 67P/Churyumov-Gerasimenko, a Jupiter family comet with a high D/H ratio. *Science* 347 (6220): 1261952.

23 Altwegg, K., Balsiger, H., Bieler, A. et al. (2017). D2O and HDS in the coma of comet 67P/Churyumov-Gerasimenko. *Phil. Trans.* 375 (2097): 20160253.

24 Bernard, M., Altwegg, K., Balsiger, H. et al. (2017). Xenon isotopic abundances in 67P/Churyumov-Gerasimenko show that comets contributed to Earth's atmosphere. *Science* 356 (6342): 1069–1072.

25 Balsiger, H., Altwegg, K., Bar-Nun, A. et al. (2015). Detection of argon in the coma of comet 67P/Churyumov-Gerasimenko. *Science Advances* 1 (8): e1500377.

26 Rubin, M., Altwegg, K., Balsiger, H. et al. (2015). Molecular nitrogen in comet 67P/Churyumov-Gerasimenko indicates a low formation temperature. *Science* 348 (6231): 232–235.

27 Bieler, A., Altwegg, K., Balsiger, H. et al. (2015). Abundant molecular oxygen in the coma of comet 67P/Churyumov-Gerasimenko. *Nature* 526 (7575): 678–681.

28 Altwegg, K., Balsiger, H., Berthelier, J.J. et al. (2017). Organics in comet 67P–a first comparative analysis of mass spectra from ROSINA–DFMS, COSAC and Ptolemy. *Monthly Notices of the Royal Astronomical Society* 469 (Suppl 2): S130–S141.

29 Sagdeev, R.Z., Kissel, J., Bertaux, J.L. et al. (1986). The element composition of comet Halley dust particles-preliminary results from the VEGA PUMA analyzers. *Soviet Astronomy Letters* 12: 254.

30 Kissel, J. and Krueger, F.R. (1987). The organic component in dust from comet Halley as measured by the PUMA mass spectrometer on board Vega 1. *Nature* 326: 755–760.

31 Fray, N., Bardyn, A., Cottin, H. et al. (2016). High-molecular-weight organic matter in the particles of comet 67P/Churyumov–Gerasimenko. *Nature* 538 (7623): 72–74.

32 Barabash, S., Sauvaud, J.A., Gunell, H. et al. (2007). The analyzer of space plasmas and energetic atoms (ASPERA-4) for the Venus Express mission. *Planetary and Space Science* 55 (12): 1772–1792.

8.6 Mass Spectrometry in the Study of Art and Archaeological Objects

Giuseppe Spoto

Dipartimento di Scienze Chimiche, Università di Catania, Catania, Italy

8.6.1 Introduction

Objects like textiles, paintings, books, furniture, and some archaeological remains constitute a complex mixture of inorganic and organic molecular systems whose identification requires powerful analytical approaches. The long-term exposure of artistic and archaeological objects to the environment often triggers degradation processes that further increase the chemical complexity of the objects. For these reasons, more robust analytical tools are required to study artifacts and archaeological objects [1].

Analytical methods used to examine art and archaeological objects should be able to provide a wealth of information to help conservation scientists to identify the chemical composition of the constituting materials. Analytical methods are also used to determine the degradation processes triggered by the long-term exposure of the objects to the environment and, sometimes, to identify changes in the original composition introduced by the procedures aimed at preserving the conservation status of the object. The complex variety of chemical information should be obtained by avoiding alteration of the chemical and physical structure of valuable pieces of art or archaeological remains [2, 3]. In this context, mass spectrometry (MS)-based methods play an essential role particularly in identifying organic and biomolecular components of art and archaeological objects [4]. MS methods offer the possibility to identify both inorganic and organic components of the objects with high sensitivity and reproducibility. Moreover, MS methods may provide new opportunities for micro-destructive and, in a future, completely nondestructive analyses, thus opening up new diagnostic approaches for the study of objects of artistic or archaeological importance [5].

8.6.2 MS Methods for the Study of Inorganic Components of Art and Archaeological Objects

Inorganic MS has been mainly used to fingerprint, on a chemical basis, archaeological material. The provenance of archaeological material can be inferred by evaluating results from trace element analysis or quantitative determination of the relative amount of selected isotopes [6]. Inorganic MS is also used to identify pigments in paints or main constituents of glass and ceramic objects, enamels, and glazes [7].

Laser ablation linked to the inductively coupled plasma (LA-ICP) MS (LA-ICP-MS) is used to investigate the elemental and isotopic compositions of a variety of artistic or archaeological samples with spatial resolution. LA-IC-MS exploits the ablated material observed upon irradiation of condensed phases with laser pulses at high laser irradiances ($>10^6$ W cm^{-2}) to perform ICP-MS elemental analysis of solid samples [8]. Metallic objects, ceramics, and glasses have been investigated for fingerprinting purposes using either line-profiling or drilling-at-one-spot modes [9]. LA-ICP-MS line-profile analysis is preferred when flat samples are going to be analyzed and benefits from the reduced elemental fractionation and the enhanced sensitivity. It has been used to evaluate the elemental distribution across the surface of written documents [10] and also to analyze skeletal remains supposedly connected to Mozart and suspected relatives [11]. The drilling-at-one-spot mode minimizes sample alterations and maximizes spatial resolution. Applications include the analysis of ancient coins [12], glazed ceramics [13], and pigments used in prehistoric rock art [14].

Secondary ion (SI) MS (SIMS) is capable of chemically analyzing the surface of both inorganic and organic components of artistic or archaeological solid samples [15]. However, only limited examples of the SIMS identification of organic components of art objects are available [16, 17]. Different pros and cons of using SIMS in analyzing the samples mentioned above are discussed elsewhere [5, 18]. A significant drawback is caused by the ultra-high vacuum environment required for the analysis (10^{-9} to 10^{-11} Torr).

Metals and alloys are ideal samples to be investigated by using SIMS, and for this reason, metal objects of archaeological and artistic interest have been studied with the aid of this technique. These include bronze artifacts [19], antique iron music wires [20], and silver coins [21]. SIMS has been also used to investigate deterioration processes affecting ancient glass objects [22, 23]. Glass degradation mechanisms involve an attack by water combined with environmental and biological agents. Water corrosion is triggered by an ion exchange between the alkali ions present in the glass and hydrogen from the environment. SIMS is also used for glass provenance [22] and dating purposes [24]. SIMS can even image pigment distributions in paint cross sections, thus contributing to the identification of the different paint techniques used in traditional painting [25–27].

8.6.3 MS Methods for the Study of Organic Components of Art and Archaeological Objects

The raw materials used in prehistorical period and by ancient artists are, for the most part, the natural products and hence composed of complex mixtures of organic and biomolecular systems. Intentional mixing of pigments and organic binders to produce painting materials and the repeated use of cooking vessels to prepare a range of different foodstuffs in prehistorical time are two examples of activities that further introduce complexity in the composition of the starting materials. The transformation of the original raw materials generated by degradation processes also contributes to the complexity of the chemical content of the objects to be studied. Proteinaceous materials used in traditional tempera painting, for instance, are subject to condensation and crosslinking reactions with other components, such as lipids, while the pH variation of paint layers contributes to the hydrolysis of peptide bonds and partial dehydration of serine and threonine. The oxidative degradation of amino acids such as cysteine, serine, and phenylalanine in old paint samples leads to the formation of aminomalonic acid [28].

Chemical complexity of the systems under study is the main reason why the identification of organic components of works of art and archaeological objects has been a challenge since 1970s when traditional organic MS started to be used for this aim. Since then, organic MS has been used in art conservation and archaeological science with a variety of purposes [4]. Unfortunately, such a robust investigative approach requires a few micrograms of solid samples to be withdrawn from the valuable object, thus significantly limiting the possibility to analyze precious pieces of art or unique archaeological artifacts. The demand for micro-destructive MS methods has led to the development of new analytical approaches based on laser desorption ionization (LDI) MS methods [5].

Matrix-assisted laser desorption/ionization (MALDI) MS (MALDI-MS) has been shown to be applicable for the identification of organic and biomolecular components of materials used in traditional painting, including drying oils, varnishes, and proteinaceous binders [29]. Drying oils are mixtures of triacylglycerols and are among the oldest binders used in paints [3]. In particular, linseed oil is one of the most popular drying oils used in traditional paints, and for this reason, processes activated during its aging have been studied in detail [30] also by using MALDI-MS [31]. MALDI-MS has also been exploited to identify proteinaceous binders used to paint Renaissance and modern paintings [32, 33] and for identification of biological samples more than 5300 years old, taken from the coat and leggings of the Tyrolean mummy (called Iceman or Oetzi) [34].

Possibilities offered by atmospheric pressure (AP)-MALDI-MS for analyzing ancient objects directly at ambient atmosphere have been shown for identification of the organic components of a variety of traditional materials, such as iron gall [35] and carbonaceous [36] inks, as well as organic dyes and pigments used to print books dated between 1911 and 1920 [37]. The latter studies demonstrated the applicability of AP-MALDI to the *in situ* and micro-destructive studies of organic components of historical objects at spatial resolution of about 400 μm. Additionally, it has been demonstrated that AP-MALDI-MS could detect polycyclic aromatic hydrocarbons (PAH) formed in the preparation of the carbonaceous components of inks used in handwritten parts of ancient books.

Questions

- What is the use of analytical methods to examine art and archaeological objects? Why mass spectrometry is especially suitable for this purpose?
- Describe the different MS methods used to analyze the inorganic and organic parts of art and archaeological objects.

References

1 Ciliberto, E. and Spoto, G. (eds.) (2000). *Modern Analytical Methods in Art and Archaeology*. New York: Wiley.
2 Spoto, G., Torrisi, A., and Contino, A. (2000). Probing archaeological and artistic solid materials by spatially resolved analytical techniques. *Chemical Society Reviews* 29: 429–439.
3 Spoto, G. (2002). Detecting past attempts to restore two important works of art. *Accounts of Chemical Research* 35: 652–659.
4 Colombini, M.P. and Modugno, F. (eds.) (2009). *Organic Mass Spectrometry in Art and Archaeology*. New York: Wiley.
5 Spoto, G. and Grasso, G. (2011). Spatially resolved mass spectrometry in the study of art and archaeological objects. *TrAC Trends in Analytical Chemistry* 30: 856–863.
6 Vanhaecke, F., Balcaen, L., and Malinovsky, D. (2009). Use of single-collector and multi-collector ICP-mass spectrometry for isotopic analysis. *Journal of Analytical Atomic Spectrometry* 24: 863.
7 Dussubieux, L., Robertshaw, P., and Glascock, M.D. (2009). LA-ICP-MS analysis of African glass beads: laboratory inter-comparison with an emphasis on the impact of corrosion on data interpretation. *International Journal of Mass Spectrometry* 284: 152–161.
8 Georgiou, S. and Koubenakis, A. (2003). Laser-induced material ejection from model molecular solids and liquids: mechanisms, implications, and applications. *Chemical Reviews* 103: 349–394.
9 Nevin, A., Spoto, G., and Anglos, D. (2012). Laser spectroscopies for elemental and molecular analysis in art and archaeology. *Applied Physics A: Materials Science and Processing* 106: 339–361.
10 Wagner, B. and Bulska, E. (2004). On the use of laser ablation inductively coupled plasma mass spectrometry for the investigation of the written heritage. *Journal of Analytical Atomic Spectrometry* 19: 1325–1329.
11 Stadlbauer, C., Reiter, C., Patzak, B. et al. (2007). History of individuals of the 18th/19th centuries stored in bones, teeth, and hair analyzed by LA-ICP-MS -a step in attempts to confirm the authenticity of Mozart's skull. *Analytical and Bioanalytical Chemistry* 388: 593–602.

12 Sarah, G., Gratuze, B., and Barrandon, J.N. (2007). Application of laser ablation inductively coupled plasma mass spectrometry (LA-ICP-MS) for the investigation of ancient silver coins. *Journal of Analytical Atomic Spectrometry* 22: 1163–1167.

13 Pérez-Arantegui, J., Resano, M., Garcìa-Ruiz, E. et al. (2008). Characterization of cobalt pigments found in traditional Valencian ceramics by means of laser ablation-inductively coupled plasma mass spectrometry and portable X-ray fluorescence spectrometry. *Talanta* 74: 1271–1280.

14 Resano, M., García-Ruiz, E., Alloza, R. et al. (2007). Laser ablation-inductively coupled plasma mass spectrometry for the characterization of pigments in prehistoric rock art. *Analytical Chemistry* 79: 8947–8955.

15 Spoto, G. (2000). Secondary ion mass spectrometry in art and archaeology. *Thermochimica Acta* 365: 157–166.

16 Sanyova, J., Cersoy, S., Richardin, P. et al. (2011). Unexpected materials in a Rembrandt painting characterized by high spatial resolution cluster-TOF-SIMS imaging. *Analytical Chemistry* 83: 753–760.

17 Voras, Z.E., de Ghetaldi, K., Baade, B. et al. (2016). Comparison of oil and egg tempera paint systems using time-of-flight secondary ion mass spectrometry. *Studies in Conservation* 61: 222–235.

18 Adriaens, A. and Dowsett, M.G. (2006). Applications of SIMS to cultural heritage studies. *Applied Surface Science* 252: 7096–7101.

19 Allen, G.C., Brown, I.T., Ciliberto, E., and Spoto, G. (1995). Scanning ion microscopy (SIM) and secondary ion mass spectrometry (SIMS) of early Iron Age bronzes. *European Mass Spectrometry* 1: 493–497.

20 Goodway, M. (1987). Phosphorus in antique iron music wire. *Science* 236: 927–932.

21 Kraft, G., Flege, S., Reiff, F., and Ortner, H.M. (2000). Investigation of contemporary forgeries of ancient silver coins. *Microchimica Acta* 145: 87–90.

22 Rutten, F.J.M., Briggs, D., Henderson, J., and Roe, M.J. (2009). The application of time-of-flight secondary ion mass spectrometry (TOF-SIMS) to the characterization of opaque ancient glasses. *Archaeometry* 51: 966–986.

23 Melcher, M., Wiesinger, R., and Schreiner, M. (2010). Degradation of glass artifacts: application of modern surface analytical techniques. *Accounts of Chemical Research* 43: 916–926.

24 Liritzis, I. and Laskaris, N. (2009). Advances in obsidian hydration dating by secondary ion mass spectrometry: World examples. *Nuclear Instruments and Methods in Physics Research Section B* 267: 144–150.

25 Van Ham, R., Van Vaeck, L., Adams, F., and Adriaens, A. (2004). Systematization of the mass spectra for speciation of inorganic salts with static secondary ion mass spectrometry. *Analytical Chemistry* 76: 2609–2617.

26 Keune, K. and Boon, J.J. (2004). Imaging secondary ion mass spectrometry of a paint cross section taken from an early Netherlandish painting by Rogier van der Weyden. *Analytical Chemistry* 76: 1374–1385.

27 Keune, K., Hoogland, F., Boon, J.J. et al. (2009). Evaluation of the "added value" of SIMS: a mass spectrometric and spectroscopic study of an unusual Naples yellow oil paint reconstruction. *International Journal of Mass Spectrometry* 284: 22–34.

28 Colombini, M.P., Modugno, F., Fuoco, R., and Tognazzi, A. (2002). A GC-MS study on the deterioration of lipidic paint binders. *Microchemical Journal* 73: 175–185.

29 Calvano, C.D., van der Werf, I.D., Palmisano, F., and Sabbatini, L. (2016). Revealing the composition of organic materials in polychrome works of art: the role of mass spectrometry-based techniques. *Analytical and Bioanalytical Chemistry* 408: 6957–6981.

30 Bonaduce, I., Carlyle, L., Colombini, M.P. et al. (2012). New insights into the ageing of linseed oil paint binder: a qualitative and quantitative analytical study. *PLoS One* 7: e49333.

31 Van Den Berg, J.D.J., Vermist, N.D., Carlyle, L. et al. (2004). Effects of traditional processing methods of linseed oil on the composition of its triacylglycerols. *Journal of Separation Science* 28: 181–199.

32 Tokarski, C., Martin, E., Rolando, C., and Cren-Olivé, C. (2006). Identification of proteins in renaissance paintings by proteomics. *Analytical Chemistry* 78: 1494–1502.

33 Kuckova, S., Hynek, R., and Kodicek, M. (2007). Identification of proteinaceous binders used in artworks by MALDI-TOF mass spectrometry. *Analytical and Bioanalytical Chemistry* 388: 201–206.

34 Hollemeyer, K., Altmeyer, W., Heinzle, E., and Pitra, C. (2008). Species identification of Oetzi's clothing with matrix-assisted laser desorption/ionization time-of-flight mass spectrometry based on peptide pattern similarities of hair digests. *Rapid Communications Mass Spectrometry* 22: 2751–2767.

35 D'Agata, R., Grasso, G., Parlato, S. et al. (2007). The use of atmospheric pressure laser desorption mass spectrometry for the study of iron-gall ink. *Applied Physics A: Materials Science and Processing* 89: 91–95.

36 Grasso, G., Calcagno, M., Rapisarda, A. et al. (2017). Atmospheric pressure MALDI for the noninvasive characterization of carbonaceous ink from Renaissance documents. *Analytical and Bioanalytical Chemistry* 409: 3943–3950.

37 Giurato, L., Candura, A., Grasso, G., and Spoto, G. (2009). In situ identification of organic components of ink used in books from the 1900s by atmospheric pressure matrix assisted laser desorption ionization mass spectrometry. *Applied Physics A: Materials Science and Processing* 97: 263–269.

8.7 Application of ICP-MS for Trace Elemental and Speciation Analysis

Aleksandra Pawlaczyk and Małgorzata Iwona Szynkowska

Institute of General and Ecological Chemistry, Faculty of Chemistry, Lodz University of Technology, Łódź, Poland

8.7.1 Introduction

Inductively coupled plasma mass spectrometry (ICP-MS) is uniquely suited for performing the analysis of liquid samples (wet chemical analysis), as well as solids or gases after application of other sampling methods, such as laser ablation (LA) or hydride generation. The ability to precisely identify and determine most of the elements in the periodic table at ultra-trace levels and with acceptable sensitivity opened up new possibilities in the analysis of complex matrices and solved numerous chemical analysis problems in many new applications. What is more, ICP-MS technique is successfully employed to obtain precise information regarding isotope ratio (IR) for elements that have stable isotopes. This, in turn, advanced the application of isotope dilution (ID) as an option to generate more accurate quantitative results [1–2]. During the last decades, many technological improvements have been introduced, which even more reinforced this technique as one of the most versatile tools for elemental detection and quantification. When compared with other techniques, such as MALDI-MS, the information gathered by ICP-MS seems to be poor because its molecular specificity is lost due to specific properties of the plasma ion source (very high temperature). However, molecule-specific information can be acquired when ICP-MS is coupled to various separation techniques, such as high performance liquid chromatography (HPLC) or gas chromatography (GC) [1, 3, 4]. Higher requirements in relation to the limits of detection with a complex matrix composition, along with a huge technical progress over the last years, have become the basis for the development of new methodological and instrumental solutions used in analytical chemistry. The introduction of hybrid or hyphenated techniques into the market, that is, two or more analytical techniques combined in one instrument, turned out to be very useful in the detection and quantification of multicomponents in complicated mixtures [2–5]. The coupling of chromatographic techniques at the first stage of analysis (separation step) with spectrometric techniques into one integral measurement system, as the most popular variant of hyphenated techniques, is now successfully used to study various forms of elements in speciation analysis, particularly in the fields of quantitative environmental and biological analyses [2, 3, 5]. On the other hand, an increasing popularity of LA coupled to ICP-MS system, together with imaging software, capable of processing of a vast amounts of raw data, has received significant attention in biology, paleontology, or biomedicine where an information about spatial distribution of chemical elements on the relatively small areas seems to be crucial [4]. This chapter summarizes examples of specific applications in the fields of qualitative and quantitative environmental and biological analysis including characteristics of current trends in this area. The unique capabilities of ICP-MS coupled to other techniques are demonstrated, and few selected recent developments in this field are discussed in more details.

8.7.2 Speciation Analysis by ICP-MS: Examples of Applications

Speciation analysis itself focuses on the identification of different forms (chemical and physical) of elements and their quantitative determination in the examined objects. The use of hybrid techniques, such as gas chromatography with inductively coupled plasma atomic emission spectroscopy (GC-ICP-AES), GC-MS, and HPLC-ICP-MS, allows the sample to be separated into individual components and then unambiguously identified. Thus, the generated analytical information can gain an additional dimension – a feedback about the structure of studied compounds – apart from their detection [1, 2]. The benefits and possibilities of these approaches for quantitative analysis can easily be demonstrated with two examples, namely, chromium speciation in natural waters and methylmercury and inorganic mercury (Hg^{2+}) determination in biological material.

As it is a well-known fact, quantification of the total chromium does not deliver complete information about its negative impact, because toxicity of Cr is strongly dependent on its chemical form; anionic hexavalent form of Cr is toxic, while chromium as a cation at trivalent oxidation state is an essential element of human nutrition. Therefore, the methods of choice to deal with this challenging task are limited to those able to obtain low detection limits (DLs) for Cr species and to separate and detect the most typical Cr species in water, such as chromate CrO_4^{2-} (an anion form) for Cr(VI) and the chromic ion Cr^{3+} (cationic) for Cr(III). Another challenge is the stability of these forms under normal laboratory conditions. Cr(III) remains at a quite stable oxidation state in water, whereas Cr(VI) ions, as strong oxidizing agents, can be easily reduced to Cr(III) by acids or organic matter. This suggests that some extra care should be taken at every step of the analytical methodology, including sample collection, storage, and preparation, to secure and preserve unchanged speciation of Cr species in the original sample. Application of ICP-MS system with quadrupole analyzer equipped with H_2 cell gas to remove the signal overlap of ^{52}Cr by $^{52}ArC^+$ and $^{52}ClOH^+$ interferences allows Cr species to be measured with a high accuracy and good sensitivity (achieved DLs for individual Cr species was $<20\,ng\,l^{-1}$). In this analytical methodology the samples, prior to the analysis, were stabilized by their incubation at $40\,°C$ with ethylenediaminetetraacetic acid (EDTA), which forms an anionic complex with Cr(III). This procedure guarantees that a single chromatographic method can be applied to separate the Cr(III)EDTA complex from Cr(VI).

Analysis of methylmercury (MeHg$^+$) and inorganic mercury (Hg^{2+}) in biological tissue by isotopic dilution GC-ICP-MS (GC-ID-ICP-MS) is another good example of the huge potential of ICP-MS technique as a combined system, where ICP mass spectrometer is used together with GC for speciation of Hg. From the toxicity of mercury point of view, its organic species are far more dangerous than inorganic ones. Elemental mercury, which is the major form of this element, can be easily oxidized to inorganic mercury and converted from volatile HgO into soluble Hg^{2+}. Afterward, highly reactive mercuric ion in natural waters can be partly transformed into methylmercury (CH$_3$Hg$^+$) by bacteria in anoxic environments. Finally, methylmercury can be easily accumulated in the biota via the food chain, resulting in human exposure through consumption of the seafood products. Apart from the fish and occupational environments, other main sources of Hg exposure include dental amalgams, skin-lightening cosmetic creams, and some beauty products (e.g. thimerosal is a mercury-based preservative used in eye shadows). In the analytical methodology, some amount of oxygen can be introduced into the plasma gas to burn off carbon deposits on the cones. Additionally, dry plasma conditions can be achieved via desolvation of sample aerosol (either via membrane desolvation or solvent evaporation techniques) and applied to generate highly specific and sensitive results. According to some reports, ID can be accomplished by spiking MeHg$^+$ and Hg^{2+} standards with ^{202}MeHg$^+$ and ^{199}Hg^{2+}, derivatized with NaBPr$_4$, and finally sonically extracted into the pesticide-grade hexane. It was proven that GC-ID-ICP-MS technique can be successfully applied for efficient, selective, and sensitive separation and subsequent detection of volatile and semi-volatile organometallic compounds of Hg. Additionally, ID can ensure the quality control and highly improves the final recovery. Achieved DLs in this case can be even up to 20 times better than those obtained for HPLC technique [6].

8.7.3 Single-Particle and Single-Cell Analysis by ICP-MS: Examples of Applications

Apart from numerous applications in speciation analysis, ICP-MS can also be employed individually for characterizing and sizing metal-based nanoparticles (NPs). In the case of samples containing NPs, which are introduced into the plasma at a low flow rate, their number in the solution is provided small enough, that it is possible to measure signal intensity for the single particle (SP), which can be linked to the size and specific fraction of the NPs. The particle size influences the transient signal intensity (peak height), and the concentration of particles is related to the frequency of the transient signal (the number of peaks in time). This analytical technique used for NP characterization is called single-particle ICP-MS (SP-ICP-MS) and has the potential to quantitate the difference between ionic and particulate signals, to evaluate particle concentration (particles per milliliter), and to assess the particle size and size dispersion. Moreover, SP-ICP-MS enables to study the particle clustering and dissolution. As a consequence, this modern approach may be a perfect tool for the analysis of the fate, behavior, and distribution of engineered nanomaterials (ENMs) in various sample matrices. In order to successfully distinguish data gathered for individual NPs, the following key issues should be considered: proper selection of the acquisition time for a given mass (dwell time) and proper dilution of the sample (the number of particles in the solution). Acquisition time should be long enough to allow the entire signal to be collected from one NP and also short enough to avoid simultaneous measurement of two NPs during one integration time. The SP-ICP-MS technique provides the information on the concentration and mass of the metal that forms NPs. Additional coupling of an appropriate separation technique such as flow fractionation technique in a field (gravitational or central) can also be very useful. The so-called field flow fractionation (FFF) with the detection technique like ICP-MS offers the possibility to separate particles according to their hydrodynamic diameter and to determine the total metal concentration associated with a particular fraction of NPs. Due to the separation stage carried out by the FFF module, the dissolved component can be fully removed from the particulate form. This feature can be beneficial for the SP-ICP-MS analysis of the non-dissolved component, which can play an important role in toxicological examination of NPs where differentiation between the dissolved and particulate components is a key factor. It should be kept in mind that direct connection of SP-ICP-MS to the FFF system still does not provide information on the content of studied metal in an SP, i.e. the unambiguous identification of NPs is difficult. The use of an HR-ICP-MS can significantly improve the signal-to-noise ratio and enables the detection of smaller particles, even those below 10 nm [7–10].

The new possibilities in data acquisition capabilities offered by SC-ICP-MS methodology opened new doors for development of a brand new concept of a single-cell ICP-MS (SC-ICP-MS), where individual cells are rapidly analyzed for their metal content (monitoring the uptake rate at the cellular level). The idea of the measurement is similar to the SP methodology – cell suspension is delivered through the SC-ICP-MS sample introduction system, and, as soon as each cell reaches the plasma, it is ionized, and the generated ion burst from the intrinsic metal is detected by the ICP-MS. Thus, each cell is recognized as an individual entity, which will produce its own ion burst. This new perspective can extend our knowledge regarding cellular interactions with metal-containing species by tracing metal content within SC and monitoring of the uptake of ionic and/or nanoparticulate contaminants. Similarly to SP-ICP-MS, certain criteria in terms of the cell and dissolved micronutrient concentrations have to be met in order to get reliable results, e.g. desirable cell concentration should be about $100\,000\,cells\,ml^{-1}$ to minimize coincidence. For example, nowadays approach includes the addition of NPs into a culture medium containing cells and then the control of the NP uptake into the cells. One of the key studies is monitoring of the commonly used compounds for cancer chemotherapy, like platinum-containing molecules (e.g. cisplatin (SP-4-2)-diamminedichloroplatinum(II)). The effectiveness of these drugs in cancer treatment is associated with their cellular uptake, which means that Pt content in individual cancer cells can be an indicator of the treatment success. The conventional approach presents many limitations. Neither the determined total Pt content in the digested entire cell population allows to assess Pt distribution, nor the number of cells containing Pt within the cell population can be estimated. All these information are essential to understand cell resistance to cisplatin treatment and for the assessment of the real degree of exposure to NPs in the organism. In a broader perspective, this might be helpful in designing strategies for anticancer medications. Such data can be beneficial for the customized design of better targeted drugs and for delivering real data about the actual risk of NPs in the environment [11, 12].

8.7.4 Imaging by LA-ICP-MS Technique

Over the last 10 years, elemental imaging of solid samples using laser ablation inductively coupled plasma mass spectrometry (LA-ICP-MS) has proved its versatile applications in many scientific fields, such as medicine, biology, and geology. The biggest advantage of LA-ICP-MS technique, when compared with other approaches, is the ability to produce bioimages at very high sensitivity. LA-ICP-MS instrument can be perceived as a hyphenated technique, in which a laser unit is coupled to ICP mass spectrometer [13–18]. The most frequently used lasers are Nd:YAG devices. This neodymium-doped yttrium aluminum garnet $Nd:Y_3Al_5O_{12}$ crystal is employed as a lasing medium for the solid-state lasers. Some portions of Y ions are replaced by Nd(III) ions of similar size. The laser is most often operated at 266 or 213 nm. Ablated material, as an aerosol in the ablation cell, is transported by a carrier gas directly into the ICP spectrometer and injected into the ICP torch. The ICP source is responsible for ions' production that are subsequently separated in the mass analyzer. The recorded data are then converted into pixels, which enables the creation of an image of a sample microarea (spatial visualization) [3, 4]. The exemplary LA-ICP-MS spectra obtained for ceramic samples are presented in Figures 8.26 and 8.27.

In 2008, this system was applied to determine and monitor the level of Pt along with a single strand of hair taken from a patient who had been treated with D,D-cisplatin for metastasized ovarian cancer. Patient was treated with the same four cycles of cisplatin/cyclophosphamide at three-week intervals, while the hair sample was taken on day 120. In order to increase the sensitivity, the total ablation of the hair cross section was performed instead of the partial ablation. The signal from [195]Pt collected along the hair strand showed

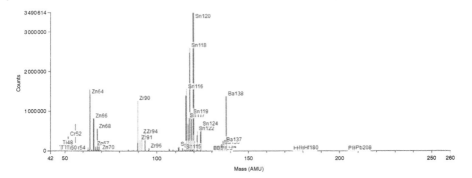

Figure 8.26 The distribution of Ti, Cr, Zn, Cd, Ba, Hf, and Pb isotopes in the laser ablation inductively coupled plasma mass spectrometry (LA-ICP-MS) spectra collected from the surface of ceramic cup sample in the mass-to-charge (*m/z*) range (42–260 AMU).

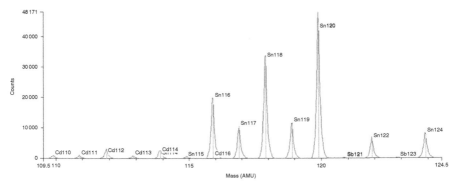

Figure 8.27 The distribution of Cd, Sn, and Sb isotopes in the LA-ICP-MS spectra collected from the surface of ceramic cup sample in the *m/z* range (110–124 AMU).

variations in Pt concentration corresponding to each cisplatin dose at time intervals, which were easily correlated with a hair growth rate. Each of the four sharp Pt signals associated with four cis-Pt doses was preceded by the much smaller Pt peaks, indicating that absorption of Pt by hair undergoes in two stages: first is an external deposition of Pt dissolved in the sweat at the orifice of the hair follicle, and then the incorporation of Pt into the hair strand takes place within the growth zone, deep in the hair follicle [19]. Almost 10 years later, new possibilities have been opened up for the analysis by LA-ICP-MS system, which was used to study cisplatin pharmacokinetics in the cochleae of mice and humans. It was shown that compared with other organs in which cisplatin was detected one hour after injection and eliminated during the next days or weeks, this anticancer drug is still present in the cochlea for months to years after treatment in both mice and humans. The map of Pt distribution within the cochlea indicated that *stria vascularis* is the region of the cochlea, in which the accumulation of cis-Pt is the highest, and this region may serve as a therapeutic target for preventing cisplatin ototoxicity [20]. This example demonstrates the progress, during only one decade, in both application and technological fields.

Nowadays, bioimaging using LA-ICP-MS enables to quantify trace elements and isotopes within different materials including tissue sections with an achievable spatial resolution, typically ranging from 10 to 100 µm at a high sensitivity. This kind of analysis performed using LA-ICP-MS can be very helpful in understanding many issues connected with the role of various elements in biomedical research. Moreover, it can even deliver some crucial information in terms of bioaccumulation and bioavailability for ecological and toxicological risk assessment in humans, animals, and plants. The main area of interest in environmental science is still covered by hyper-accumulating plants and pollution-indicating organisms [15, 17, 21–23]. The usefulness of IR combined with bioimaging for predicting geographical origin, habitat, migrations of subjects, diet, and lifestyle has been shown in some papers as well [21, 23]. For instance, LA-ICP-MS was applied to discriminate specimens during all stages of the fish life, and this methodology delivered information on metabolic studies on barium. Moreover, this technique was used to assess the isotopic composition of Sr, which can be treated as a stable signature over years and can serve in the salmon habitat/migration studies or as a migration indicator to identify ancient population mobility. In other research, the levels of K, Mg, Mn, P, and S were monitored as a consequence of Cu stress after ^{65}Cu accumulation in the petiole and main veins of the leaves. The literature survey over the last 10 years suggests that LA-ICP-MS proves its usefulness and is constantly exhibiting its huge potential in bioimaging of metals and metalloids in biomedical research. Apart from the animal models and plant sciences, many achievements of mass spectrometry imaging (MSI) and metallomics have been reported in brain and cancer research (such as studies of Fe, Cu, and Zn distribution in rodent brain) or drug development (distribution and delivery of metallodrugs in animals). For example, the quantitative image of metallodrugs (Sr/Pt based) were generated and magnetic resonance imaging (MRI) contrast agents (Gd based) were additionally employed. It was shown that in the rat tumor tissue Pt was distributed in the left brain hemisphere, which may indicate that some regions of the tumor had become necrotic, resulting in the accumulation of Gd in the occluded vasculature. Moreover, some areas in the hard tissues with elevated Sr levels were detected. In other studies, sections of rat brain samples were investigated, and images of P, S, Fe, Cu, and Zn distribution were obtained. It was discovered that P, S, and Fe distributions were depleted in the tumor tissue, whose shape could have been easily reproduced and distinguished from the surrounding healthy tissue. Similar research was carried out for thin tissue sections of primary human brain tumor, in which tumor mass region and tumor invasion zones were discriminated. The visualization of the spatial distribution of Cu, Zn, Pb, and U was performed, which enabled the localization and indication of elements belonging to the tumor and its invasion zones [23].

8.7.5 Improvements of LA-ICP-MS Technique

Recently, all efforts are focused on the design of the LA cells to achieve more efficient washout times, thus reducing duration of single signals generated by the LA device. One of the implemented solutions is minimization of the internal volume of the cell and tubing, which is responsible for aerosol delivery to the ICP. The goal is a spot-resolved imaging, which requires very high level of synchronization between the LA unit and ICP-MS system (typically with TOF analyzer) where every shot laser will be assigned to the particular pixel in the generated image. The detection of individual laser pulses of ablated sample as sharp peaks, instead of a slowly decaying peak, leads to the improved resolution of individual LA events by reducing the pulse-to-pulse intermixing. In this case, higher repetition rates up to 100 Hz can be applied. As the result of the signal-to-noise ratio improvement, together with the application of the much smaller laser spots, and much better spatial resolution can be obtained, which significantly enhances the capability of the spot-resolved LA-ICP-MS imaging performance [15]. All these developments led to the markedly improved spatial resolutions down to 1 μm. For instance, it was possible to construct high-resolution (HR) (6 μm^2) quantitative imaging of ^{56}Fe and ^{57}Fe in the mouse brain that corresponds to the dimensions of SC [24]. Still challenging is the issue of heterogeneity of the samples being ablated by LA-ICP-MS systems since no homogeneously distributed internal standard can be found, which is necessary to perform quantitative analysis. For some matrices, the solution to such problem can be subsequent 100 wt.% normalization of the results for the entire matrix. However, this concept can be successfully realized only if all elements within a matrix are detected. Some new problems may arise if highly interfering elements in the matrix are present at high amounts. The future belongs to new ICP-MS developments leading to the design of HR time-of-flight MS (TOF-MS) systems or improvements in the reaction cell manufacturing, in combination with a TOF-MS system. Both possibilities are still commercially unavailable, but those challenges are already undertaken [18]. The use of ultra-short pulse lasers (fs-LA) opened new doors for elemental imaging, as the reduction of target heating during femtosecond laser pulses is achieved, which eliminated significant limitations in the studies of volatile samples along with assured highly improved precision. In consequence, fs-LA-ICP-MS already found numerous applications in elemental characterization of diverse samples and may soon become a standard in many medical, biological, environmental, or geological laboratories [5].

8.7.6 LA-ICP Mass Spectrometer with LIBS

A yet another powerful approach should be mentioned. Incorporation of the laser-induced breakdown spectroscopy (LIBS), being an atomic emission technique working under ambient conditions, into LA-ICP-MS resulted in tandem LIBS–LA-ICP-MS setup. In LIBS instrument, the characteristic photons are generated during relaxation of the excited atomic and ionic species, resulting in generation of the spectral signature of the elements. So far, all different types of mass spectrometers (quadrupole, TOF, magnetic sector) have been applied, and also fs-LA unit has been incorporated. The biggest outcome of this unit is the ability to cover a wider concentration range, as both techniques used can generate distinct information about elemental and isotopic compositions. In the case of LIBS, typical DLs are in micrograms-per-gram ($\mu g\,g^{-1}$) levels, while for LA-ICP-MS, the achievable DLs are in the range of sub-microgram-per-gram levels [5].

Questions

- How is it possible to obtain molecular-specific information by performing ICP-MS?
- What are the main analytical challenges in detecting chromium toxicity in samples?
- How is it possible to perform the analysis of methylmercury ($MeHg^+$) and inorganic mercury (Hg^{2+}) in biological tissues?
- What are the key experimental issues to be considered in the SP-ICP-MS analysis?
- What is field flow fractionation and what is its application?
- What are the potential applications of single-cell ICP-MS?
- How is it possible to perform ICP-MS on solid samples?
- Give some examples of LA-ICP-MS applications.
- Why the synchronization between the LA unit and the ICP-MS system is important to obtain a well-resolved image?
- What is the main advantage of fs-LA-ICP-MS?

References

1 Pröfrock, D. and Prange, A. (2012). Inductively coupled plasma–mass spectrometry (ICP-MS) for quantitative analysis in environmental and life sciences: a review of challenges, solutions, and trends. *Applied Spectroscopy* 66 (8): 843–868.
2 Ammann, A.A. (2007). Inductively coupled plasma mass spectrometry (ICP MS): a versatile tool. *Journal of Mass Spectrometry* 42: 419–427.

3 Nelms, S.M. (ed.) (2005). *ICP Mass Spectrometry Handbook*. Oxford: Blackwell Publishing, CRC Press.

4 López-Fernández, H., de S. Pessôa, G., Arruda, M.A.Z. et al. (2016). LA-iMageS: a software for elemental distribution bioimaging using LA–ICP–MS data. *Journal of Cheminformatics* 8 (65): 1–10.

5 Balaram, V. (2018). Recent advances and trends in inductively coupled plasma–mass spectrometry and applications. *Chromatography Online* 16 (2 (38)): 8–13.

6 (2007). *Handbook of Hyphenated ICP-MS Applications*. USA: Agilent Technologies.

7 Tadjiki, S. (2015). *Field-Flow Fractionation with Single Particle ICP-MS as an Online Detector*. Utah: Agilent Technologies.

8 Merrifield, R., Amable, L., and Stephan, C. (2018). Laser ablation and inductively coupled plasma–time-of-flight mass spectrometry: a powerful combination for high-speed multielemental imaging on the micrometer scale. *Spectroscopy* 32 (5): 14–20.

9 Lee, S., Bi, X., Reed, R.B. et al. (2014). Nanoparticle size detection limits by single particle ICP-MS for 40 elements. *Environmental Science and Technology* 48 (17): 10291–10300.

10 Wilbur, S., Yamanaka, M., and Sannac, S. (2017). *Characterization of Nanoparticles in Aqueous Samples by ICP-MS*, White Paper. USA: Agilent Technologies.

11 Amable, L., Stephan, C., Smith, S., and Merrifield, R. (2017). *An Introduction to Single Cell ICP-MS Analysis*, vol. 2016. Waltham: PerkinElmer.

12 Corte Rodríguez, M., Álvarez-FernándezGarcía, R., Blanco, E. et al. (2017). Quantitative Evaluation of Cisplatin Uptake in Sensitive and Resistant Individual Cells by Single-Cell ICP-MS (SC-ICP-MS). *Analytical Chemistry* 89 (21): 11491–11497.

13 Limbeck, A., Galler, P., Bonta, M. et al. (2015). Recent advances in quantitative LA-ICP-MS analysis: challenges and solutions in the life sciences and environmental chemistry. *Analytical and Bioanalytical Chemistry* 407: 6593–6617.

14 Sylvester, P.J. and Jackson, S.E. (2016). A brief history of laser ablation inductively coupled plasma mass spectrometry (LA–ICP–MS). *Elements* 12 (5): 307–310.

15 Bussweiler, Y., Borovinskaya, O., and Tanner, M. (2017). Laser ablation and inductively coupled plasma–time-of-flight mass spectrometry: a powerful combination for high-speed multielemental imaging on the micrometer scale. *Spectroscopy* 32 (5): 14–20.

16 Becker, S., Zoriy, M., Becker, S.J. et al. (2007). Laser ablation inductively coupled plasma mass spectrometry (LA-ICP-MS) in elemental imaging of biological tissues and in proteomics. *Journal of Analytical Atomic Spectrometry* 22: 736–744.

17 Becker, J.S., Matusch, A., and Wu, B. (2014). Bioimaging mass spectrometry of trace elements—recent advance and applications of LA-ICP-MS: a review. *Analytica Chimica Acta* 835: 1–18.

18 Spectroscopy Editors (2016) Laser-ablation ICP-MS imaging of geological samples, *Spectroscopy Online*. http://www.spectroscopyonline.com/laser-ablation-icp-ms-imaging-geological-samples (accessed 30 October 2019).

19 Pozebon, D., Dressler, V.L., Matusch, D.A., and Becker, S. (2008). Monitoring of platinum in a single hair by laser ablation inductively coupled plasma mass spectrometry (LA-ICP-MS) after Cisplatin treatment for cancer. *International Journal of Mass Spectrometry* 272 (1): 57–62.

20 Breglio, A.M., Rusheen, A.E., Shide, E.D. et al. (1654). (2017) Cisplatin is retained in the cochlea indefinitely following chemotherapy. *Nature Communications* 8: 1–9.

21 Pozebon, D., Scheffler, G.L., Dressler, V.L., and Nunes, M.A.G. (2014). Review of the applications of laser ablation inductively coupled plasma mass spectrometry (LA-ICP-MS) to the analysis of biological samples. *Journal of Analytical Atomic Spectrometry* 29 (12): 2204–2228.

22 Van Malderen, S.J.M., Managh, A.J., Sharpb, B.L., and Vanhaecke, F. (2016). Recent developments in the design of rapid response cells for laser ablation-inductively coupled plasma-mass spectrometry and their impact on bioimaging applications. *Journal of Analytical Atomic Spectrometry* 31: 423–439.

23 Pozebon, D., Schefflera, G.L., and Dressler, V.L. (2017). Recent applications of laser ablation inductively coupled plasma mass spectrometry (LA-ICP-MS) for biological sample analysis: a follow-up review. *Journal of Analytical Atomic Spectrometry* 32: 890–919.

24 Lear, J., Fryer, D.F., Hare, D.J., and Doble, P. (2012). High-resolution elemental bio-imaging of Ca, Mn, Fe, Co, Cu and Zn employing LA-ICP-MS and hydrogen reaction gas. *Analytical Chemistry* 84 (15): 67076714.

8.8 Mass Spectrometry in Forensic Research

Marek Smoluch[1] and Jerzy Silberring[1,2]

[1] Department of Biochemistry and Neurobiology, Faculty of Materials Science and Ceramics, AGH University of Science and Technology, Kraków, Poland
[2] Centre of Polymer and Carbon Materials, Polish Academy of Sciences, Zabrze, Poland

8.8.1 Introduction

This chapter describes, in brief, basic concepts and future prospects of mass spectrometry applications in forensic research. Not only this subject includes typical investigations on the crime scene but also anti-doping research (described in Section 8.9), forgery in art, anti-terror actions, and counterfeit medicines, among others. All these explorations demand specific types of instrumentation and sample preparation. Also, time of analysis, precision, and accuracy may play an important role, depending on the sample type.

8.8.2 Forgery in Art

Art objects and old manuscripts were always copied, which is presently referred to as plagiarism. Even Michelangelo was doing that by creating a marble statue "Sleeping Eros" in 1496, presented as a Roman art, which was quite quickly recognized to be a fake. This falsification made the artist famous, and his reputation became prosperous, because "only true master could fake so well" [1].

Even more spectacular was the work done by John Myatt, also known as the Master Forger, together with his assistant John Drewe who counterfeit over 200 various artists, among them Giacometti, Monet, Renoir, and Matisse. They succeeded to cheat Sotheby's and Christie's, but in contrast to Michelangelo, their career ended up in jail. Despite this unpleasant episode, their work, now labeled on the back of the canvas as "genuine fake," reaches quite high revenue, going from c. £500 to 5000 [2].

A yet another example of almost perfect forgery was the case known as Getty Kouros. A couros (κοῦρος) denotes for a nude standing youth that first appears in the Archaic period in Greece, c. seventh century BCE. This marble sculpture was purchased by the J.P. Getty Museum for 10 million USD. Before final decision, the museum has commissioned many expertises, including mass spectrometry. Despite such detailed number of analyses, the item has been considered as a fake. Nevertheless, the Getty Kouros Colloquium organized in Athens in 1992 with a number of renowned experts still raises a debate on its authentication.

Isotope ratio mass spectrometry, in combination with ICP, was applied to study authenticity of the Vermeer's "Saint Praxedis." The painting was primarily attributed to another artist, but a discovery of signature "Meer-1655" raised speculations that this might be Vermeer's work. Lead isotopes, particularly the ratio $^{206}Pb/^{204}Pb$, can be used to point out location of the lead ore used to prepare a popular pigment, lead white applied in the painting. This technique needs, however, comparison with other unquestionable artwork of Vermeer, and the sample has been taken from "Diana and her companions." The analysis, together with other observations, allowed to unambiguously assign the work to Vermeer.

Apart from the forgeries, many studies are performed to reveal painting techniques of famous artists. This is usually achieved during art conservation where tiny scrapes of paints or groundings can be acquired without unnecessary dismantling of the artwork at other occasions. The vast number of binders contains proteinaceous components of natural origin, such as whole egg, egg yolk, egg white, casein, bone glue, fish glue, etc. To reveal identity of such binders, proteomics can be applied. An example of such work was performed on the artwork "Sitting nude and grotesque masque" by Edvard Munch [4]. Whole egg was identified as a binder using peptide mass fingerprinting and MALDI-TOF instrument.

Another application concerns unexpected identification of starch in a Rembrandt painting, "Portrait of Nicolaes van Bambeeck," where the layered analysis by TOF-SIMS was proven very useful [5]. From their discovery, authors concluded that there was a replacement of a part of the lead white by starch or flour to increase the transparency of the second ground layer.

More fascinating stories can be found in recent books written by J. Ragai [6] and by P. Craddock [7].

8.8.3 Psychoactive Substances and Narcotics

Apart from psychoactive substances, the novel psychoactive substances comprise the young generation of drugs of abuse. Their chemical structures and psychoactive properties are similar to those already existing and prohibited. In other words, these compounds are a "legal" alternative to those already outlawed by a majority of countries. Many categories of such compounds have been synthesized, for example, psychoactive bath salts (PABS), and two major groups should be mentioned. One is named "legal highs" and the other is dubbed "designer drugs." Legal highs are compounds that can easily be synthesized or used in an unmodified form, from the over-the-counter medicines. This can be done by unexperienced persons such as kids and are difficult to monitor or size by authorities. Designer drugs denote for the more advanced compounds, and their production requires arrangement of laboratory equipment and skillful person in chemical synthesis. Having in mind that only in 2014, over 100 new psychoactive compounds appeared on the global market, this raises an unsolved yet problem – how to identify those molecules and their metabolites in human body to reduce potential harm to the patients in the emergency room. Here, we will focus on some technical problems associated with sampling of biological material, detection of metabolites, and legal regulations.

Taking into consideration that daily marijuana use in 2017 exceeded daily cigarette use among 8th, 10th, and 12th graders [8], nations are well aware that drug trafficking, selling, and manufacturing comprise a serious health problem to the entire societies. Another recently discovered issue concerns opioid epidemic [9, 10], due to the substantial increase, in the 1990s, in the use of prescription opioids – among them fentanyl, strong and addictive painkiller. This resulted in a report released by the US Drug Enforcement Administration stating (in 2015!!) that situation with overdose deaths and number of addicts have reached epidemic level [11]. Many actions are undertaken on a global scale and coordinated by the Interpol. One of such operations devoted to narcotics was the Lionfish III, aiming at disrupting illicit flows and narco groups in Latin America and West Africa, responsible for trafficking drugs to Europe [12]. During a 10-day period, 52 tons of cocaine, cannabis, and heroin was seized. Moreover, 20 clandestine laboratories were dismantled, and 3 tons of precursor chemicals was seized.

Besides these numbers, there is a yet another problem with vehicle drivers being, apart from the alcohol, upon influence of various psychoactive drugs (predominantly stimulants), and patients admitted to the emergency room (ER), often unconscious that are unable to describe what they took. In a majority of cases, they are unaware of the content as it is impossible for a layperson to reveal the composition of a pill purchased on the street. Another issue concerns the unknown content of the substances, which can contain a variety of adulterants, other compounds than those offered (e.g. MDMA replaced by cheaper, longer-onset, and more toxic para-methoxyamphetamine or 4-methylthioamphetamine – Yaba), or additives, such as talcum, powdered glass, or psychoactive compounds (caffeine, quinine, lidocaine, synthetic cannabinoids, etc.). Toxicology laboratories routinely detect four main groups of psychoactive compounds, which are cocaine, opiates, amphetamines, and cannabinoids, and ELISA method is the most common. The rest, in particular "designer drugs" and "legal highs," comprising a vast number of chemical structures and contaminants, often unrelated to these four groups, requires additional monitoring, not accessible in ER. Therefore, fast and unambiguous monitoring methods should be available to address such problems, and mass spectrometry certainly is the central technique in such an approach. The primary goal is proper biosampling and preparation of the material for further analysis. Here, several sources of biological material are available, namely, plasma, urine, dried blood spots, saliva, and hair. All these sources are used in forensic research and for medical diagnosis; however sample withdrawal and its processing vary significantly, including its assignment. Saliva seems to be an ideal material, as its sampling is noninvasive and can be performed anywhere by an unqualified person with a help of a cotton swab. On the other hand, saliva content undergoes many variations, depending on food ingestion, drinks, salivation itself, etc. Luckily, the exact quantitation is not important at this stage, and a simple answer "yes" or "no" is sufficient to choose further procedure. From this point of view, this material may serve as indicative measure to be further confirmed by the more specific and quantitative techniques using blood/urine samples. The key issue is time from patient's admission to complete analysis, often decisive for the person's life. More information can be found in Ref. [13]. The ideal situation would be to eliminate as many steps as possible to avoid time-consuming sample handling. New trends point out clearly application of ambient mass spectrometry as a suitable tool for such purposes. These devices utilize a variety of ion sources, such as DART, DESI, FAPA, and DBDI (Sections 4.1.3.4 and 4.1.4) capable of analyzing substances at ambient conditions or without a need for sample preparation. An example of such applications may be the rapid and direct analysis of the cigarette smoke by FAPA-MS (Figure 8.28) or methcathinone (a typical "legal high") produced from pseudoephedrine-containing pills (Figure 8.29). Such recipe for its preparation is freely available on the Internet. The instruments constantly undergo miniaturization down to a true portability, which was possible thanks to miniature components, such as vacuum pumps, analyzers, and power supplies [14–18], and find their applications in many disciplines from forensic research, monitoring of chemical synthesis to space exploration, and environmental protection. Forensic analysis should also fulfill judicial requirements, which are more restrictive than the methodologies developed for basic research purposes.

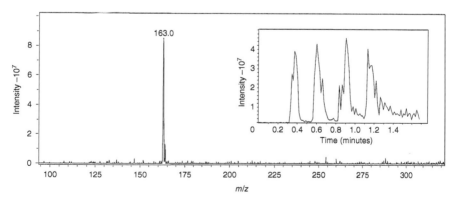

Figure 8.28 Direct analysis of the exhaled cigarette smoke. Exhaled air was blown by the smoker directly to the space between FAPA source and MS inlet. Signal at *m/z* 163.0 corresponds to the protonated molecule of nicotine. Air was exhaled four times in the frame of one minute (see insert for the extracted ion chromatogram (EIC) for *m/z* 163.0).

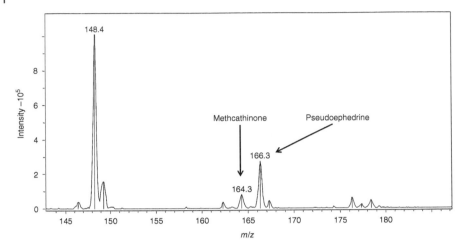

Figure 8.29 FAPA-MS analysis of the crude mixture resulting from the synthesis of methcathinone from the pill containing pseudoephedrine. The substance in a liquid form was introduced from the glass slide. Signal at *m/z* 164.3 corresponds to the protonated methcathinone. The presence of methcathinone was also confirmed by MS/MS experiment (data not shown).

Identification of a single ion belonging to a given substance may not be sufficient, and for example, one or two metabolites should also be detected to prove the identity of a molecule.

8.8.4 Counterfeit Drugs and Generation of Metabolites

An international action called "Operation Pangea" is regularly arranged, and the last one was 10th in a row, where 123 countries participated in 2017 [19, 20]. For only one week, fake and illicit medicines were sized, 3 000 adverts were suspended, 3 584 websites were taken offline, and 470 000 packages were sized. Pangea is not exclusively focused on narcotics, but broadly deals with all illegal or counterfeit substances and products, such as implants, prophylactics, syringes, etc. Counterfeit medicines, defined as fake medicines, according to FDA definition, "may be contaminated or contain the wrong or no active ingredient. They could have the right active ingredient but at the wrong dose" [21]. They are easily available over the Internet or delivered directly to the buyers, and the commonly distributed medicines are antibiotics, cancer medicines (Avastin, Herceptin), Viagra or analogues for treatment of erectile dysfunctions and cardiovascular diseases, and antimalarial. According to the World Health Organization, 1 in 10 medical products in developing countries is substandard or falsified. This gives a huge profit of c. 30 billion USD [22]. Such uncontrolled medicines may also cause serious health problems, which was a

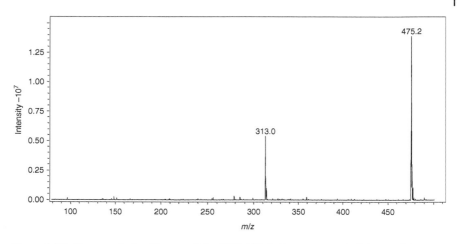

Figure 8.30 Direct analysis of Viagra pill by FAPA-MS. Active compound of the medicine sildenafil can be seen at *m/z* 475.2.

case in Bangladesh. It has been concluded that paracetamol elixirs with diethylene glycol as a diluent were responsible for a large outbreak of fatal renal failure in Bangladesh [23]. Many actions are undertaken at both the global and local levels to countermeasure manufacturing and distribution of bogus drugs.

With a quite advanced equipment for printing labels and holograms, also packages resemble those original, and, as such, they are indistinguishable to the majority of unexperienced persons. Visual examination of boxes and packages, as well as labels and leaflets, can be supported by the analysis utilizing Raman or infrared spectroscopy, X-ray fluorescence spectroscopy, scanning electron microscopy, and computer tomography. Mass spectrometry, in particular imaging options, can be used here, together with other methods, such as HPLC or electrophoresis, thus providing clear answer to the authenticity problem. NMR and spectroscopic methods are also commonly applied. Other methods for authentication of chemical composition include simpler techniques that can be applied in field, such as thin layer chromatography or colorimetry. In addition to the analysis of medicines, also packages are inspected for their authenticity. An example of the fast and unambiguous identification of the content of a pill is shown in Figure 8.30.

8.8.5 Terrorism/Explosives/Chemical Warfare

Mass spectrometry has already established itself in security protection as an unambiguous technique. There are still many obstacles to be solved, such as true portability of the instruments, sensitivity, sampling, time of analysis, and simplicity. An additional asset of mass spectrometry is that almost all potential agents can be simultaneously detected. There are c. 200 explosives listed by the ATF Bureau of Alcohol, Tobacco, Firearms and Explosives to be in hands of terrorists. Among them are cyclotrimethylenetrinitramine (RDX), Semtex, nitroglycerin, and nitrates [24]. Poisonous agents comprise one of the elements of terrorist attacks. As a recent example, a tragic terror attack in Japan (Tokyo Subway Attack in 1994) has been performed using sarin, a warfare gas. Thorough analysis of blood, water, formalin-fixed brain tissues, and wipe samples by GC-MS revealed the presence of sarin together with its hydrolysis products. A definite presence of the nerve agent was confirmed by its presence in erythrocytes of four victims of the attack [25]. In this work, sarin-bound acetylcholinesterase (AChE) was solubilized from erythrocyte membranes of sarin victims and digested with trypsin. The hydrolysis products bound to AChE were released by alkaline phosphatase digestion, and the digested sarin hydrolysis products were subjected to TMS derivatization and detected by GC-MS. Sarin, together with a newly applied *Novichok*, are organophosphates acting as cholinesterase inhibitors [26]. This and other attacks triggered research and development of reliable methods for detection of such compounds.

Not only at the battlefield the instruments for on-site analysis are necessary, with a possibility for operating in a continuous mode, but also constant monitoring of borders, airports, large event venues, executive facilities, and water supplies is necessary to protect people. A similar issue concerns rapid detection of psychoactive substances at public places, such as dance scenes, public meetings, etc. Current equipment involves ion mobility spectrometers, gas chromatography, and various sensors [27].

Besides low molecular mass agents that are quite easy to be identified by portable instruments, a more complex problem arises when microbial warfare needs to be detected. A plethora of screening methods, mainly involving MALDI-TOF, are routinely used in hospitals and diagnostic centers to detect bacterial strains being a threat to the patients. Other harmful molecules such as ricin or botulinum toxin can be identified in the specialized laboratories, with all necessary safety precautions undertaken by the personnel. In such cases, on-field analysis is practically impossible due to technical limitations of the available equipment and a need for the well-trained operators. It can be concluded here that portable mass spectrometers equipped with ambient pressure ionization are capable of detecting low molecular weight molecules but face difficulties to analyze larger species.

8.8.6 Future Prospects

Ambient mass spectrometry is a rapidly evolving technique, with a good perspective to release commercial fieldable instruments. Such instruments will certainly gain broader popularity offering capabilities suitable to various disciplines, such as chemical laboratories, customs, defense, art authentication, and forensic research.

Each domain has its own regulations and demands. For instance, art authentication requires minimal destruction of precious art objects, but the speed of analysis is not important. In contrary, identification of psychoactive substances as a part of medical diagnostics and detection of explosives or warfare agents should be immediate and with high sensitivity and unambiguity. Here, exact quantitation may be of lower importance. Moreover, continuous mode of operation is vital to protect public against terror attack. Design and construction of such instruments, from truly portable to fieldable models, are appearing in the literature, and the number of publications of this topic shows good perspectives for the development of novel concepts in mass spectrometry.

Questions

- Give some examples of the forgery in art that could be unveiled by applying mass spectrometric investigation.
- List the four main groups of psychoactive compounds.
- What sources of biological material are available for the screening of psychoactive substances and narcotics?
- What are the advantages and limitations of portable mass spectrometers equipped with the ambient pressure ionization?
- Provide examples of the requirements that MS analysis should follow, depending on the type of samples available and the forensic investigation to be carried out.

References

1 Charney, N. (2015). *The Art of Forgery*. London: Phaidon.
2 Honigsbaum, M. (2005) The master forger. *The Guardian* (8 December 2005). https://www.theguardian.com/artanddesign/2005/dec/08/art (accessed 30 January 2019).
3 Honigsbaum, M. (2005) The master forger. *The Guardian* (8 December 2005). https://www.theguardian.com/artanddesign/2005/dec/08/art (accessed 30 January 2019).
4 Kuckova, S., Hynek, R., and Kodicek, M. (2007). Identification of proteinaceous binders used in artworks by MALDI-TOF mass spectrometry. *Analytical and Bioanalytical Chemistry* 388: 201–206.
5 J. Sanyova, S. Cersoy, P. Richardin, O. Laprévote, P. Walter, and A. Brunelle. Unexpected materials in a Rembrandt painting characterized by high spatial resolution cluster-TOF-SIMS imaging. *Analytical Chemistry* 2011;83(3):753–60.

6 Ragai, J. (2015). *The Scientist and the Forger*. London: Imperial College Press.

7 Craddock, P. (2009). *Scientific Investigation of Copies, Fakes and Forgeries*. Amsterdam: Elsevier.

8 National Institute for Drug Abuse. NIDA DrugFacts'2017. https://www.drugabuse.gov/publications/drugfacts/monitoring-future-survey-high-school-youth-trends (accessed 30 January 2019).

9 Volkow, N.D. and Collins, F.S. (2017). The role of science in addressing the opioid crisis. *New England Journal of Medicine* 377 (4): 391–394.

10 Shipton, E.A., Shipton, E.E., and Shipton, A.J. (2018 Jun). A review of the opioid epidemic: what do we do about it? *Pain and Therapy* 7 (1): 23–36.

11 U.S. Department of Justice, Drug Enforcement Administration (2015). National Drug Threat Assessment Summary; DEA-DCT-DIR-008-16. https://www.dea.gov/sites/default/files/2018-07/2015%20NDTA%20Report.pdf (accessed 30 January 2019).

12 United Nations Office on Drugs and Crime, UNODC. Operation Lionfish III highlighted transnational organized crime threat: supported by CRIMJUST. https://www.unodc.org/unodc/en/drug-trafficking/crimjust/news/operation-lionfish-iii-highlighted-transnational-organized-crime-threat--supported-by-crimjust.html (accessed 30 January 2019).

13 Mercolini, L. and Protti, M. (2016). Biosampling strategies for emerging drugs of abuse: towards the future of toxicological and forensic analysis. *Journal of Pharmaceutical and Biomedical Analysis* 130: 202–219.

14 Jiang, T., Xu, Q., Zhang, H. et al. (2018). Improving the performances of a "brick mass spectrometer" by quadrupole enhanced dipolar resonance ejection from the linear ion trap. *Analytical Chemistry* 90 (19): 11671–11679.

15 Guo, M., Li, D., Cheng, Y. et al. (2018). Performance evaluation of a miniature magnetic sector mass spectrometer onboard a satellite in space. *European Journal of Mass Spectrometry (Chichester)* 24 (2): 206–213.

16 Pulliam, C.J., Bain, R.M., Osswald, H.L. et al. (2017). Simultaneous online monitoring of multiple reactions using a miniature mass spectrometer. *Analytical Chemistry* 89 (13): 6969–6975.

17 Jjunju, F.P., Maher, S., Li, A. et al. (2015). Analysis of polycyclic aromatic hydrocarbons using desorption atmospheric pressure chemical ionization coupled to a portable mass spectrometer. *Journal of The American Society for Mass Spectrometry* 26 (2): 271–280.

18 Costanzo, M.T., Boock, J.J., Kemperman, R.H.J. et al. (2017). Portable FAIMS: applications and future perspectives. *International Journal of Mass Spectrometry* 422: 188–196.

19 INTERPOL (2017). Millions of medicines seized in largest INTERPOL operation against illicit online pharmacies. https://www.interpol.int/News-and-media/News/2017/N2017-119 (accessed 30 January 2019).

20 EUROPOL (2017). Millions of medicines seized in largest operation against illicit online pharmacies. https://www.europol.europa.eu/newsroom/news/

millions-of-medicines-seized-in-largest-operation-against-illicit-online-pharmacies (accessed 30 January 2019).

21 FDA (2016). Counterfeit Medicine. https://www.fda.gov/drugs/resourcesforyou/consumers/buyingusingmedicinesafely/counterfeitmedicine (accessed 30 January 2019).

22 WHO (2017). *News Release*, Geneva (28 November). http://www.who.int/news-room/detail/28-11-2017-1-in-10-medical-products-in-developing-countries-is-substandard-or-falsified (accessed 30 January 2019).

23 Hanif, M., Mobarak, M.R., Ronan, A. et al. (1995). Fatal renal failure caused by diethylene glycol in paracetamol elixir: the Bangladesh epidemic. *British Medical Journal* 311: 88–91.

24 Forbes, T.P. and Sisco, E. (2018). Recent advances in ambient mass spectrometry of trace explosives. *The Analyst* 143 (9): 1948–1969.

25 Nagao, M., Takatori, T., Matsuda, Y. et al. (1997). Definitive evidence for the acute sarin poisoning diagnosis in the Tokyo subway. *Toxicology and Applied Pharmacology* 144 (1): 198–203.

26 Chai, P.R., Hayes, B.D., Erickson, T.B., and Boyer, E.W. (2018). Novichok agents: a historical, current, and toxicological perspective. *Toxicology Communications* 2 (1): 45–48.

27 Seto, Y. (2014). On-site detection as a countermeasure to chemical warfare/terrorism. *Forensic Science Review* 26 (1): 23–51.

8.9 Doping in Sport

Dorota Kwiatkowska

Polish Anti-Doping Laboratory, Warsaw, Poland

The earliest reports of anti-doping tests go as far back as the sixth century BCE and the games in Thebes, where participants were not allowed to drink wine. A priest posted at the entrance to the arena used his nose as detector, checking the breath of competing athletes.

One of the pioneers of modern anti-doping research was a Pole, the Warsaw-based pharmacist Alfons Bukowski, who performed the first anti-doping tests in 1910. He tested the saliva of racehorses and found evidence of alkaloid use. Unfortunately, the exact method he utilized is not known [1].

Efforts to combat doping have been, and still are, closely associated with the advances in analytical methods that are, in turn driven by the demands posed on anti-doping laboratories. One of such methods is mass spectrometry. It is worth mentioning here that only laboratories accredited by the World Anti-Doping Agency (WADA) can test samples collected from athletes. Today, there are 32 such facilities in the world, among them the laboratory in Warsaw [2].

The need to detect new and constantly emerging doping substances and prohibited methods coupled with the increasingly strict requirements of WADA is the driving force behind the development of analytic techniques introduced in anti-doping testing. Since 1972, when Professor Manfred Donike first used gas chromatograph coupled with a mass spectrometer in anti-doping tests during the Olympic Games in Munich, the GC-MS system has become the staple of anti-doping laboratories. The next step was tandem mass spectrometry (MS/MS) and high-resolution mass spectrometry (HRMS), followed by the introduction of liquid chromatographs coupled with spectrometers, initially with tandem mass spectrometers (LC-MS/MS) and then with ultra-high performance liquid chromatographs combined with a tandem mass spectrometer (UHPLC-MS/MS).

The anti-doping laboratories of today, however, use multiple device methods, such as gas chromatography combined with isotope ratio mass spectrometry (GC-C/IRMS), liquid chromatography with a time-of-flight detector (LC-Q-TOF), electrophoresis, chemiluminometry, fluorometry, colorimetry, immunochemistry, luminometry, flow cytometry, impedance, and cumulative electrical impulse counting. The use of such a wide array of methods is necessary due to the variety of substances and methods banned in sports, published in one of the *WADA International Standards, List of Prohibited Substances and Methods (Prohibited List)*, which is updated annually and published along with the *WADA Technical Documents* and *WADA Guidelines* that contain information on recommended testing methods and minimum required performance levels (MRPL) [3–14].

The implementation of new techniques at a testing laboratory goes hand in hand with amendments to the *Prohibited List* and WADA requirements. In 1967, the International Olympic Committee (IOC) established the Medical Commission to organize and supervise the fight against doping. The Committee published the first list of substances prohibited by IOC, which was originally limited to stimulants and illicit drugs. Through the years, the list has been significantly expanded. At that time, analysis was predominantly done using gas chromatographs and, since 1972, gas chromatographs combined with mass spectrometers. Since 2004, when WADA took over from the IOC Medical Commission as the main coordinating body of the international anti-doping system, it has been responsible for the preparation and publication of the *Prohibited List*. By lowering the limits of detection [12], WADA forced the laboratories to bring in MS/MS and HRMS detectors coupled with gas chromatographs and, by including, in the 2004, the *List* of glucocorticosteroids and two steroid substances, gestrinone and tetrahydrogestrinone (THG), induced them to perform testing using liquid chromatography combined with MS/MS [15]. Later additions of new substance classes obliged laboratories to implement additional methods, though different than previously introduced mass spectrometry.

The *Prohibited List*, updated annually with new substance classes and methods, is also open ended [16]. It contains, along with examples of specific substances in the anabolic agents' class, anabolic androgenic steroids (S.1.1), peptide hormones, growth factors, related substances and mimetics (S.2), diuretics and masking agents (S.5); stimulants: specified stimulants (S.6.b), and other substances with a similar chemical structure or similar biological effect(s). The *Prohibited List* also includes non-approved substances (S.0): "Any pharmacological substance which is not addressed by any of the subsequent sections of the List and with no current approval by any governmental regulatory health authority for human therapeutic use (e.g. drugs under pre-clinical or clinical development or discontinued, designer drugs, substances approved only for veterinary use)." In addition, some groups are appended with a note that reads "(...) including, but not limited to (...)." Such wording poses a huge challenge to laboratories, forcing them to constantly seek out new solutions. In most cases, the answer is mass spectrometry.

Systems with mass spectrometer as one of the components make it possible to detect and test for multiple compounds at the same time. This is a particularly important aspect of anti-doping, where efficiency, time, and costs are of the essence. Systems with mass spectrometers enable us to detect and test for both the free and conjugated forms of compounds. To that end, samples (usually urine) are prepared using either very simple methods, such as dilute and shoot, or more complex ones, involving various extraction procedures and hydrolysis.

The fight against doping often requires the development of new models for interpreting results, finding new metabolites of known substances, e.g. in order

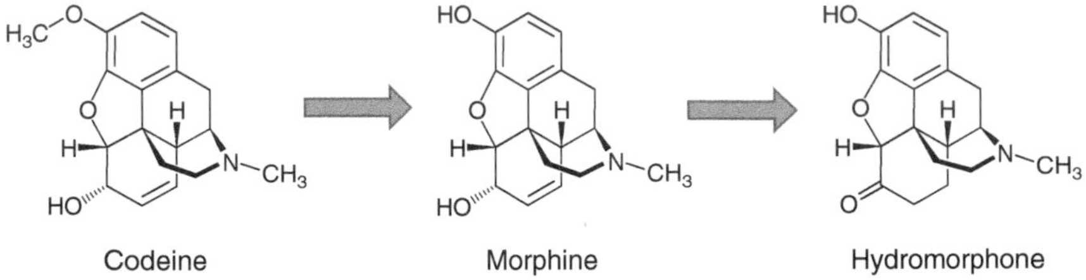

Codeine   Morphine   Hydromorphone

Figure 8.31 A simplified model of codeine metabolism.

to expand detection windows (detecting xenobiotics for a longer time following ingestion), and identifying new compounds, some of which are modifications of the already existing ones. All this is made possible by systems including mass spectrometers.

One example where the interpretation of findings requires knowledge of the compound's metabolism is morphine, which is included on the *Prohibited List*. The presence of morphine in a urine sample may result from the ingestion of codeine, which the current regulations allow [6]. Additionally, the sample may also contain hydromorphone, which is banned. The interpretation of morphine findings is precisely where the GC-MS/MS system comes in handy (Figures 8.31 and 8.32).

The most spectacular examples of substances whose detection windows were successfully extended thanks to mass spectrometry are metandienone and dehydrochlormethyltestosterone or DHCMT. By discovering new metabolites, we were able to prolong the detection time for these compounds from several days up to 19 days for metandienone (5 mg single dose) and 50 days for DHCMT (40 mg single dose) [17–18].

In addition, for some substances the simultaneous monitoring of original compounds and their metabolites also extends the so-called detection windows. A good example is clomifene, shown in Figure 8.33 [19].

An example of a new compound developed on the basis of an existing one is the aforementioned tetrahydrogestrinone (THG) and the associated BALCO scandal [20]. It was synthesized by hydrogenation of gestrinone (active substance of a drug used to treat, among others, endometriosis). Another interesting example of compound modification is delta-6-methyltestosterone, a derivative of methyltestosterone [21]. In this case finding out which specific substance was ingested is only possible by analyzing its metabolism (Figure 8.34). It turned out that the new compound had two metabolites in common with methyltestosterone and a third one that was unique to it.

Mass spectrometry makes it possible to monitor all those compounds simultaneously, which is crucial for the interpretation of findings.

Another challenge is presented by the derivatives of phenethylamine (PEA). WADA banned the use of all substituted phenethylamines [22]. Examples

Compounds at a glance

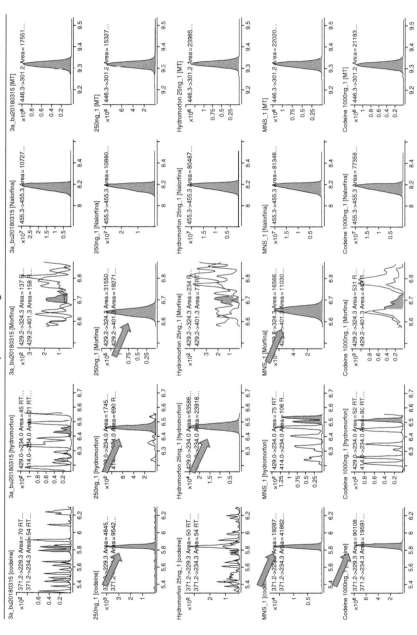

Figure 8.32 Example of the GC-MS/MS (7000D Agilent Technologies) system used to interpret morphine results.

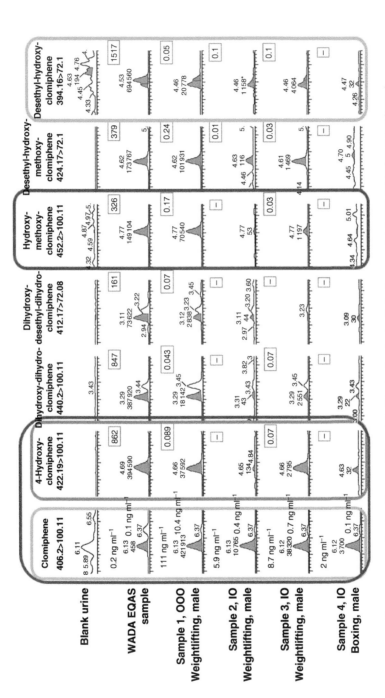

Figure 8.33 Chromatograms for clomifene and its metabolites in standard anti-doping samples (UPLC-MS/MS Premier Waters). Chromatogram in selected column 2: previously recommended, chromatograms in selected columns 1 and 4: recommended by Polish Laboratory, chromatograms in selected columns 1, 2, and 3: currently recommended by WADA.

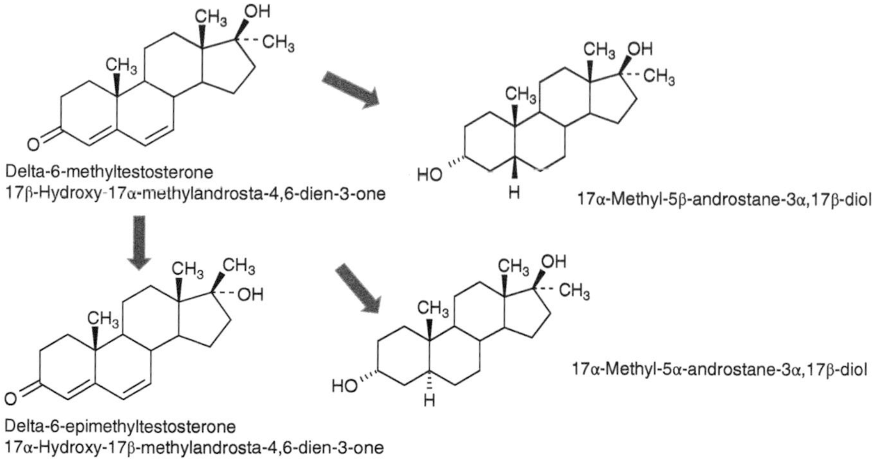

Delta-6-methyltestosterone
17β-Hydroxy-17α-methylandrosta-4,6-dien-3-one

17α-Methyl-5β-androstane-3α,17β-diol

Delta-6-epimethyltestosterone
17α-Hydroxy-17β-methylandrosta-4,6-dien-3-one

17α-Methyl-5α-androstane-3α,17β-diol

Figure 8.34 Metabolism of delta-6-methyltestosterone. The metabolites 17α-methyl-5β-androstane-3α,17β-diol and 17α-methyl-5α-androstane-3α,17β-diol are common to both delta-6-methyltestosterone and methyltestosterone.

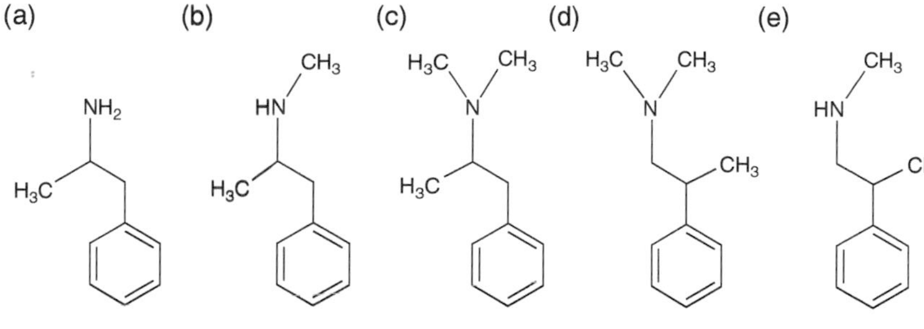

(a) (b) (c) (d) (e)

Figure 8.35 The chemical structure of amphetamine (a) and its analogues: methylamphetamine (b), dimethylamphetamine (c), *N,N*-dimethyl-2-phenylpropan-1-amine (d), and phenpromethamine (e).

include amphetamine and its beta isomer. One of the problems in interpreting results involves the fact that the spectra of those compounds are identical. However, with correct sample preparation and the use of mass spectrometry, the isomers can be differentiated [23].

With mass spectrometry, new compounds for which there are no reference materials can be discovered, such as the PEA derivatives pictured in Figure 8.35. This is accomplished by examining mass spectra that may hint toward a new compound or its metabolite whose reference material can then be synthesized. A case in point is *N,N*-dimethyl-2-phenylpropan-1-amine, which is similar to phenpromethamine (Section S6 of the *Prohibited List*) [24].

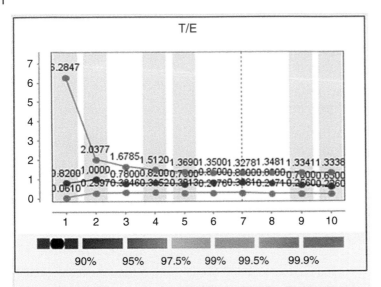

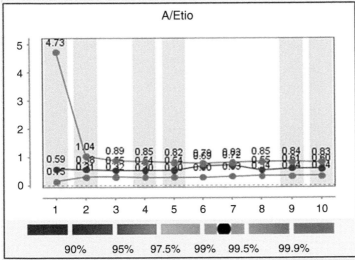

Figure 8.36 Sample steroidal passport.

The use of standardized methods specified in relevant WADA regulations means the analytical findings are consistent across different laboratories. This makes it possible for the so-called Athlete Biological Passport (ABP) to be developed. The new development is the result of WADA and, by extension, anti-doping laboratories being always on the lookout for indirect evidence of the use of prohibited substances and methods [25]. ABP is a system for the

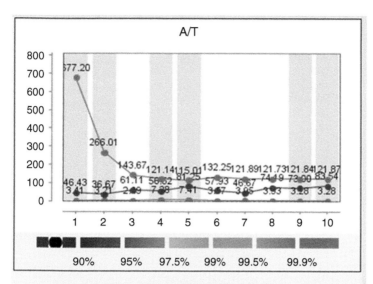

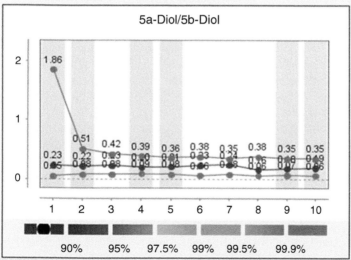

Figure 8.36 (*Continued*)

long-term assessment of biological markers in athletes. It is based on a statistical model, the so-called adaptive model, that allows for calculating quantitative ranges of analyzed markers and exceeded limits that suggest the use of doping substances or methods. One of the modules used is the steroidal passport (Figures 8.36 and 8.37). It is analyzed in the GC-MS or GC-MS/MS system and helps track multiple steroidal parameters of an athlete. Any apparent changes warrant further investigative action [7, 11]. There is another emergent

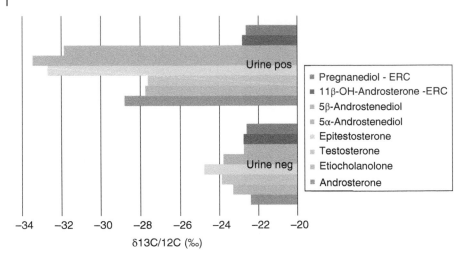

Figure 8.37 δ13C/12C [‰] values for selected compounds (steroids) in two quality control samples: urine neg (negative QC – sample contains endogenous steroids) and urine pos (sample with exogenous steroids) [26].

challenge here: to differentiate between endogenous and exogenous compounds, in this case S.1.1.b class steroids. The GC-C/IRMS method is used for that purpose [11]. The test uses gas chromatograph coupled with a combustion chamber and isotope mass spectrometer to observe the values of the 13C/12C isotope ratio (Figures 8.36 and 8.37).

As has been discussed above, mass spectrometry is a vital aspect in the operation of any anti-doping laboratory. It helps detect the majority of substances banned by WADA, often quite different in structure and properties. Without it, the efforts to stamp out the pathology that is doping would be impossible.

It should also be mentioned that anti-doping labs use the above methods to test for doping not only in humans but also in animals (horses, dogs, camels, pigeons) and in toxicology tests ordered by prosecutors, courts, military, police, and other similar authorities. Testing can be performed on a variety of materials, biological as well as others [27].

Questions

- Why mass spectrometry is an especially useful and advantageous tool of analysis in sport doping?
- Why is it important to detect metabolites of a substance by mass spectrometry in sport doping?
- What is the Athlete Biological Passport, and why is important in the battle against doping in sport?

References

1 Pokrywka, A., Gorczyca, D., Jarek, A., and Kwiatkowska, D. (2010). In memory of Alfons Bukowski on the centenary of anti-doping research. *Drug Testing and Analysis* 2: 538–541.

2 WADA. Laboratories. https://www.wada-ama.org/en/what-we-do/science-medical/laboratories (accessed 1 May 2018).

3 WADA. Prohibited list 2018. https://wada-main-prod.s3.amazonaws.com/resources/files/wada-2018-prohibited-list-en.pdf (accessed 1 May 2018).

4 WADA. Technical Document TD2017BAR Blood Analytical Requirements for the Athlete Biological Passport Ver. 1.0 January 2017. https://www.wada-ama.org/sites/default/files/resources/files/wada_td2017bar_blood_analysis_requirements_en.pdf (accessed 1 May 2018).

5 WADA. Technical Document TD2018CG/LH Reporting & Management of Urinary Human Chorionic Gonadotrophin (hCG) and Luteinizing Hormone (LH) Findings in Male Athletes Ver. 1.0 March 2018. https://www.wada-ama.org/sites/default/files/resources/files/td2018cglh_v1_en.pdf (accessed 1 May 2018).

6 WADA. Technical Document TD2018DL Decision Limits for the Confirmatory Quantification of Threshold Substances Ver. 1.0 March 2018. https://www.wada-ama.org/sites/default/files/resources/files/td2018dl_v1_en.pdf (accessed 1 May 2018).

7 WADA. Technical Document – TD2016EAAS Endogenous Anabolic Androgenic Steroids Measurement and Reporting Ver. 1.0 January 2016. https://www.wada-ama.org/sites/default/files/resources/files/wada-td2016eaas-eaas-measurement-and-reporting-en.pdf (accessed 1 May 2018).

8 WADA. Technical Document TD2014EPO Harmonization of Analysis and Reporting of Erythropoiesis Stimulating Agents (ESAs) by Electrophoretic Techniques. Ver. 1.0 September 2014. https://www.wada-ama.org/sites/default/files/resources/files/WADA-TD2014EPO-v1-Harmonization-of-Analysis-and-Reporting-of-ESAs-by-Electrophoretic-Techniques-EN.pdf (accessed 1 May 2018).

9 WADA. Technical Document TD2015GH Human Growth Hormone (hGH) isoform differential immunoassays for doping control analyses Ver. 1.0 September 2015. https://www.wada-ama.org/sites/default/files/resources/files/wada_td2015gh_hgh_isoform_diff_immunoassays_en.pdf (accessed 1 May 2018).

10 WADA. Technical Document TD2015IDCR Minimum Criteria for Chromatographic-Mass Spectrometric Confirmation of the Identity of Analytes for Doping Control Purposes Ver. 1.0 September 2015. https://www.wada-ama.org/sites/default/files/resources/files/td2015idcr_-_eng.pdf (accessed 1 May 2018).

11 WADA. Technical Document TD2016IRMS Detection of synthetic forms of Endogenous Anabolic Androgenic Steroids by GC-C-IRMS Ver. 1.0 January 2016. https://www.wada-ama.org/sites/default/files/resources/files/wada-td2016irms-detection_synthetic_forms_eaas_by_irms-en.pdf (accessed 1 May 2018).

12 WADA. Technical Document TD2018MRPL Minimum Required Performance Levels for Detection and Identification of Non-Threshold Substances Ver. 1.0 January 2018. https://www.wada-ama.org/sites/default/files/resources/files/td2018mrpl_v1_finaleng.pdf (accessed 1 May 2018).

13 WADA. Technical Document TD2017NA Harmonization of Analysis and Reporting of 19-Norsteroids Related to Nandrolone Ver. 1.0 September 2017. https://www.wada-ama.org/sites/default/files/resources/files/wada-td2017na-en_0.pdf (accessed 1 May 2018).

14 WADA. Guidelines: Human Growth Hormone (hGH) Biomarkers Test. Ver. 2.0 April 2016. https://www.wada-ama.org/sites/default/files/resources/files/wada-guidelines-for-hgh-biomarkers-test-v2.0-2016-en.pdf (accessed 1 May 2018).

15 WADA. Prohibited list 2004. https://wada-main-prod.s3. amazonaws.com/resources/files/WADA_Prohibited_List_2004_EN.pdf (accessed 1 May 2018).

16 Pokrywka, A., Kwiatkowska, D., Kaliszewski, P., and Grucza, R. (2010). Some aspects concerning modifications of the List of Prohibited Substances and Methods in sport. *Biology of Sport* 27: 307–314.

17 Schänzer, W., Geyer, H., Fußhöller, G. et al. (2006). Mass spectrometric identification and characterization of a new long-term metabolite of metandienone in human urine. *Rapid Communications in Mass Spectrometry* 20 (15): 2252–2258.

18 Sobolevsky, T. and Rodchenkov, G. (2012). Detection and mass spectrometric characterization of novel long-term dehydrochloromethyltestosterone metabolites in human urine. *The Journal of Steroid Biochemistry and Molecular Biology* 128 (3–5): 121–127.

19 Chołbinski, P., Wicka, M., Turek-Lepa, E. et al. (2013). Clomiphene: how about including the parent compound in screening procedures? In: *Recent Advances in Doping Analysis (21)* (ed. W. Schänzer, M. Thevis, H. Geyer and U. Mareck), 94–97. Köln: Sportverlag Strauss.

20 Catlin, D.H., Sekera, M.H., Ahrens, B.D. et al. (2004). Tetrahydrogestrinone: discovery, synthesis, and detection in urine. *Rapid Commun Mass Spectrom* 18: 1245–1249.

21 Parr, M.K., Pokrywka, A., Kwiatkowska, D., and Schänzer, W. (2011). Ingestion of designer supplements produced positive doping cases unexpected by the athletes. *Biology of Sport* 28: 153–157.

22 WADA. Prohibited List 2015. https://www.wada-ama.org/sites/default/files/resources/files/wada-2015-prohibited-list-en.pdf (accessed 1 May 2018).

23 Chołbiński, P., Wicka, M., Kowalczyk, K. et al. (2014). Detection of β-methylphenethylamine, a novel doping substance, by means of UPLC/MS/MS. *Analytical and Bioanalytical Chemistry* 406: 3681–3688.

24 Kwiatkowska D., Wójtowicz M., Jarek A., Goebel C., Chajewska K., Turek-Lepa E., Pokrywka A., Kazlauskas R.: N,N-dimethyl-2-phenylpropan-1-amine: new designer agent found in athlete urine and nutritional supplement. *Drug Testing and Analysis* 2015,7(4): 331–335.

25 WADA. Athlete Biological Passport Operating Guidelines and Compilation of Required Elements Ver. 6.0 January 2017. https://www.wada-ama.org/sites/default/files/resources/files/guidelines_abp_v6_2017_jan_en_final.pdf (accessed 1 May 2018).

26 Bulska, E., Gorczyca, D., Zalewska, I. et al. (2015). Analytical approach for the determination of steroid profile of humans by gas chromatography isotope ratio mass spectrometry aimed at distinguishing between endogenous and exogenous steroids. *Journal of Pharmaceutical and Biomedical Analysis* 106: 159–166.

27 ISL. International Standard for Laboratories Ver. 9.0 June 2016. https://www.wada-ama.org/en/resources/laboratories/international-standard-for-laboratories-isl (accessed 1 May 2018).

8.10   Miniaturization in Mass Spectrometry

Marek Smoluch[1] and Jerzy Silberring[1,2]

[1]Department of Biochemistry and Neurobiology, Faculty of Materials Science and Ceramics, AGH University of Science and Technology, Kraków, Poland
[2]Centre of Polymer and Carbon Materials, Polish Academy of Sciences, Zabrze, Poland

Since mass spectrometry has become one of the most important analytical techniques, constant progress has been made in improving instrument's parameters and decreasing their size. On the one hand, it is important to maintain the ability to analyze the smallest amount of substances and on the other to perform such analyses not only in a specialized laboratory but also outside in the field as portable devices. Implementing these expectations would greatly increase the potential for effective usage of mass spectrometry in many areas. The construction of a portable mass spectrometer has become a matter of time. The biggest problem to overcome when designing this type of instrument is to provide a high vacuum. This requirement implies a need for efficient pumps that do not fit to lightweight devices. Recently developed miniature turbopumps from Creare Inc. weighs only 0.5 kg and requires below 18 W to work. The second essential issue is to provide continuous supply of gas (helium or nitrogen), which also involves additional cylinders with compressed gas. All other elements of portable mass spectrometer have to be miniaturized. This includes mass analyzers, which, in the case of miniaturized systems, are usually time of flight or ion trap. For a device to be considered as portable, its total weight should not exceed approximately 20 kg, which is the mass that allows the device to be handled by one person. For several years, more and more of these types of spectrometers appear on the market, which are used in many interesting applications described below.

The miniaturized 25 kg Mini 12 mass spectrometer (earlier versions of this instrument were even below 10 kg) was used for clinical trials (blood analysis) [1]. The design of the instrument allows the use of nanoelectrospray ionization, APCI, discontinuous atmospheric pressure interface (DAPI), or paper spray (Figure 8.38). The latter technique involves soaking the chromatographic paper with the analyte and its direct ionization (Figure 8.39). By using the ion trap as an analyzer, the spectrometer also provided the possibility of multistage fragmentation. The maximum scan range was 900 m/z units. A very good detection limit of 7.5 ng ml^{-1} for this instrument was reported.

Paper spray was used for blood analysis applied from the silica-coated paper and dried *in situ* [2]. Both qualitative and quantitative analyses were performed. The subjects of this research involved drugs, such as verapamil, citalopram, lidocaine, and amitriptyline. The limit of detection of 10–20 ng ml^{-1} was reached, which is sufficient for this type of analysis. Another device with a total weight of only 12 kg and low temperature plasma (LTP) ionization source was

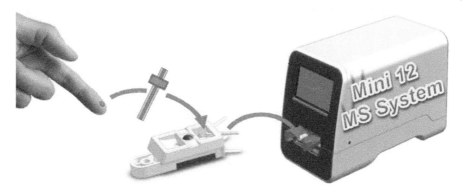

Figure 8.38 Analysis of drugs in blood by Mini 12 system. Blood quantitatively taken up with capillary, transferred to a sample cartridge and inserted to MS system for automatic analysis. *Source*: Reprinted with permission from American Chemical Society [1].

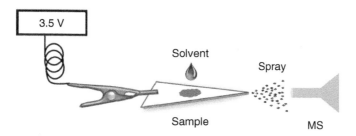

Figure 8.39 Ionization of the sample applied and separated on chromatographic paper (paper spray).

used for detection of warfare agents, explosives, or illegal substances [3]. Nanogram quantities of the substance were detected directly on the test surfaces. The maximum range of the analyzed masses is 925 Da, and the operating time of the instrument is 1.5 hours as a stand-alone battery-driven device. Traces of explosives were also successfully detected on the surface by means of another LTP plasma source spectrometer. To increase the sensitivity of the analyses, the samples were heated using a halogen lamp. As a result, using this device, it was possible to detect substances at the level of 10–20 pg. LTP ionization is relatively common in portable spectrometers. Such a moderately inexpensive device can continuously hold the plasma source for about two hours using, for example, a 7.4 V lithium battery. A miniature helium cylinder (discharge gas) is sufficient for eight hours of operation. Warfare agents such as phosgene, mustard gas, and hydrogen cyanide were detected using a portable mass spectrometer equipped with electron cyclotron resonance (ECR) ion source. Identification was made based on the characteristic isotopic pattern and identification of specific fragment ions. Sensitivity at ppb level was sufficient to obtain quantitative results. This level of sensitivity is well below the potential lethal dose of those warfare agents, which guarantees the safety of the persons exposed. Direct analysis of quaternary ammonium salts in oil was made using the previously mentioned paper spray method. Salts of this type are commonly used as active ingredients to manufacture corrosion inhibitors. In this case, the detection limit of $1\,ng\,ml^{-1}$ was reached. A triple quadrupole mass spectrometer and the electrospray ion source with the unique sample inlet system were used for the analysis of trace amounts ($ppt\,ppb^{-1}$) of substances (chlorophenols, triclosan, etc.) from complex matrices as beer, natural water, sewage, or plant tissue [4]. The portable spectrometer can also be used to verify the progress and correctness of chemical synthesis. For this purpose, the instrument was combined with a preparative flow-through chemical synthesis system. The portable spectrometer was also used in combination with a gas chromatograph to analyze a mixture of polychlorinated biphenyls. At a high resolution on the order of 10 000, a detection limit of 1 ppb was reached. For pheromone analysis in the air, a portable system consisting of a sample inlet system, a gas chromatograph, a mass spectrometer, and an electroantennographic detector was used. Pheromones were detected in the air at concentrations of $3\,ng\,m^{-3}$. The portable mass spectrometer was used for the quantitative analysis of a substance of abuse in the urine [5]. A small pump was applied for wetting the previously dried sample, which was then analyzed by nanoelectrospray ionization. In such a way, samples were quantified in just 15 minutes, and for cocaine the analysis has reached the limit of quantitation of $40\,ng\,ml^{-1}$. The entire machine weighs 25 kg and can be successfully used as a portable analytical instrument. A miniature version of DESI has been used for the analysis of illicit synthetic cathinones and other psychoactive substances. Verification of structures was achieved by fragmentation in the ion trap. The portable time-of-flight spectrometer with laser ablation was designed for chemical analysis of the planetary surface. It can perform elemental and isotope analysis on solid surfaces with high sensitivity and sampling resolution. Taking into account the rigorous parameters control such as laser characteristics, ion source optics, and sample positioning, measurements can be performed with high repeatability. Another instrument was designed in such a way that the length of the time-of-flight analyzer was only 12 cm. This ion source allowed for analysis of substances at atmospheric pressure. Another miniaturized spectrometer of impressive size (20 cm length) with time-of-flight analyzer was used for the study of polycyclic aromatic hydrocarbons, which are an important class of compounds in the Earth and other planets [6]. Probably the lightest portable instrument currently present on the market is palm portable MS from Samyang Chemical Co. (South Korea) with total weight of only 1.5 kg. However this instrument has limited applicability – no MS/MS option, mass range up to 300 m/z, and resolution 150. Miniature mass

Table 8.7 Portable mass spectrometer producers.

Company name	Website	Ionization
1st Detect (United States)	www.1stdetect.com	EI
908 Devices	908devices.com	Internal ionization
Advion (United States)	advion.com	ESI, APCI
Aston Labs (Purdue University)	aston.chem.purdue.edu	MIMS, EI, APCI, ESI, DESI, LTP
BaySpec (United States)	www.bayspec.com	ADI compatible
INFICON (Switzerland)	www.inficon.com/en	GCEI
FLIR Systems	www.flir.com	GCEI
Microsaic (United Kingdom)	www.microsaic.com	SPME, EI
PerkinElmer	www.perkinelmer.com	SPME, GCEI

spectrometers producers are rapidly growing due to the continuously increasing need for such instruments in the market. The list of some leading producers is shown in the Table 8.7.

Questions

- What are the purposes and needs to have miniaturization in mass spectrometry?
- What are the main problems encountered in designing a miniaturized mass spectrometer?
- What are the limits of a miniaturized mass spectrometer?
- Give some examples of applications of miniaturized mass spectrometers.

References

1 Li, L., Chen, T.C., Ren, Y. et al. (2014). Mini 12, miniature mass spectrometer for clinical and other applications-introduction and characterization. *Analytical Chemistry* 86: 2909–2916.
2 Zhang, Z., Xu, W., Manicke, N.E. et al. (2012). Silica coated paper substrate for paper-spray analysis of therapeutic drugs in dried blood spots. *Analytical Chemistry* 84: 931–938.
3 Wiley, J.S., Shelley, J.T., and Cooks, R.G. (2013). Handheld low-temperature plasma probe for portable "point-and-shoot" ambient ionization mass spectrometry. *Analytical Chemistry* 85: 6545–6552.

4 Duncan, K.D., Willis, M.D., Krogh, E.T., and Gill, C.G. (2013). A miniature condensed-phase membrane introduction mass spectrometry (CP-MIMS) probe for direct and on-line measurements of pharmaceuticals and contaminants in small, complex samples. *Rapid Communications in Mass Spectrometry* 27: 1213–1221.

5 Kirby, A.E., Lafrenière, N.M., Seale, B. et al. (2014). Analysis on the go: quantitation of drugs of abuse in dried urine with Digital microfluidics and miniature mass spectrometry. *Analytical Chemistry* 86: 6121–6129.

6 Getty, S.A., Brinckerhoff, W.B., Cornish, T. et al. (2012). Compact two-step laser time-of-flight mass spectrometer for in situ analyses of aromatic organics on planetary missions. *Rapid Communications in Mass Spectrometry* 26: 2786–2790.

9

Appendix

Kinga Piechura and Marek Smoluch

Department of Biochemistry and Neurobiology, Faculty of Materials Science and Ceramics, AGH University of Science and Technology, Kraków, Poland

9.1 Pressure Units

Unit	Unit symbol	Comments
Pascal	Pa	SI unit
Bar	bar	105 Pa
Inch of mercury	mm Hg	133 322 Pa
Torr	Tr	1 Tr = 1 mm Hg (at 0 °C)
Atmosphere	atm	101 325 Pa = 760 mm Hg
Technical atmosphere	at	98 066.5 Pa
Pound per square inch	psi	6 895 Pa

9.2 Most Commonly Detected Fragments Generated by Electron Impact (EI) Ionization

No.	Fragment mass (*m/z*)	Fragment
1	1.0078	$H^{\bullet}$
2	15.0235	$CH_3^{\bullet}$
3	17.0027	$HO^{\bullet}$

(Continued)

Mass Spectrometry: An Applied Approach, Second Edition. Edited by Marek Smoluch, Giuseppe Grasso, Piotr Suder, and Jerzy Silberring.
© 2019 John Wiley & Sons, Inc. Published 2019 by John Wiley & Sons, Inc.

(Continued)

No.	Fragment mass (*m/z*)	Fragment
4	18.0106	H_2O
5	18.9984	$F^{\bullet}$
6	20.0062	HF
7	26.0031	$^{\bullet}C\equiv N$
	26.0157	$CH\equiv CH$
8	27.0109	HCN
	27.0235	$CH_2=CH^{\bullet}$
9	27.9949	CO
	28.0061	N_2 (from the air)
	28.0187	$(HCN+H)$
	28.0313	$CH_2=CH_2$
10	29.0027	$^{\bullet}CHO$
	29.0391	$CH_3CH_2^{\bullet}$
11	29.9980	$NO^{\bullet}$
	30.0106	CH_2O
	30.0344	$NH_2CH_2^{\bullet}$
12	31.0184	$^{\bullet}OCH_3, ^{\bullet}CH_2OH$
	31.0422	CH_3NH_2
13	31.9898	O_2 (from the air)
	32.0262	CH_3OH
14	32.9799	$HS^{\bullet}$
	33.0340	$(^{\bullet}CH_3$ and $H_2O)$
15	33.9877	H_2S
16	34.9689	$Cl^{\bullet}$
17	35.9767	HCl
18	36.9845	$(HCl+H)$
19	39.0235	C_3H_3
20	40.0313	$CH_3C\equiv CH$
21	41.0391	$CH_2=CHCH_2^{\bullet}$
22	42.0106	$CH_2=C=O$
	42.0218	$NCNH_2$
	42.0470	$\triangle, CH_2=CHCH_3$
23	43.0058	HCNO
	43.0184	$CH_3C^{\bullet}=O$
	43.0548	$C_3H_7^{\bullet}$

No.	Fragment mass (*m/z*)	Fragment
	43.9898	CO_2
	44.0011	N_2O
24	44.0136	$^\bullet CONH_2$
	44.0262	$CH_2=CHOH$
	44.0500	$^\bullet NHCH_2CH_3$
	44.9977	$^\bullet COOH$
25	45.0340	$CH_3CH^\bullet OH$, $CH_3CHO + H^\bullet$, $CH_3CH_2O^\bullet$
	45.0578	$CH_3CH_2NH_2$
26	45.9929	$^\bullet NO_2$
	46.0419	$(H_2O$ and $CH_2=CH_2)$, CH_3CH_2OH
27	46.9955	$CH_3S^\bullet$
28	48.0034	$(CH_3S^\bullet + H^\bullet)$
29	48.9845	$^\bullet CH_2Cl$
30	52.0313	C_4H_4
31	53.0391	$C_4H_5^\bullet$
32	54.0470	$CH_2=CH-CH=CH_2$
33	55.0184	$CH_2=CHC^\bullet=O$
	55.0548	$CH_2=CH-CH^\bullet CH_3$
34	55.9898	$2CO$
	56.0626	$CH_2=CHCH_2CH_3$, $CH_3CH=CHCH_3$
35	57.0340	$C_2H_5C^\bullet=O$
	57.0704	$C_4H_9^\bullet$
	57.9751	$NCS^\bullet$
36	57.9929	$(NO + CO)$
	58.0419	CH_3COCH_3
	58.0657	$C_2H_5C^\bullet HNH_2$
37	58.9955	![structure: H-S• on cyclopropane ring]
	59.0133	$CH_3OC^\bullet=O$
	59.0371	CH_3CONH_2
38	60.0086	$^\bullet CH_2ONO$
	60.0575	$(CH_3OH + C_2H_4)$

(Continued)

(Continued)

No.	Fragment mass (m/z)	Fragment
39	61.0112	$^{\bullet}CH_2SCH_3$, , $CH_3CH_2S^{\bullet}$
40	62.0190	(H_2S and $CH_2{=}CH_2$)
41	63.0002	$^{\bullet}CH_2CH_2Cl$
42	63.9441	S_2
	63.9619	SO_2
	64.0313	C_5H_4
43	67.0548	$C_5H_7^{\bullet}$
44	68.0626	$CH_2{=}C(CH_3){-}CH{=}CH_2$
45	68.9952	$CF_3^{\bullet}$
	69.0704	$C_5H_9^{\bullet}$
46	71.0497	$C_3H_7C^{\bullet}{=}O$
	71.0861	$C_5H_{11}^{\bullet}$
47	73.0290	$CH_3CH_2O{-}C^{\bullet}{=}O$
48	74.0732	($C_2H_5OH + C_2H_4$)
49	75.0235	$C_6H_3^{\bullet}$
50	75.9441	CS_2
	76.0313	C_6H_4
51	77.0391	$C_6H_5^{\bullet}$
52	78.0344	$^{\bullet}C_5H_4N$
	78.0470	C_6H_6
53	78.9183	Br
	79.0422	C_5H_5N
54	79.9262	HBr
55	81.0704	, $C_6H_9^{\bullet}$
56	82.9455	$^{\bullet}CHCl_2$
	82.9955	
57	84.9657	$^{\bullet}CClF_2$
	85.1017	$C_6H_{13}^{\bullet}$
58	90.0191	$^{\bullet}CH_2CH_2ONO_2$
59	93.0704	$^{\bullet}C_7H_9$

No.	Fragment mass (*m/z*)	Fragment
60	97.0112	
61	99.9936	$CF_2{=}CF_2$
	100.1126	$C_5H_{11}C^{\bullet}HNH_2$
62	105.0340	
63	118.9920	$CF_3{-}CF_2^{\bullet}$
64	121.1017	$C_9H_{13}^{\bullet}$
65	122.0368	C_6H_5COOH
66	126.9045	$I^{\bullet}$
67	127.9123	HI

9.3 Trypsin Autolysis Products

Monoisotopic mass (Da)	Fragment	Missing cleavages
5191.4975	1–49	1
4551.1318	1–43	0
4417.1980	50–89	1
3686.8040	100–136	1
3670.5916	150–186	1
3338.6467	92–125	1
3195.4795	140–170	1
2910.3178	171–200	1
2803.4220	44–69	1
2552.2483	100–125	0
2514.3384	70–91	1
2321.2070	187–208	1
2273.1594	70–89	0
2193.9943	150–170	0
2163.0564	50–69	0
1999.0469	201–217	1

(Continued)

(Continued)

Monoisotopic mass (Da)	Fragment	Missing cleavages
1725.8628	209–223	1
1497.7617	126–139	1
1495.6151	171–186	0
1433.7205	187–200	0
1364.6912	137–149	1
1153.5735	126–136	0
1111.5605	209–217	0
1046.5953	90–99	1
1020.5030	140–149	0
906.5043	201–208	0
805.4162	92–99	0

Active trypsin autolysis products (Swiss-Prot database for bovine trypsin – Swiss-Prot no. P00760; PeptideMass tool was used for virtual digestion and fragment mass calculations).

9.4 Proteolytic Enzymes for Protein Identification

		Amino acid residue		
Enzyme name	EC number	Digestion place after following amino acid	Interference	Side of bond digestion
1	2	3	4	5
Trypsin	3.4.21.4	Arg, Lys	Pro	C-terminal
Arg-C endoproteinase (clostripain)	3.4.22.8	Arg	Pro	C-terminal
Asp-N endoproteinase (metalloendopeptidase peptidyl-Asp endoproteinase)	3.4.24.33	Asp, Glu	—	N-terminal
Chymotrypsin	3.4.21.1	Phe, Tyr, Trp, Leu	Pro	C-terminal
Lys-C endoproteinase	3.4.21.50	Lys	Pro	C-terminal
Pepsin	3.4.23.1	Phe, Leu, Tyr, Trp		C-terminal
Glu-C endoproteinase (*Staphylococcus* protease V8)	3.4.21.19	Glu, Asp	Pro	C-terminal
Cyanogen bromide (CNBr)	—	Met	—	C-terminal

The fifth column indicates on which side, from the amino acid residue indicated, the enzyme cleaves the peptide bond (C-terminus = carboxyl side of the indicated amino acid, N-terminal = amino side of the indicated amino acid). In the fourth column, amino acids have been entered, which are adjacent to the digesting sites and interfere with the digestion of the protein chain by a given enzyme.

9.5 Molecular Masses of Amino Acid Residues

			Mass (Da)	
Name	Code	Amino acid residue	Monoisotopic	Average
Alanine	Ala (A)	CH_3 —NH—CH—CO—	71.037 11	71.0790
Arginine	Arg (R)	H_2C—$(CH_2)_2$—NH—C—NH_2 —NH—CH—CO— NH	156.101 11	156.1881
Asparagine	Asn (N)	CH_2—$CONH_2$ —NH—CH—CO—	114.042 93	114.1041
Aspartic acid	Asp (D)	CH_2—COOH —NH—CH—CO—	115.026 94	115.0887
Cysteine	Cys (C)	CH_2—SH —NH—CH—CO—	103.009 18	103.1434
Phenylalanine	Phe (F)	H_2C —NH—CH—CO—	147.068 41	147.1771
Glycine	Gly (G)	—NH—CH_2—CO—	57.021 46	57.0520
Glutamine	Gln (Q)	CH_2—CH_2—$CONH_2$ —NH—CH—CO—	128.058 58	128.1310
Glutamic acid	Glu (E)	CH_2—CH_2—COOH —NH—CH—CO—	129.042 59	129.1157

(Continued)

(Continued)

Name	Code	Amino acid residue	Mass (Da)	
			Monoisotopic	Average
Histidine	His (H)		137.058 91	137.1415
Isoleucine	Ile (I)	H₃C—CH—CH₂—CH₃ / —NH—CH—CO—	113.084 06	113.1600
Leucine	Leu (L)	H₂C—CH(CH₃)₂ / —NH—CH—CO—	113.084 06	113.1600
Lysine	Lys (K)	H₂C—(CH₂)₃—NH₂ / —NH—CH—CO—	128.094 96	128.1747
Methionine	Met (M)	H₂C—CH₂—S—CH₃ / —NH—CH—CO—	131.040 48	131.1974
Proline	Pro (P)	—N—HC—CO—	97.052 76	97.1170
Serine	Ser (S)	CH₂—OH / —NH—CH—CO—	87.032 03	87.0783
Threonine	Thr (T)	HO—CH—CH₃ / —NH—CH—CO—	101.047 68	101.1053
Tryptophan	Trp (W)	H₂C / —NH—CH—CO—	186.079 31	186.2139
Tyrosine	Tyr (Y)	H₂C——OH / —NH—CH—CO—	163.063 33	163.1764
Valine	Val (V)	H₃C—CH—CH₃ / —NH—CH—CO—	99.068 41	99.1330

9.6 Molecular Masses of Less Common Amino Acid Residues

Name	Code	Amino acid residue	Mass (Da)	
			Monoisotopic	Average
2-Aminobutyric acid	2-Aba	$-NH-CH-CO-$; CH_2-CH_3	85.05276	85.1060
Aminoethylcysteine	AECys	$-NH-CH-CO-$; $CH_2-S-(CH_2)_2-NH_2$	146.05138	146.2121
2-Aminoisobutyric acid	Aib	$-NH-C-CO-$; CH_3 ; CH_3	85.05276	85.1060
Carboxymethylcysteine	Cmc	$-NH-CH-CO-$; CH_2-S-CH_2-COOH	161.01466	161.1800
Cysteic acid	Cys(O$_3$H)	$-NH-CH-CO-$; CH_2-SO_3H	150.99393	151.1413
Dehydroalanine	Dha	$-NH-C-CO-$; CH_2	69.02146	69.0631
2-Dehydro-2-aminobutyric acid	Dhb	$-NH-C-CO-$; $CH-CH_3$	83.03711	83.0901

(Continued)

(Continued)

Name	Code	Amino acid residue	Mass (Da)	
			Monoisotopic	Average
4-Carboxyglutamic acid	Gla	COOH | CH₂—CH—COOH | —NH—CH—CO—	173.032 42	173.1253
Homocysteine	Hcy	CH₂—CH₂—SH | —NH—CH—CO—	117.024 83	117.1704
Homoserine	Hse	CH₂—CH₂—OH | —NH—CH—CO—	101.047 68	101.1053
5-Hydroxylysine	Hyl	OH | H₂C—CH₂—CH—CH₂—NH₂ | —NH—CH—CO—	144.089 88	144.1740
4-Hydroxyproline	Hyp	OH —N—HC—CO—	113.047 68	113.1163
Isovaline	Iva	H₂C—CH₃ | —NH—C—CO— | CH₃	99.068 41	99.1330

Norleucine	Nle	$-NH-CH-CO-$ $CH_2-(CH_2)_2-CH_3$	113.08406	113.1600
Norvaline	Nva	$-NH-CH-CO-$ $H_2C-CH_2-CH_3$	99.06841	99.1330
Ornithine	Orn	$-NH-CH-CO-$ $H_2C-(CH_2)_2-NH_2$	114.07931	114.1477
2-Piperidinecarboxylic acid	Pip	$-N-CH-CO-$ (piperidine ring)	111.06841	111.1440
Pyroglutamic acid	pGlu	$-N-CH-CO-$ (pyrrolidinone ring)	111.03203	111.1004
Sarcosine	Sar	$-N-CH_2-CO-$ CH_3	71.03711	71.0790

9.7 Internet Databases

9.7.1 Literature Databases

National Center for Biotechnology Information (www.ncbi.nlm.nih.gov)
The NCBI is involved in disseminating of broadly defined biomedical information. At www.ncbi.nlm.nih.gov, there is, among others, an information retrieval system that uses several related databases. Among them, the database of biochemical literature (PubMed), online books, a digital archive of full-text articles from scientific journals, and many others can be found. Highly recommendable is Coffee Break tab, which includes the set of short, written inaccessible language articles, which are focused on the latest biological discoveries.

National Library of Medicine (www.nlm.nih.gov)
The library service allows to access to databases, such as MedlinePlus, MEDLINE/PubMed, Entrez, TOXNET, and many others. On the website, information about history of medicine can be found, as well as online exhibition and interesting projects, such as Visible Human Project that allows to study the anatomical structure of the human body.

Virtual Library ICM (vls.icm.edu.pl)
It is a website of Virtual Library of Interdisciplinary Centre for Mathematical and Computer Modelling where full-text articles of journals published by Elsevier, Springer, American Chemical Society, Ovid Biomedical Collections, and Kluwer can be found. This library includes also abstracts of articles from more than 5000 journals in the area of mathematics and natural sciences. Moreover it contains Beilstein (CrossFire plus Reactions) database of organic and bioorganic compounds, Gmelin database of nonorganic and organometallic compounds, and many others databases. Both articles and databases are mostly available only for licensed users.

Jagiellonian Library (www.bj.uj.edu.pl)
It contains full-text digital journals. Some of them are available only from computers connected to the Jagiellonian University's network. This library allows to use databases such us Medline, ScienceDirect, Science Citation Index, Springer Link, Scopus, Chemical Abstracts, Inspec, and many others.

Google Print (print.google.com)
Google Print is a base of book from greater libraries. The item of interest can be ordered, or the library of searched item can be found. Moreover Google Print offers a search of books reviews and information related to the item of interest. Some of books are available in full view (those out of copyright or that the publisher has made fully viewable). Access to the copyrighted books is subject to certain restrictions.

Google Scholar (scholar.google.com)
The server offers a simple access to scientific literature, searching of journals, articles, review of articles, citations, etc.

www.scirus.com
Science-specific search engine of sources in subject areas as mathematics and natural sciences. It is possible to searches databases, university websites, texts of scientific papers, etc. In addition to websites, files in other format, such as pdf, poscript, and Word, can be searched.

www.scopus.com
Database of abstracts and citations from peer-reviewed publications: magazines, scientific books and post-conference materials. Contains titles from Nature, Biomed Central, Springer-Verlag, and many others (even non-English titles). Provides a comprehensive overview of global research results in technology, medicine, and social sciences.

www.ingentaconnect.com
Enables searching articles in various subject areas, such as medicine, chemistry, biology, mathematics, economy, psychology, philosophy, and the arts.

9.7.2   Scientific Journals

Nature (www.nature.com)
The website of the world-renowned publisher. It offers a search of the *Nature* publications database by keywords. Some of the articles can be downloaded free of charge.

Science (www.sciencemag.org)
The "Science Magazine" website allows to search the archives of the journal. Some of the articles are available free of charge. Recently, this magazine is subscribed by the Jagiellonian University.

Proceedings of the National Academy of Sciences USA (www.pnas.org)
This website offers search of the articles published by *Proceedings of the National Academy of Sciences* (PNAS). The access of these articles is free of charge from the Internet university network.

ScienceDirect (www.sciencedirect.com)
The website allows to access to over 2000 Elsevier journals in technology and medicine science.

BioMed Central (www.biomedcentral.com)
Provides free access to specialist biological and medical journals. Registration is required.

Springer Link (www.link.springer.com)
Access to articles related to chemistry, technology, medicine, physics, mathematics, technical sciences, computational techniques, economy, law, and sociology, published by the Springer-Verlag. Full version of articles is available from the internal university network with a valid license. For users without license, only abstracts of articles are available.

9.7.2.1 Journals Related to Mass Spectrometry

Journals of Mass Spectrometry
https://www.onlinelibrary.wiley.com/journal/10969888c

Mass Spectrometry Reviews
https://www.onlinelibrary.wiley.com/journal/10982787

RCMS
https://www.onlinelibrary.wiley.com/journal/10970231

International Journal of Mass Spectrometry
http://ees.elsevier.com/ijms

Analytical Chemistry
http://pubs.acs.org/journals/ancham

9.7.3 Bioinformatics Databases

9.7.3.1 Protein Databases

Swiss-Prot, https://web.expasy.org/docs/swiss-prot_guideline.html
It provides access to a variety of data, dedicated to proteins. This database contains information about protein sequence, structural motifs, protein domain, and posttranslational modifications of function of the protein. All data is reviewed and verified manually.

TrEMBL, www.expasy.org/sprot
Database generated by computer translation of genetic information (nucleotide sequences) from EMBL database into amino acid sequences. It also contains additional information and descriptions, generated automatically by computer programs.

PIR, http://pir.georgetown.edu
Protein Information Resource (PIR), database of Georgetown University. It is a resource of databases and tools needed for their analysis, among others, the Protein Sequence Database (PSD), the first international database created from the Atlas of Protein Sequence and Structure, published from 1965 to 1978.

UniProt, http://www.uniprot.org
Universal Protein Resource (UniProt), a base of protein sequences and functions. The database is assembly of several other databases such as Swiss-Prot, TrEMBL, and PIR-PSD.

NCBI nr, ftp://ftp.ncbi.nih.gov/blast/db
Database of NCBI, which contains non-redundant amino acid sequences from other significant protein databases.

IPI, ftp://ftp.ebi.ac.uk/pub/databases/IPI
International Protein Index (IPI), a source of data describing the proteome of higher eukaryotic organisms.

9.7.3.2 Database of Structures and Functions of Protein
Protein Data Bank, www.rcsb.org/pdb/home/home.do
Database of three-dimensional structures of proteins, nucleic acids, protein–nucleic acid complexes, and viruses. It contains data related to structure of interest, such as structural analogues, visualizations of the structures, or links to external sources.

InterPro, www.ebi.ac.uk/interpro
Database of protein families, domains, structural motifs, and active sites of proteins. It is a source that provides prediction of the function or structure of newly developed protein, based on the analogy of its sequence to the sequence of proteins already known.

SignalP, www.cbs.dtu.dk/services/SignalP
One of the CBS server that allows to predict the presence and location of signal peptides in specific amino acid sequences.

Human Protein Reference Database, www.hprd.org
It contains information about interactions between proteins found in human body.

PROSITE, prosite.expasy.org
Database that defines the function of protein and identify which family of protein the new one belongs to. The basis for identification is the sequence obtained directly from the translation of nucleic acid sequence into amino acid sequence.

MEROPS, merops.sanger.ac.uk
Protein database that includes data about proteases and their inhibitors. It is a source of information about classification and protein naming. Moreover,

it contains external links to literature sources or structure of searched protein, if such structure is known. It is also possible to find the gene coding for the protease or enzyme that is the most suitable to digest specific substrate.

9.7.3.3 Other Databases

Entrez, www.ncbi.nlm.nih.gov
Data search system that integrates the databases subjected to NCBI, containing protein and nucleotide sequences, macromolecules structures, genomes, and MEDLINE database.

METLIN, metlin.scripps.edu/index.php
Metabolite database that supports research and metabolite identification, based on data received from mass spectrometry, liquid chromatography–mass spectrometry system, and Fourier transform mass spectrometry. Each metabolite is linked to the Kyoto Encyclopedia of Genes and Genomes.

BRENDA Enzymes, www.brenda-enzymes.org
It contains data about all known enzymes with their detailed characteristics. The database can be searched with many parameters. It uses unified enzymes names accordingly to the EC record.

ChemSpider, www.chemspider.org
A chemical structures database that contains information on more than 63 million molecules (data from 2018). Compounds can be searched by mass, chemical composition, or partial structure. Highly useful tool for the chemists.

9.7.4 Bioinformatics Tools

Mascot, www.matrixscience.com/search_form_select.html
Search engine that provides identification of peptides and proteins using mass spectrometry data. It uses protein sequence databases (MSDB, Swiss-Prot, NCBI nr, and others)

BLAST, FASTA, ClustalW
Programs for analyzing and comparing the nucleotide and amino acid sequences, also finding homology or sequence similar to the given one. These programs are available free of charge on many websites, among others (www.abi.ac.uk/services).

ProteinProspector, prospector.ucsf.edu/prospector/mshome.htm
Server that allows to search bioinformatic databases where many tools specifically designed for mass spectrometry can be found.

ExPASy, www.expasy.org
Tools for analyze the amino acid sequences. Compute pJ/MW, tools for calculating molecular mass and isoelectric point. PeptideMass, tool for "theoretical" digestion of a protein. Also ProtParam, tool that allows to determine amino acid composition, molecular weight, half-life, hydrophobicity factor, and other parameters of searched protein or its fragment.

9.7.5   Useful Websites

www.abrf.org
The Association of Biomolecular Resource Facilities. The website with interesting discussion forum, which was set up by scientists who work within biotechnology and related sciences. Moreover, it contains Delta Mass Database of posttranslational modifications.

www.spectroscopynow.com
Subfolder in a large portal dedicated to spectroscopy. Site that offers access to interesting articles focused on proteomics, mass spectrometry, educational facts, and news.

www.separationsnow.com
Site that comprises description of analytical methods for separations. Contains information about electrophoresis, chromatography, and proteomics.

www.unimod.org
Contains comprehensive data about protein modifications, particularly for mass spectrometry applications.

www.rockefeller.edu
The site is a source of bioinformatics tools for biological macromolecules handling (e.g. ProFound, ProteinInfo, PeptideMap).

www.ionsource.com
Mass spectrometry educational resource where some noteworthy tutorials can be found.

www.phosida.de
Phosphorylation site database.

www.peptide.ucsd.edu/MassSpec
Site comprising tools for mass spectra interpretations of *de novo* sequenced peptides.

10

Abbreviations

Kinga Piechura and Marek Smoluch

Department of Biochemistry and Neurobiology, Faculty of Materials Science and Ceramics, AGH University of Science and Technology, Kraków, Poland

2D-PAGE	two-dimensional polyacrylamide gel electrophoresis
ABS	acrylonitrile butadiene styrene
AC	alternating current
ACE	angiotensin I-converting enzyme
ADI	ambient desorption/ionization
ADI-MS	ambient desorption/ionization mass spectrometry
AED	atomic emission detector
AgTFA	silver trifluoroacetate
AP-MALDI	atmospheric pressure matrix-assisted laser desorption/ionization
APB	Antimicrobial Peptide Database
APCI	atmospheric pressure chemical ionization
APGD	atmospheric pressure glow discharge
API	atmospheric pressure ionization
APPI	atmospheric pressure photoionization
ASAP	Atmospheric Solids Analysis Probe
B	magnetic analyzer
BIRD	blackbody infrared radiative dissociation
BSE	backscattered electron
CAD	collisionally activated dissociation
CAF	chemically activated fragmentation
CCD	charge-coupled device
CCT	collision cell technology
CD	conductivity detector
CE	capillary electrophoresis

CHCA	α-cyano-4-hydroxycinnamic acid
CI	chemical ionization
CID	collision-induced dissociation
CL	cathodoluminescence
CM	carboxymethyl
CMC	critical micelle concentration
CMOS	complementary metal–oxide–semiconductor
CN	casein
CZE	capillary zone electrophoresis
DAD	diode array detector
DAPCI	direct atmospheric pressure chemical ionization
DAPI	discontinuous atmospheric pressure interface
DAPPI	desorption atmospheric pressure photoionization
DART	direct analysis in real time
DBDI	dielectric barrier discharge ionization
DC	direct current
DCBI	desorption corona beam ionization
DCP	direct current plasma
DDA	data-dependent acquisition
DEAE	diethylaminoethyl
DESI	desorption electrospray ionization
DHB	2,5-dihydroxybenzoic acid
DIOS	desorption/ionization on porous silicon
DMF	dimethylformamide
DMSO	dimethyl sulfoxide
DNPH	dinitrophenylhydrazine
E	electrostatic analyzer
EC	electron capture
ECD	electron capture detector
ECD	electrochemical detector
ECR	electron cyclotron resonance
EDD	electron detachment dissociation
EI	electron impact
EI	electron ionization
EIC	extracted ion chromatogram
ELISA	enzyme-linked immunosorbent assay
EMP	electron microprobe
EOF	electroosmotic flow
ESI	electrospray ionization
EST	expressed sequence tag
ETD	electron transfer dissociation
ETV	electrothermal evaporation
FA	formic acid

FAB	fast atom bombardment
FAIMS	high field asymmetric waveform ion mobility spectrometer
FAMIS	high field asymmetric waveform ion mobility spectrometer
FAPA	flowing atmospheric pressure afterglow
FD	field desorption
FDR	false discovery rate
FID	flame ionization detector
FLD	fluorescence detector
FPD	flame photometric detector
fs-LA	femtosecond laser ablation
FTICR	Fourier transform ion cyclotron resonance
FWHM	full width at half maximum
GC	gas chromatography
GC-MS	gas chromatography–mass spectrometry
GCEI	gas chromatography/electron ionization
GD	glow discharge
GFP	green fluorescent protein
GO	Gene Ontology
HCD	high-energy collisional dissociation
HERS-IMS	high-energy range sensor of the ion mass spectrometer
HG	hydride generation
HILIC	hydrophilic interaction chromatography
HMM	hidden Markov model
HMW-GS	high molecular weight glutenin subunit
HMX	1,3,5 7-tetranitro-1,3,5,7-tetraazacyclooctane
HPLC	high performance liquid chromatography
HR/AM	high resolution and accurate mass
HRMS	high-resolution mass spectrometry
HTD	high-throughput mass spectrometry data
HTP	hydroxyapatite
ICAT	isotope-coded affinity tag
ICP	inductively coupled plasma
ICR	ion cyclotron resonance
ID	isotope dilution
IEC	ion exchange chromatography
IM	ion mobility
IMS	ion mobility spectrometry
IMMS	ion mobility mass spectrometry
IRMPD	infrared multiphoton dissociation
IRZ	initial radiation zone
IT	ion trap
ITO	indium tin oxide glass
KNN	*k*-nearest neighbors

LA	laser ablation
LAESI	laser ablation electrospray ionization
LASS	laser ablation split stream
LC	liquid chromatography
LC-MS	liquid chromatography–mass spectrometry
LDA	linear discriminant analysis
LIBS	laser-induced breakdown spectroscopy
LID	laser induced dissociation
LIT	linear ion trap
LLSD	evaporative light scattering detector
LMW-GS	low molecular weight glutenin subunit
LTP	low temperature plasma
LTQ-FTICR	hybrid linear trap/Fourier transform ion cyclotron resonance
MALDI	matrix-assisted laser desorption/ionization
MCAT	mass-coded abundance tagging
MDA	malondialtonaldehyde
MECK	micellar electrokinetic chromatography
MFGDP	micro-fabricated glow discharge plasma
MHCD	microhollow cathode discharge
MIM	multiple ion monitoring
MIP	microwave-induced plasma
MIPDI	microwave-induced plasma desorption/ionization
MMPcd	matrix metalloprotease catalytic domain
MPT	microwave plasma torch
MRM	multiple reaction monitoring
MS	mass spectrometry
MS/MS	tandem mass spectrometry
MSEA	Metabolite Set Enrichment Analysis
MSI	mass spectrometry imaging
MudPIT	multidimensional protein identification technology
MW	molecular weight
NALDI	nanostructure-assisted laser desorption/ionization
nanoESI	nanoelectrospray
nanoLC	nano-liquid chromatography
NAZ	normal analytical zone
NETD	negative electron transfer dissociation
nePTM	nonenzymatic posttranslational modification
NIST	National Institute of Standards and Technology
NMR	nuclear magnetic resonance
NMS	neutral mass spectrometer
NPD	nitrogen phosphorus detector
oaTOF	orthogonal acceleration time of flight
OES	optical emission spectrometry

P2VP	poly(2-vinylpyridine)
PABS	psychoactive bath salts
PADI	plasma-assisted desorption ionization
PAH	polycyclic aromatic hydrocarbons
PBS	phosphate buffered saline
PCA	principal component analysis
PCB	polychlorinated biphenyl
PDO	protected by denomination of origin
PEG	polyethylene glycol
PENT	pentrite
PET	polyethylene terephthalate
PET	positron emission tomography
PFF	peptide fragmentation fingerprinting
PHZ	preheating zone
pI	isoelectric point
PI	photoionization
PID	photoionization detector
PMF	peptide mass fingerprinting
PMMA	poly(methyl methacrylate)
PPI	protein–protein interaction
ppm	parts per million
PQD	pulsed Q collision-induced dissociation
PRM	parallel reaction monitoring
PS	polystyrene
PSD	post-source decay
PTFE	Teflon
PTM	posttranslational modification
PTR	proton transfer reaction
Q	quadrupole analyzer
QDA	quadratic discriminant analysis
QIT	quadrupole ion trap
QMS	quadrupole mass spectrometry
QqQ	triple quadrupole
qTOF	quadrupole time of flight
RDA	retro-Diels–Alder
RDX	hexogen
RF	radio frequency
RID	refractive index detector
RMS	root-mean-square
RSD	relative standard deviation
SA	sinapinic acid
SAX	strong anion exchange
SCD/NCD	sulfur/nitrogen chemiluminescence detector

SCX	strong cation exchange
SD	standard deviation
SDS	sodium dodecyl sulfate
SEC	size exclusion chromatography
SELDI	surface-enhanced laser desorption/ionization
SID	surface-induced dissociation
SILAC	stable isotope labeling in culture
SILAM	stable isotope labeling of mammals
SIM	selected ion monitoring
SIMS	secondary ion mass spectrometry
SORI	sustained off-resonance irradiation
SPME	solid-phase microextraction
SRM	selected reaction monitoring
STRIG	stacked-ring ion guide
SWATH-MS	Sequential Windowed Acquisition of All Theoretical Fragment Ion Mass Spectra
TCD	thermal conductivity detector
TFA	trifluoroethanoic acid
THB	sodium tetraborohydride
THF	tetrahydrofuran
TIC	total ion current
TIMS	thermal ionization mass spectrometry
TLC	thin layer chromatography
TMA	tissue microarrays
TNT	trotyl
TOF	time of flight
TWIG	traveling wave ion guides
U	potential permanent
UPLC	ultra performance liquid chromatography
UV	ultraviolet detector
UV/Vis	ultraviolet–visible
V	potential variable
VDAC	voltage-dependent anion-selective channel
VOC	volatile organic compound
VUV-SP	vacuum UV single photon

Index

Mass Spectrometry: An Applied Approach, Second Edition. Edited by Marek Smoluch,
Giuseppe Grasso, Piotr Suder, and Jerzy Silberring.
© 2019 John Wiley & Sons, Inc. Published 2019 by John Wiley & Sons, Inc.